BIBLIOGRAPHY
of the
HISTORY OF TECHNOLOGY

This is Number 5 in a series of monographs in the history of technology and culture published jointly by the Society for the History of Technology and The M.I.T. Press. The members of the editorial board for the Society for the History of Technology Monograph Series are Melvin Kranzberg, Cyril S. Smith, and R. J. Forbes.

Publications in the series include:

History of the Lathe to 1850
Robert S. Woodbury

English Land Measuring to 1800: Instruments and Practices
A. W. Richeson

The Development of Technical Education in France 1500–1850
Frederick B. Artz

Sources for the History of the Science of Steel 1532–1786
Cyril Stanley Smith

Bibliography of the History of Technology
Eugene S. Ferguson

BIBLIOGRAPHY of the HISTORY OF TECHNOLOGY

EUGENE S. FERGUSON

Published jointly by

The Society for the History of Technology

and

The M.I.T. Press

Massachusetts Institute of Technology
Cambridge, Massachusetts, and London, England

FERNALD LIBRARY
COLBY-SAWYER COLLEGE
NEW LONDON. N. H. 03257

Copyright © 1968 by
The Massachusetts Institute of Technology

Set in Linotype Caledonia and printed by The Heffernan Press Inc.
Bound in the United States of America by The Colonial Press, Inc.

All rights reserved. No part of this book may be
reproduced or utilized in any form or by any means,
electronic or mechanical, including photocopying,
recording, or by any information storage and retrieval
system, without permission in writing from the publisher.

Library of Congress catalog card number: 68–21559

To

EARLE DUDLEY ROSS

*who, with the sure touch of a master teacher,
encouraged me to find my own proper direction*

"In executing this task I shall aim at more than the character of a diligent collector; for to bring together information of this kind, to arrange it, and to make it useful, requires no less readiness of thought than the labours of those who assume the character of original thinkers, and who imagine that they render others inferior to themselves when they bestow on them the appellation of compilers."

—JOHANN BECKMANN,
A History of Inventions
(4th ed., London, 1846),
Vol. 1, pp. 435–436.

"I, too, am versed in *title pages*."

—JOHN RANDOLPH
Quoted in George R. Taylor,
The Great Tarrif Debate, 1820–1830
(Boston, 1953), p. vii.

PREFACE

The purpose of this book is to provide a reasonably comprehensive introduction to primary and secondary sources in the history of technology. Although more than a third of the book is devoted to monographs and articles that will answer specific questions on a given subject, I have put particular emphasis upon the kinds and classes of information that may be available and upon specialized bibliographies and finding aids. I hope that some readers who have specific questions may find just the title they seek—to that end I have provided an index—but such readers are not my central concern.

The reader that I expect to be helped by this book is the student who is trying to get his bearings in a new and largely uncharted field. I have tried to direct him to the tools and resources of the scholar. I assume that he will take time to study and become familiar with the guides and other finding aids that are listed here. I hope he will recognize that many works of scholarly guidance offer wisdom as well as direction. I hope also that he will come to appreciate the richness and complexity of the resources that a large library can offer him if he will study with care its structure, responding with sensitivity to its nuances and ambiguities.

The history of technology is viewed here as a strand of cultural history. The relationships between technology and the culture in which it exists are of primary importance, but most of the historical works that have appeared thus far are concerned with the narration of how things were made and how they were used. While full and exact information on "what" and "how" are necessary if we are to understand how we are related to technology, the questions starting with "why" should also be asked. As Lynn White, Jr., has observed, it is time that the historian of technology began to explore the jungle of meaning.[1]

I have followed implicitly, if not consistently, Brooke Hindle's concept of technology, which encompasses activities of man that result

[1] *Isis*, 55 (June 1964), 228–230.

in artifacts. "Technology seeks means for making and doing things," he writes. "It is a question of process, always expressible in terms of three-dimensional 'things.'" Often the products as well as the tools and means are "things," but even when the product is not, as in the telegraph or the computer, the means are.[2]

The history of science has been omitted because excellent bibliographies and guides in that field already exist. Economic, business, and entrepreneurial history have been tapped for titles that bear upon technology, but the coverage of these subjects is in no sense comprehensive; and I have not even tried to deal with sociological works. The limits of subject matter can be gauged by scanning Parts XII, XIII, and XIV of the table of contents.

Every specialist will see immediately the deficiencies in his own field. Those who are familiar with works in languages other than English will recognize my provincialism and limited view. Because I have not put my hands on every work listed, it is possible that "ghosts" (titles of books that do not exist) have crept in, and it is certain that some of my evaluations are faulty if not dead wrong. The discovery of historical information so often depends upon chance references of the most tenuous sort that I have followed the rule that an incomplete and untidy something is more useful than a judiciously omitted nothing. Some titles that sounded promising but that, on inspection, were not particularly helpful have been included so that other readers may avoid repeating the process.

For all these reasons I shall welcome corrections and additions. I must ask, however, that additions include critical evaluation as well as reasonably complete bibliographic data. Every conversion of an incomplete note to a finished entry has to be made by me; there is no staff to which I can turn.

A list of corrections and additions will appear in *Technology and Culture* about two years after the appearance of this book, if present plans are carried out.

I have taken a considerable number of recent titles from Jack Goodwin's "Current Bibliography in the History of Technology," which has appeared annually since 1964 in *Technology and Culture*. Nevertheless, any thorough search must include a combing of that bibliography. The task will be greatly simplified by Goodman's cumulative subject index covering the first four years, published in 8 (Spring 1967), 291–309, and other name and subject indexes that appear in the annual numbers. The possibilities of consolidating and reprinting the series, perhaps at ten-year intervals, is being considered.

If my book is used to guide the formation of a collection of books and serials, its selective nature ought to be recognized. Just as the student is expected to go well beyond my listing, so will the book

---

[2] Brooke Hindle, *Technology in Early America* (Chapel Hill: North Carolina University Press, 1966), pp. 4–5.

buyer have to consult the bibliographies in the books that are listed in order to learn of additional titles in the same or allied fields. This is particularly true in Part XIV.

Earlier versions of parts of this book were published serially in *Technology and Culture*, 3 (1962), 73–84, 167–174, 298–306; 4 (1963), 318–330; 5 (1964), 416–434, 578–594; 6 (1965), 99–107. All parts have been extensively revised and augmented. Most of Part XII and all of Parts XIII and XIV are published here for the first time.

EUGENE S. FERGUSON

Iowa State University
Ames, Iowa
October 1967

ACKNOWLEDGMENTS

Robert P. Multhauf, Director of the Museum of History and Technology, warned me of some of the hazards of this undertaking, but he generously encouraged me to proceed with it while I was at the Smithsonian Institution.

Melvin Kranzberg, in his capacity as editor in chief of *Technology and Culture*, supplied crucial encouragement by agreeing to serial publication of parts of this bibliography as they were completed over a period of several years.

George R. Town, Dean of Engineering, and Henry M. Black, Head of the Department of Mechanical Engineering, made it possible for me to pursue the work after I returned to Iowa State University.

A grant from the Iowa State University Engineering Research Institute, awarded upon the recommendations of David R. Boylan, Director, and George K. Serovy, Professor of Mechanical Engineering, enabled me to put my manuscript into a final form much sooner than would otherwise have been possible.

The Eleutherian Mills Historical Library, in making a grant that enabled me to study there during the summer of 1964, contributed substantially to my knowledge of resources.

Mrs. Robert Garland helped me incalculably by correcting very numerous errors and inconsistencies as she checked and typed the final draft of the work.

Each of the following persons has helped me in some particular way, many over a considerable period of time, in supplying information, advice, corrections, and encouragement. To those who should be included but through oversight have been omitted, I add apologies to my thanks.

Collamer M. Abbott, White River Junction, Vt.; Philip M. Arnold, Bartlesville, Okla.; Karlos I. Ators, Philadelphia, Pa.; Silvio A. Bedini, Washington, D.C.; Robert Brittain, London; Lynwood Bryant, Cambridge, Mass.; Logan O. Cowgill, Washington, D.C.; Jacques Denavit, Evanston, Ill.; H. Drubba, Hannover, Germany; Meyer H.

Fishbein, Washington, D.C.; Arthur H. Frazier, Madison, Wis.; Hugh R. Gibb, Greenville, Del.; Jack Goodwin, Washington, D.C.; Richard S. Hartenberg, Evanston, Ill.; M. W. Hill, London; Kurt Mauel, Düsseldorf, Germany; Joseph A. Miller, New Haven, Conn.; Warren G. Ogden, Jr., North Andover, Mass.; Jacques Payen, Paris; Frank D. Prager, Abington, Pa.; Nathan Reingold, Washington, D.C.; Ladislao Reti, Los Angeles, Cal.; Joseph Rossman, Philadelphia, Pa.; John D. Stanitz, University Heights, Ohio; Sigvard Strandh, Stockholm; Edward L. Towle, Rochester, N.Y.; Peter C. Welsh, Washington, D.C.; Norman B. Wilkinson, Greenville, Del.; and Evelyn Wimersberger, Ames, Iowa.

In addition to all of those named, there is a host of people in libraries from Maine to California (literally) who have served me over the years. Almost without exception they have given me to believe that my problems were as important as any they had been faced with. Particularly helpful has been the entire staff of the Iowa State University Library.

I thank all of these individuals and institutions and also those whose names I failed to make note of, for this book has been greatly enriched by their contributions. I hasten to take full credit for the mistakes and failures that remain.

Jo, my wife, has helped most of all, as usual.

E.S.F.

NEEDS AND OPPORTUNITIES

The present shape of the field of history of technology and its possibilities in the future can be deduced from Brooke Hindle, *Technology in Early America: Needs and Opportunities for Study* (p. 22), which is of much wider significance than its title indicates. My own opinions regarding various aspects of the field will be found throughout the pages that follow.

The comments here are not based upon a systematic study but merely reflect the notes I have made from time to time as I have noticed particular lacunae in the body of work that has been published. Some of the gaps will be easily bridged by a single work; others will call for a generation of scholarly effort.

One of the principal difficulties in the history of technology is in finding out what actually happened. On many topics there are more conclusions than data. In few if any subjects do we have thorough and systematic studies, quantitative as well as qualitative. A few of the topics calling for study are the following:

Bridge design: evaluation of the structural analyses of, for example, Squire Whipple and Herman Haupt.
Canning of food. Incredibly little has been written about either the processes or the industry.
Manufactured gas systems in American cities.
Electrical power networks: their growth and occasional collapse.
Celestial navigation since about 1800.
Railroad permanent way: problems of adhesion, grades, track, ties and ballast.
Internal combustion engine: theories of combustion in the twentieth century, building on the work of Bryant (p. 245).
Printing presses. Very little has been written about the modern printing press (after 1800).
Papermaking machines in the United States.
Weighing devices: modern precision balances and scales for trade.

Machine-building shops in the United States: development before 1860.
The Civil War (U.S.) and its effect upon technical capabilities. Most studies have approached this question from an economic viewpoint and have been vague about "hardware."
Gas and electric welding: its development and acceptance in boiler and structural work.
Nuclear power development in England and the United States.
Metric-English system controversies. The history of these runs out into the present.
The assembly line. This is but one aspect of the nearly untouched field of industrial materials handling.
Prison contract labor; for example, in quarrying of blocks for railroad foundations and in making of carpenters' molding planes.
Horsepower and other units: the development of mechanical and electrical measurement systems.
Engineering education in the United States. The effect of l'Ecole Polytechnique on engineering education has been often asserted but little studied.
Handbook information: its origin, availability, and use by engineers and technicians. This might also include steam tables and boiler codes.
Quality control: development of inspection systems and philosophy and their relationship to statistical quality-control doctrine and practice.
American system of manufacture. It is quite remarkable that so little thorough work has been done on a topic so central to industrial development in the United States.
European sources of American techniques. The tide of American visitors to Europe, bent upon learning specific details of European practice, has been only dimly apprehended.

Every scholar owes his successors at least one solid piece of bibliography. The absence of systematic bibliographies in subject fields should be evident throughout this book. Article-length annotated bibliographies, based upon intimate knowledge of a particular field, would be welcomed in dozens of subjects: for example, hand tools, the steam engine, industrial ceramics, textile machinery, technology and culture, various subdivisions of industrial chemistry, industrial management, quality control, technical societies, technical education, sixteenth- and seventeenth-century machine books.

Such bibliographies should, I think, meet the needs of (a) the reader who wants to learn something of the field but who does not want to become trapped in it and (b) the reader who will eventually contribute to knowledge within the field. Because good bibliographies have a very long life, their publication in *Technology and Culture* would ensure their retrievability by readers a generation hence. As a model for individual entries—informative, dispassionate, not quirky —I recommend **Larson**, *Guide* (p. 23).

In addition to subject bibliographies, there is a need for intelligently compiled lists and indexes. For example, a biographical index of technologists, which would guide a reader to existing obituaries and other short sketches, would immediately became a standard work; an annotated bibliography of technical museum publications could be of great value; and the investment of perhaps two years of a young man's (or young woman's) time in compiling a descriptive list of museum artifacts, based upon personal visits, could I think yield another standard work that would live longer than its author.

Inquiry into the pathology of engineering would include studies of failures of structures, such as bridges, buildings, and dams; of machines and pressure vessels; and also of systems—for example, transportation systems, management systems, and extensive control systems, such as those for railroad and air traffic.

Biographies of a few technical people have been written, but the surface has barely been uncovered. No biographies exist for such important and influential men as George Corliss, John Smeaton, Henry Maudslay, Joseph Whitworth, John Jervis, Frank Sprague, and many others. For the one who is willing to wrestle with the meaning as well as minutiae of an individual human being, the challenge and rewards of biography are very considerable.

Historiography is dealt with by Brooke Hindle in the context of early American technology (p. 22), and the discussion of the philosophy of technology (p. 208) elicits mention of philosophical assumptions of historians, but as yet no work has been centrally and systematically concerned with historiography.

Finally, the social and cultural effects of "Scientific Management" (Taylorism) are beginning to be studied. This is a subject that is extremely pertinent to the present as successive waves of young men entering American industry adopt methods whose philosophy they do not comprehend and whose underlying assumptions they are completely unaware of.

E.S.F.

CONTENTS

"So I hit upon the thought of dividing the little book into four chapters, named after the seasons. Like all classifications, it is imperfect, but 'twill serve."
—George Gissing
The Private Papers of Henry Ryecroft
(New York: Dutton Everyman Library Ed., 1927), p. xiii.

    Preface   *vii*
    Acknowledgments   *x*
    Needs and Opportunities   *xii*

I. **GENERAL WORKS**   1
    A. Encyclopaedic Works in the History of Technology   *1*
    B. General Histories of Technology
       (more than one chronological period)   *3*
    C. Histories of Technology of Particular Periods   *8*
       1. *Ancient and Classical Periods* (to ca. 500 A.D.)   *8*
       2. *Middle Ages and Renaissance* (ca. 500–1600 A.D.)   *10*
       3. *Modern Period* (ca. 1600 to present)   *11*
          a. American   *11*
          b. Other   *13*
    D. Histories of Technology Published before 1900   *14*

II. **GENERAL BIBLIOGRAPHIES AND LIBRARY LISTS**   17
    (More than one subject. For single subjects, see Part XIV.)
    A. General Bibliographies   *17*
    B. Library Lists   *25*
    C. Miscellaneous   *29*

III. **DIRECTORIES OF TECHNICAL, ACADEMIC, AND BUSINESS ORGANIZATIONS**   35
    (Histories of societies will be found in Part XII.)

IV. **EARLY SOURCE BOOKS AND MANUSCRIPTS** 39
(before 1750)
A. Bibliographies, Guides, and Descriptions of Books 39
B. Translations and Editions of Books 44
C. Descriptions and Editions of Early Manuscripts 49

V. **ENCYCLOPAEDIAS, COMPENDIA, AND DICTIONARIES OF TECHNOLOGY** 53
A. Encyclopaedias 53
  1. *Bibliographies of Encyclopaedias* 54
  2. *Encyclopaedias in English* 54
  3. *Encyclopaedias in French* 57
  4. *Encyclopaedias in German* 60
B. Handbooks and Compendia of Technology 60
  1. *Compendia (mechanical "dictionaries" and "guides")* 61
  2. *Handbooks* 65
  3. *Trade Catalogues* 67
C. Glossaries 69
D. Bi- and Multilingual Dictionaries 70

VI. **BIOGRAPHY** 73
A. Biographical Dictionaries 73
  1. *Published Finding Aids* 73
  2. *Dictionaries of Biography* 74
  3. *Who's Whos* 77
B. Bibliographies and Indexes of Biography 77
C. Other Biographical Works 81
  1. *Collected Biographies* 81
  2. *Biographies in Local Histories* 83
  3. *Individual Autobiographies* 84
D. Portraits 87

VII. **GOVERNMENT PUBLICATIONS AND RECORDS** 91
A. Guides to the Records 91
  1. *United States* 91
  2. *Other Countries* 94
B. Departmental and Other Publications (U.S.) 95
  1. *Census Records and Publications* 95
  2. *Smithsonian Institution Publications* 97
  3. *Army, Corps of Engineers* 99
  4. *Consular Reports* 100

CONTENTS   xvii

      5. *Other Agencies*   100
      6. *Miscellaneous*   101
   C. Patents   102
      1. *Finding Aids (U.S.)*   102
      2. *Manuscript Patent Records (U.S.)*   105
      3. *Patent Models (U.S.)*   106
      4. *Searching of Patents (U.S.)*   106
      5. *Foreign Patents*   107
   D. U.S. National Archives   109

VIII. MANUSCRIPTS   113
   (See also Early Manuscripts, Part IV.C.)
   A. United States   113
   B. Other Countries   120

IX. ILLUSTRATIONS   123
   A. General Guides   123
   B. Individual Collections   124
   C. Other Sources   125
   D. Motion Pictures   127

X. TRAVEL AND DESCRIPTION   129
   A. Finding Aids and Bibliographies   129
   B. Individual Travels   132
      1. *United States*   132
      2. *Other Countries*   135
   C. Travel Guides and Descriptive Works   139

XI. PERIODICAL AND SERIAL PUBLICATIONS   141
   A. Finding Aids and Lists of Periodicals   142
   B. Indexes of Articles in Periodicals   144
   C. Current Serials Having Articles on Technical, Scientific, and Economic History   146
   D. Independent Technical and General Magazines (chiefly nineteenth century)   155
   E. Technical and Scientific Society Publications   161
      1. *United States*   162
      2. *Great Britain*   166
      3. *Canada*   168
      4. *France*   169

F. Newspapers   169
G. Miscellaneous Series   170

XII. TECHNICAL SOCIETIES, EDUCATION, AND EXHIBITIONS   173
    A. Technical and Scientific Societies   173
    B. Technical Colleges and Institutes   176
    C. Technical Museums   181
        1. Museums, Objects, and Lists of Museums   181
        2. Museum Publications   186
        3. Industrial Archaeology   191
    D. Exhibitions   192
        1. Exhibitions in General   193
        2. International Exhibitions   194
        3. Local and Regional Exhibitions   199
    E. Diffusion of Technology (geographic transfer of techniques)   200

XIII. TECHNOLOGY AND CULTURE   203
    A. Interrelationships between Technology and Its Cultural Matrix   204
    B. Philosophy of Technology   208
    C. Human Ecology and Natural Resources   209
        1. Ecologic Problems   209
        2. Use and Conservation of Natural Resources   210

XIV. SUBJECT FIELDS (Monographs, Articles, and Bibliographies)   213
    A. Food Production, Preservation, and Preparation   214
        1. Agricultural Machinery and Grain Milling   214
        2. Food Preservation and Preparation
        (See also Refrigeration, Part XIV.D.7)   217
    B. Civil Engineering   219
        1. General   219
        2. Buildings and Similar Structures   221
        3. Bridges and Tunnels   223
        4. Hydraulic Engineering, Water Supply, and Dams   225
        5. Surveying and Mapping   227
    C. Transportation   228
        1. General   228
        2. Ships and Boats, Navigation and Charting   228

3. Canals, Rivers, and Harbors   232
4. Roads and Vehicles   233
5. Railroads and Vehicles   236
6. Aircraft and Spacecraft   240

D. Energy Conversion   243
1. General   243
2. Horsepower and Manpower   243
3. Waterwheels, Hydro-power, Windmills, and Pumps   244
4. Internal-Combustion Engines   245
5. Steam Engines, Turbines, and Boilers   247
6. Atomic Energy   249
7. Refrigeration and Heating   249
8. Lighting   250

E. Electrical and Electronic Arts   250
1. General   250
2. Power Generation and Transmission, Motors, and Lighting   252
3. Communications and Electronic Arts, Including Electronic Computers   253

F. Materials and Processes   255
1. General   255
2. Manufacturing in General   256
3. Metals: Mining, Metallurgy, and Metallography   256
4. Chemical Industries, Including Photography   262
5. Glass, Ceramics and Cement   265
6. Coal, Oil, and Gas   266
7. Paper and Printing, Including Inks   269
8. Textiles and Allied Industries   272
9. Timber and Wood Industries   274
10. Testing of Materials   275

G. Mechanical Technology   276
1. Hand Tools and Machine Tools   276
2. Crafts and Craftsmen   280
3. Mechanisms and Automatic Control   284
4. Timekeepers   285
5. Scientific Instruments and Calculating Machines   287
6. Weights, Measures, and Standards   290
7. Mechanical Power Transmission, Bearings and Lubrication   292
8. Cranes, Rigging, and the Moving of Heavy Objects   294

H. Musical Instruments   294

I. Military Technology and War   295

J. Industrial Organization 297
   1. General, Including the "Labor Problem" 297
   2. "American System" of Manufacture; Assembly Line 298
   3. Scientific Management and Systems Analysis 300
   4. Quality Control 303
K. "Engineering Sciences": Thermodynamics, Hydraulics, Aerodynamics, Strength of Materials, and Kinematics 304
L. Process of Invention and Innovation 306

INDEX 307

I. GENERAL WORKS

Although it is too early in the career of the history of technology as an accepted academic discipline to look for an entirely satisfactory general work, those of us who are too busy with details to attempt the grand syntheses would be unable to see beyond our own backyards if some of the books listed in Part I had not been written. On the other hand, it is my opinion that we now have enough general works to serve the present generation and that more important contributions are to be made in monographs and in summary histories of particular periods or places. There will always be room, of course, for new and sharp insights into the relevance to our times of the whole story and particularly for fresh and imaginative books intended for classrooms, but additional general catalogues of events will be superfluous. Monographic works are listed in other parts of this volume under appropriate subject headings.

The multivolume encyclopaedic works of Daumas, Needham, and Singer are listed in the section immediately following.

A. Encyclopaedic Works in the History of Technology

Ludwig Darmstaedter, *Handbuch zur Geschichte der Naturwissenschaften und der Technik* (2nd ed., Berlin, 1908, 1,262 pp.). Reprint (New York: Kraus Reprint Corp., 1960). The book consists of about 13,000 short paragraphs noting discoveries and advances, 3500 B.C.–1908 A.D., arranged chronologically; 708 pages are devoted to the nineteenth century. No sources are given. Fully indexed.

Maurice Daumas, Ed., *Histoire générale des techniques* (4 vols., Paris: Presses universitaires de France, 1962———). Vol. 1, to 1350 A.D. (652 pp., illustrated), has been published. Three more volumes are projected: Vol. 2, to *ca.* 1700; Vol. 3, *ca.* 1700–1850; Vol. 4, *ca.* 1850 to present. As noted by Lynn White, Jr. in *Isis*, 55 (June 1964), 228–

230, this work is "a decided intellectual advance over that of Singer's (below), even though discussion of individual topics is usually more detailed in Singer's." Specifically, the treatment is global where Singer is parochial, and the history of the twentieth century will be included. On the other hand, the task of connecting technology with general history was not undertaken. White reminds us that every historian of technology must eventually begin to "explore the jungle of meaning."

Franz M. Feldhaus, *Die Technik der Vorzeit, der geschichtlichen Zeit und der Naturvölker: ein Handbuch* (2nd ed., Munich: Heinz Moos Verlag, 1965; 748 pp., 900 illustrations). The second edition of this major reference work consists of a facsimile reprint of the first edition (1914; 700 pp., 873 illustrations), to which have been added a 26-page appendix of supplementary subject matter and an appendix of some 500 corrections. References to further sources are listed at the end of nearly every article in the book. An informative review of the new edition by Otto Mayr is in *Technology and Culture,* 8 (Apr. 1967), 213–214.

Joseph Needham, *Science and Civilisation in China* (Cambridge: University Press, 1954———). A comprehensive account from earliest times to the end of the seventeenth century. Profusely illustrated. Needham has brought Chinese science and technology to the West, where it can never again be ignored as it has been throughout modern times.

Seven volumes are projected, some in parts: Vol. 1, Introductory (1954); Vol. 2, History of scientific thought (1956); Vol. 3, Mathematics and sciences of heavens and earth (1959); Vol. 4, Physics and physical technology: pt. 1, Physics (1962); pt. 2, Mechanical engineering (1965). Not yet published are Vol. 4, pt. 3, Engineering and nautics; Vol. 5, Chemistry and chemical technology: pt. 1, Arts of peace and war; pt. 2, Chemical discovery and invention; Vol. 6, Biology and biological technology; Vol. 7, Social background. Needham's bibliographies are voluminous. For example, in Vol. 4, pt. 2, Oriental works occupy pages 611–647; books and journal articles in Western languages, pages 648–707.

The reasons for China's scientific and technical decline *ca.* 1500 A.D., after a brilliant 2000-year performance, are summarized in Joseph Needham's "Science and Society in East and West," *Centaurus,* 10, No. 3 (1964), 174–197. These arguments, greatly expanded, are to appear in Vol. 7 of Needham's work. The same article appeared in Maurice Goldsmith and Alan Mackay, Eds., *Society and Science* (New York: Simon and Schuster, 1964), pp. 127–149. The same book was published in England as *The Science of Science* (p. 205).

Charles Singer, E. J. Holmyard, A. R. Hall, and Trevor I. Williams, Eds., *A History of Technology* (5 vols., New York and London:

Oxford University Press, 1954–1958). The view of the editors is by their own admission parochial, being generally limited to the Near East, Europe, and the United States. The period covered is from the beginnings to *ca.* 1900. The illustrations are outstanding, for both their quality and the ease with which their sources may be established. The interest in this monumental series and the implicit recognition of its value are shown in the large numbers of searching reviews that it has received. An investment of time in reading the reviews will make the books even more valuable. The Fall 1960 issue of *Technology and Culture* (Vol. 1, No. 4) was devoted entirely to reviews of the set. I have noted also careful reviews of Vols. 2 and 3 in *Economic History Review*, 2nd series, *11* (1959), 506–514; *12* (1959), 120–125. Imperial Chemical Industries Limited played an important and commendable supporting role in intellectual as well as financial matters.

Just as a wise reader frequently turns first to a good encyclopaedia when he prepares to enter a subject he is unfamiliar with, so the student of the history of technology should permit individual chapters of this work to give him an overview and a point of departure.

Arturo Uccelli, *Enciclopedia Storia delle Scienze e delle loro Applicazione* (3 vols., Milan, 1941). Ambitious project considering all aspects of the history of science and technology from the earliest times to the present. Countless text illustrations and plates. Regrettably the author does not include credits for sources of illustrations, many of which appear to be unique, or at least not found in other sources. [This note by Silvio Bedini.]

Arturo Uccelli (with numerous collaborators), *Scienza e Tecnica del Tempo Nostro nei Principii e nelle Applicazioni* (Milan: Hoepli, 1958; 768 pp., 1,098 illustrations). Well-illustrated and comprehensive one-volume encyclopaedia of the history of science and technology. [This note by Silvio Bedini.]

B. General Histories of Technology (more than one chronological period)

T. K. Derry and Trevor I. Williams, *A Short History of Technology* (New York: Oxford University Press, 1961, 782 pp.). The "Epilogue: Technological and General History" (pp. 700–711) suggests the problem of showing the relevance of technological history to the present human situation. Apparently designed as a popular reference and textbook, this work is based on the five-volume **Singer**, *History of Technology* (p. 2). There are too many facts, however, and too little juice for a textbook.

P. Ducassé, *Histoire des Techniques* [Que sais-je? No. 126] (Paris: Presses universitaires de France, 1944).

U. Eco and G. B. Zorzoli, *A Pictorial History of Inventions* (London: Weidenfeld and Nicolson, 1962). Published in the United States as *The Picture History of Inventions* (New York: Macmillan, 1963). A lavishly illustrated popularization.

Franz M. Feldhaus, *Ruhmesblätter der Technik von der Urerfindungen bis zur Gegenwart* (Leipzig, 1910; 630 pp., 231 illustrations); (2nd ed., 2 vols., Leipzig, 1924–1926; 421 illustrations).

James Kip Finch, *Engineering and Western Civilization* (New York: McGraw-Hill, 1951; 397 pp., illustrated). While the book is mildly polemical as it celebrates triumphant technology, a sense of historical perspective comes through. The bibliography, pp. 331–374, is very good.

James Kip Finch, *The Story of Engineering* (Garden City: Doubleday, 1960; 528 pp., illustrated). An original paperback. Generally a paragraph per event or man; easily read if the reader is content with this level of coverage. Insufficient depth or interpretation for textbook. No index; no footnotes; no bibliography.

Robert J. Forbes, *Man the Maker* (London and New York: Abelard-Schuman, 1950; 365 pp., illustrated). Bibliography, pp. 341–348. The book is sound history, despite numerous errors of detail. The author's tendency to develop catalogues of names makes reading tedious and summarizing difficult. On the other hand, A. G. Keller, in *History of Science*, 1 (1962), 110–111, concludes that no better short survey of the history of technology exists. I am reluctantly inclined to agree.

R. J. Forbes and E. J. Dijksterhuis, *A History of Science and Technology* (2 vols., Baltimore: Penguin, 1963; 294, 242 pp.). Vol. 1, "Ancient Times to the Seventeenth Century"; Vol. 2, "Eighteenth and Nineteenth Centuries." Bibliographies, pp. 261–274, 511–516. Biology and medicine have been omitted; what remains is so compressed that it can have meaning only to those who already know the story. In my opinion this work is totally unsuitable as a textbook, despite its wide-ranging scope.

H. J. Fyrth and M. Goldsmith, *Science, History and Technology* (2 vols., London: Cassell, 1966——). The first volume, 800 A.D.–1840's, 260 pp., illustrated, is an overambitious synthesis marred by numerous errors that suggest unfamiliarity with subject, according to a note in *The Times Literary Supplement*, Sept. 1, 1966.

GENERAL WORKS 5

Max Geitel, *Der Siegeslauf der Technik* (2nd ed., 3 vols., Stuttgart, 1922). Profusely illustrated; on the triumphant advance of technology.

Richard S. Kirby, Sidney Withington, Arthur B. Darling, and Frederick G. Kilgour, *Engineering in History* (New York: McGraw-Hill, 1956; 530 pp., illustrated). Selection of material to be treated is good, but the treatment includes a vein of superiority-by-hindsight (e.g., "Today's teletypists may be amused" by the thought of using cannon to convey the word of the opening of the Erie Canal) that makes the book unsuitable as a textbook for historically unsophisticated students, such as engineers. On the other hand, see a favorable review by Carl Condit in *Isis, 48* (1957), 484–485. There is a helpful book list at the end of every chapter.

Friedrich Klemm, *Kurze Geschichte der Technik* (Freiburg im Breisgau: Herder, 1961; 190 pp., illustrated). This book provides in narrative form the substance of the author's earlier work, which follows immediately.

Friedrich Klemm, *Technik: eine Geschichte ihrer Probleme* (Munich, 1954). Translated by Dorothea W. Singer as *A History of Western Technology* (London: Allen & Unwin, 1959; 401 pp., illustrated; M.I.T. Press paperback, 1964). An intelligently selected series of readings connected by short interlocutory passages. The original work has a useful, closely printed bibliography, pp. 414–440, which was emasculated in the English edition. Students dislike the absence of dogmatic interpretation; many can see no reason to read ten pages when a sentence in a textbook might summarize the gross facts. Nevertheless, I keep imposing the book on my students because Klemm comes closer than most authors to establishing connections between technology and culture. "Quellenschriften zur Technik" in the German edition is a very closely printed list of short titles, with references to articles and books describing authors or works: Greek-Roman, 424–425; Middle Ages, 425–427; Renaissance, 427–431; Baroque, 431–434; to 1800, 435–438; to 1860, 439–440.

Carl Graf von Klinckowstroem, *Knaurs Geschichte der Technik* (Munich and Zurich: Droemersche Verlagsanstalt, 1964; 488 pp., illustrated). Popular, "pleasant reading," providing a "detached view of the development of American technology in relation to general development," according to Edwin A. Battison in *Technology and Culture, 7* (Spring 1966), 229–230.

Melvin Kranzberg and Carroll W. Pursell, Jr., Eds., *Technology in Western Civilization* (2 vols., New York: Oxford University Press, 1967; 802, 772 pp., illustrated). Vol. 1, "The Emergence of Modern Industrial Society," covers the early periods lightly and emphasizes the seventeenth through nineteenth centuries; Vol. 2 is on "Technology in the Twentieth Century." There are well-constructed basic bibliographies in both volumes: Vol. 1, pp. 745–774; Vol. 2, pp. 709–739. The latter bibliography is the best I have seen for any considerable part of twentieth-century technology.

Designed as a textbook for proposed courses in the U.S. Armed Forces Institute, this work is long and has some of the inevitable shortcomings of a collaborative work, such as repetition and uneven quality in the various chapters. Nevertheless, as I have observed the book take shape out of nothing in slightly more than three years, I have been more impressed by the very fact of its existence and its general competence than by its shortcomings. It is the first comprehensive work to make any serious attempt to deal with the social matrix in which technology developed, and it is the first attempt to bring together the history of technology in the twentieth century. In both respects I think it has succeeded remarkably well.

Samuel Lilley, *Men, Machines and History; the Story of Tools and Machines in Relation to Social Progress* (Rev. ed., 1st U.S. ed., New York: International Publishers, 1966; 352 pp., illustrated). A polemical work which advocates unlimited technical proliferation. Since "society" does not keep up with technology, society must change in order to maintain progress. Lilley would install the Russian system as the one most conducive to progress.

E. O. von Lippman, *Beiträge zur Geschichte der Naturwissenschaften und der Technik* (Berlin: Springer, 1923; 314 pp., illustrated). Bibliographical footnotes.

Lewis Mumford, *Technics and Civilization* (New York: Harcourt, Brace, 1934; Harbinger paperback, 1963; 495 pp., illustrated). Annotated bibliography, pp. 447–474. One of the best works of an important thinker and social critic. Concerned with cultural aspects of technology since medieval times, the book is full of fresh and exceedingly sharp insights. Too devastating for a first book of history to be read by a technical student, urgent though the need is for such criticism.

G. Neudeck, *Geschichte der Technik* (Stuttgart, 1923, 569 pp.). Profusely illustrated. I have not seen this book.

Pierre Rousseau, *Histoire des techniques et des inventions* (Paris: Libraire Arthème Fayard, 1958, 526 pp.). Easily read popularization, but Maurice Daumas, in a review in *Revue d'histoire des sciences,* 11 (1958), 93–95, makes the point that it is too facile, decidedly not good history.

GENERAL WORKS 7

S. W. Schuchardin, *Grundlagen der Geschichte der Technik; Versuch einer Ausarbeitung der theoretischen und methodologischen Probleme* (Leipzig: VEB Fachbuchverlag, 1963, 195 pp.). Translated from Russian; a Party view of the proper approach to the history of technology. Reviewed by K.-H. Ludwig in *Technikgeschichte*, 32, No. 1 (1965), 98.

R. Soulard, *A History of the Machine* [New Illustrated Library of Science and Invention, Vol. 11] (New York: Hawthorn, 1964; 112 pp., over 100 illustrations, 24 in color). According to Melvin Kranzberg in *Engineer*, Vol. 5 (New York, Autumn 1964), the thirty-three pages of text are free of gross errors and the pictures are most attractive, but the book is not serious history.

A. A. Sworykin, N. I. Osmawa, W. I. Tschernyschew, and S. W. Schuchardin, *Geschichte der Technik* (Leipzig: VEB Fachbuchverlag, 1964; 831 pp., 444 figs., 16 pls.). According to J. Payen in *Bulletin Signalétique*, Pt. 22, 20, No. 2 (1966), 857, this is a translation from a Russian work of 1962, which gives the Party view of the history of technology.

Albrecht Timm, *Kleine Geschichte der Technologie* (Stuttgart, Kohlhammer, 1964; 219 pp., 4 pls.). Bibliography, 49 refs. Reviewed informatively by R. S. Hartenberg in *Technology and Culture* (Summer 1965), pp. 448–449; also by Eduard Fueter in *Technikgeschichte*, 33, No. 2 (1965), 196–198.

Hidomei Tuge, Ed., *Historical Development of Science and Technology in Japan* [Series on Japanese Life and Culture, Vol. 5] (Tokyo: Kokusai Bunka Shinkokai [Society for International Cultural Relations], 1961; 200 pp., illustrated). In English. Early period, 0–1700 A.D., pp. 1–52; to 1850, pp. 53–88; since 1850, pp. 89–183. The direct influences of Western thought and Western visitors in the late nineteenth century are particularly interesting.

Arturo Uccelli, *Storia della Tecnica dal Medioevo ai Nostri Giorni* (Milan, 1944; large 4to, 933 pp., 2,717 illustrations).

Abbott Payson Usher, *A History of Mechanical Inventions* (2nd ed., Cambridge, Mass.: Harvard University Press, 1954). Reprint (1959 as Beacon paperback BP84). This pioneering work has four chapters on the place of technology in economic history and the nature and process of innovation. The bibliography, pp. 433–443, in keeping with the whole book, is carefully constructed. Difficult to read as a textbook; nevertheless desirable under some circumstances because work is remarkably accurate and has not been made obsolete by later scholarship.

Arthur Vierendeel, *Esquisse d'un Histoire de la Technique* (2 vols., Brussels, 1921).

See also
Sarton, Guide (p. 17), pp. 167–168.
Russo, Bibliographie (p. 17), pp. 32–34.

C. Histories of Technology of Particular Periods

1. *Ancient and Classical Periods (to ca. 500 A.D.)*

A large number (50–100) of general works pertaining to this period will be found in **Forbes**, *Bibliographia Antiqua* (p. 21), pt. X, titles 7,973–8,143. Note, for example, H. Blümner (7,983) and H. Diels (8,017).

Ancient Peoples and Places. A continuing series published in London by Thames and Hudson, in New York by Frederick A. Praeger, Inc. The books are sensibly made; they are written by first-rate authorities; they are well illustrated; through their bibliographies they provide an entry into a subject; and, since the focus of the series is on archaeological evidence, there is due attention paid to technology. Some forty titles are in print, and forty-five more are being prepared. See, for example, Cyril Aldred, *The Egyptians* (1961), 268 pp., 82 illustrations; Holger Arbman, *The Vikings* (1961); J. F. S. Stone, *Wessex before the Celts* (1960); David Diringer, *Writing* (1962); George F. Bass, *Archaeology Under Water* (1966).

V. Gordon Childe, *Man Makes Himself* (4th ed., London: Watts, 1964, 244 pp.). One of several works on prehistory and emerging civilizations by a British historian whose Marxist tendencies supported his emphasis upon technology. Chapter 7, "The Urban Revolution," develops the basic requirements of a civilization. Comments by Glyn Daniel on Childe's work are in the preface to this edition. Five other titles by Childe are described briefly in **A.H.A.,** *Guide* (p. 18), pp. 80, 82, 127.

Dictionnaire Archéologique des Techniques (2 vols., Paris: Editions d'Accueil, 1963–1964). Selected bibliography, pp. 1081–1084. Silvio Bedini, in a review in *Isis* (Summer 1965), calls this a compendium for ready references by amateurs and educators but not of much use to scholars because presentation is general, and some major subjects are omitted. Nevertheless, as pointed out to me by Jack Goodwin, the

GENERAL WORKS 9

comparative treatment of subjects is convenient. "Matières colorantes," for example (pp. 596–603), are traced in separate short articles: Prehistory, Pre-Columbian America, South America, Far East, India, Western Asia, Egypt, Greece, and Rome. Articles are signed.

Hermann Diels, *Antike Technik: sieben Vorträge* (3d ed., Leipzig: B. G. Teubner, 1924; 243 pp., illustrated). Chapters on Greek science and technology and ancient automata, locks, telegraph, artillery, timekeepers, and chemistry.

Franz M. Feldhaus, *Die Technik der Antike und des Mittelalters* (Potsdam, 1931, 442 pp.). Profusely illustrated. In *Speculum*, 15 (Apr. 1940), 159, Lynn White, Jr., notes: "Feldhaus must be honored as a pioneer, but [this] work is sadly defective; it has no topical organization; it is a series of undocumented notes arranged in rough chronological order."

R. J. Forbes, *Studies in Ancient Technology* (11 vols., Leiden: Brill, 1955———; 2nd revised eds. of most vols., 1964 ff.) are solid and wide ranging, though not free of errors. For nature of difficulties see review of Vol. 7 by J. R. Harris in *Isis* (Spring 1965), pp. 90–92. Abbreviated contents are given below; all volumes are illustrated; bibliographies follow each subject-matter section.

Vol. 1: Bitumen and petroleum; alchemy; water supply. 2nd ed., 1964, 199 pp.
Vol. 2: Irrigation, drainage; power, including water and windmills; land transport, road building. 2nd ed., 1965, 220 pp.
Vol. 3: Cosmetics, perfumery; food, alcohol, vinegar, and fermented beverages, 500 B.C.–1500 A.D.; crushing; salt, preservation, mummification; paint, pigment, ink, varnish. 2nd ed., 1965, 276 pp.
Vol. 4: Fibres, fabrics; washing, bleaching, fulling, felting; dyes and dyeing; spinning; sewing, basketry, weaving. 2nd ed., 1964, 263 pp.
Vol. 5: Leather; sugar and substitutes; glass. 2nd ed., 1966, 241 pp.
Vol. 6: Heat and heating; refrigeration; light. 2nd ed., 1966, 200 pp.
Vol. 7: Geology; mining and quarrying; mining techniques. 2nd ed., 1966, 259 pp.
Vol. 8: Metallurgy; tools; evolution of smithy; gold; silver and lead; zinc and brass. 1964, 288 pp.
Vol. 9: Copper; tin and bronze, antimony and arsenic; early iron. 1964, 295 pp.
Vol. 10: Pharmacy. In preparation [1967].
Vol. 11: Clay, stone, brick, tiles and pantiles. In preparation [1967].

The Legacy Series. One or more chapters on early technology will be found in each of the books of this fine series, which is published by Oxford University Press. Titles in the series: *Legacy of Greece, Legacy of Rome, Legacy of Egypt* (1942) (chapter on technology by Engelbach), *Legacy of Israel, Legacy of Islam, Legacy of India, Legacy of the Middle Ages* (1926), and *Legacy of China* (1964) (chapter on technology by Needham).

Curt Merckel, *Die Ingenieurtechnik im Alterthum* (Berlin: Springer, 1899; 4to, 658 pp., 261 illustrations). On civil engineering in the Mediterranean area and Near and Far East from the earliest empires to Roman times. Extensive bibliographies follow the principal chapters.

Albert Neuburger, *The Technical Arts and Sciences of the Ancients* (London: Methuen, 1930, 518 pp.). Translated from the German by Henry L. Brose. Through Roman times. Well illustrated, but strictly secondary and thus often in error.

2. *Middle Ages and Renaissance (ca. 500–1600 A.D.)*

Alistair C. Crombie, *Medieval and Early Modern Science* (2 vols., Garden City: Doubleday, 1959; Anchor paperbacks A167a, A167b). A revision of *Augustine to Galileo: the History of Science A.D. 400–1650* (London, 1952). See particularly "Technics and Science in the Middle Ages," 1, 175–238; see also 2, 244–254. Excellent bibliographies.

Umberto Forti, *Storia della Tecnica, dal Medioevo al Rinascimento*, (Milan: Sansoni, ca. 1958; 650 pp., 432 illustrations in text, 34 plates in black-and-white and color). Italian text. Technological inventions and scientific discoveries which have with their developments contributed to human existence. Very fine coverage of the subject in one volume; illustrations are consistently excellent, including many illustrations not generally known or easily found. [This note by Silvio Bedini.]

Bertrand Gille, *Engineers of the Renaissance* (Cambridge, Mass.: The M.I.T. Press, 1967; 256 pp., many illustrations). Published in England as *The Renaissance Engineers* (London: Lund, Humphries, 1966). Translated from the French *Les Ingénieurs de la Renaissance* (Paris: Hermann, 1964; 239 pp., illustrated). An index and list of illustrations have been added in the English edition. This is an important work, based upon thorough familiarity with the manuscript sources. The bibliography, pp. 229–239 (242–253 in English ed.), includes a "Catalogue des Manuscrits" more comprehensive than any other published to date. German mss. (codices of 20–300 ff.), thirty-odd titles; Italian (including 21 of Leonardo), about ninety titles.

GENERAL WORKS

Leonardo Olschki, *Geschichte der neusprachlichen wissenschaftlichen Literatur* (3 vols., Leipzig and Halle, 1919–1927). On science and technology in the Middle Ages and the Renaissance. Vol. 1 includes Alberti, Ghiberti, Filarete, Martini, Leonardo da Vinci, Dürer; Vol. 2 considers the rise and diffusion of science and technology in the Italian Renaissance; Vol. 3 includes Bruno, Tartaglia, and Galileo.

William B. Parsons, *Engineers and Engineering in the Renaissance* (Baltimore, 1939; 661 pp., over 200 illustrations). A spacious and generally sound book based on long and diligent study. Perhaps because the work was published posthumously, the bibliography, pp. 619–623, is quite spotty.

Lynn White, Jr., *Medieval Technology and Social Change* (Oxford: Clarendon, 1962; 194 pp., illustrated). A major work which presents fresh and deeply significant interpretations. In an enthusiastic and informative review (*Isis, 54* [Sept. 1963], 418–420), Joseph Needham properly calls this "the most stimulating book of the century on the history of technology." See also the review by R. W. Southern in *History of Science, 2* (1963), 130–135.

Lynn White, Jr., "Technology and Invention in the Middle Ages," *Speculum, 15* (Apr. 1940), 141–159. A classic paper, described by Joseph Needham in *Isis, 54* (Sept. 1963), p. 418, as "a *libellus aureus*, treasured and minutely annotated by many scholars, constantly photostated and passed from hand to hand." Extensive bibliographic notes point to origins of, for example, the sternpost rudder, spectacles, feltmaking, the ski, soap, distillation, paper, bells, the wheelbarrow, spinning wheel, cannon, and so forth. A large number of sources are critically examined.

Lynn White, Jr., "Tibet, India, and Malaya as Sources of Western Medieval Technology," *American Historical Review, 45* (1960), 515–526. Extensive bibliographic notes.

3. Modern Period (ca. 1600 to present)

a. AMERICAN

Roger Burlingame, *Engines of Democracy* (New York: Scribner, 1940). A social history of American invention from *ca.* 1865. The bibliography lists about 500 titles and has a subject index. A sequel to the following work.

Roger Burlingame, *March of the Iron Men* (New York: Scribner, 1938; Grossett paperback, 500 pp.). A social history of American invention to *ca.* 1865. The bibliography lists about 500 titles and has a subject index.

Edward W. Byrn, *The Progress of Invention in the Nineteenth Century* (New York: Munn, 1900, 476 pp.). A straightforward account, with good wood-engraved illustrations.

Siegfried Giedion, *Mechanization Takes Command* (New York: Oxford University Press, 1948; 743 pp., 501 illustrations). Deals mainly with the United States, England, and France in the nineteenth century and later. A provocative, inquiring work that promotes fresh insights. One of very few sources of information on early industrial assembly lines.

Courtney R. Hall, *History of American Industrial Science* (New York: Library Publishers, 1954, 453 pp.). Bibliography, pp. 419–423. Arranged by industries (transport, chemical, electrical, etc.). The author is anxious to show that "the American system . . . is one of relatively free enterprise," and that "to change the American economic system into a different kind would be a disaster of great national and international consequence." A generally uncritical review of "progress."

Brooke Hindle, *The Pursuit of Science in Revolutionary America, 1735–1789* (Chapel Hill: The University of North Carolina Press, 1956, 410 pp.). The chapter on "American Improvement," pp. 190–215, sets technological development in the matrix of science of the Enlightenment.

Brooke Hindle, *Technology in Early America.* See p. 22.

Waldemar Kaempffert, Ed., *A Popular History of American Invention* (2 vols., New York: Scribner, 1924). About 500 illustrations.

John W. Oliver, *History of American Technology* (New York: Ronald, 1956; 676 pp., illustrated). An uncritically enthusiastic compendium of facts, many of which are questionable. A careful review by Carl Condit is in *Isis, 48* (1957), 485–487.

Dirk J. Struik, *Yankee Science in the Making* (Boston: Little, Brown, 1948, 430 pp.). A long, generally unfocused, but not inconsiderable account of science and technology to *ca.* 1865. Bibliographic notes, pp. 359–385; Bibliography, pp. 387–416.

Mitchell Wilson, *American Science and Invention, A Pictorial History* (New York: Simon and Schuster, 1954, large 4to, 437 pp.). Many good pictures. The text is routine; numerous picture captions are misleading or wrong.

The following three volumes of economic history are in the publisher's series "The Economic History of the United States." Aside

GENERAL WORKS 13

from their having been written by very knowledgeable scholars, each book contains a lengthy and good critical bibliographical essay.

Curtis P. Nettels, *The Emergence of a National Economy 1775–1815* (New York: Holt, Rinehart & Winston, 1962). Bibliography, pp. 341–380.

George R. Taylor, *The Transportation Revolution 1815–1860* (New York: Holt, Rinehart & Winston, 1951). Bibliography, pp. 399–438.

Edward C. Kirkland, *Industry Comes of Age 1860–1897* (New York: Holt, Rinehart & Winston, 1961). Bibliography, pp. 410–436.

b. OTHER

W. H. G. Armytage, *A Social History of Engineering* (London: Faber and Faber, 1961). Although this book starts at the beginning, its emphasis is upon the modern period, since 1700, with "especial reference to Britain." Useful bibliography, pp. 335–353; list of professional institutions in Great Britain, with founding dates, pp. 354–357. The author pays more attention than most to the social matrix of technology. A second printing, 1966, is called a new edition.

Charles Ballot, *L'Introduction du machinisme dans l'industrie française* (Lille, 1923, 576 pp.). A classic work, difficult to locate. Sections on steam engine, manufacturing, machine tools, and so forth.

W. H. Chaloner and A. E. Musson, *Industry and Technology* [A Visual History of Modern Britain] (London: Vista Books, 1963; 60 pp. + 238 illustrations). The brief text, starting in medieval times, is equally divided between the early and post-1700 period. Emphasis of pictures, which are generally fresh and well selected, is heavily upon eighteenth and nineteenth centuries. The bibliographical note, pp. 63–65, is a concise, sensible guide to further reading.

G. N. Clark, *Science and Social Welfare in the Age of Newton* (2nd ed., Oxford: Clarendon, 1949, 159 pp.). A brilliant short work on the relationships between science and technology, the economic and social aspects of technology, and the rise of scientific societies.

Arthur P. M. Fleming and H. J. Brocklehurst, *A History of Engineering* (London: Black, 1925). A history of modern British engineering.

Franz Hendrichs, *Der Weg aus der Tretmühle. Ein Abriss der Geschichte der Technik der neueren Zeit.* (2nd ed., Düsseldorf: VDI Verlag, 1958; 236 pp., illustrated). A brief and general history, starting with seventeenth-century steam-engine development.

Thomas P. Hughes, *The Development of Western Technology since 1500* [Main Themes in European History] (New York: Macmillan, 1964, 149 pp.). Designed for the classroom; consists of two or three contemporaneous accounts and several interpretive articles by good twentieth century historians. No reading list, but a few footnote references to further information.

Friedrich Klemm, Ed., *Die Technik der Neuzeit* (Potsdam, 1941– ca. 1950). An unfinished collaborative work, well illustrated. No more published; no index. Vol. 1, Sec. 1, "Von der mittelalterlichen Technik zum Maschinenzeitalter" (48 pp.); Vol. 2, Sec. 1–5, "Rohstoffgewinnung und Verarbeitung" (240 pp.); "Verkehrs- und Bautechnik" (240 pp.).

Paul Mantoux, *The Industrial Revolution in the Eighteenth Century* (rev. ed., London: Cape, 1961; Harper Torchbook ed., 1962). In the paperback edition the bibliography occupies pp. 478–516. The bibliography of Mantoux, revised in 1928, has been supplemented by one compiled in 1958 by A. J. Bourde.

Abraham Wolf, *A History of Science, Technology, and Philosophy in the 16th and 17th Centuries* (2nd ed., London: Allen & Unwin, 1950; 692 pp., 316 illustrations). A monumental compendium based upon the author's study of a vast number of printed works of the period covered. In Chapters 20–23, on technology, two dozen books are treated briefly, in the fields of agriculture, textiles, structures, mining and metallurgy, glassmaking, and mechanical engineering. Extensive treatment of Agricola. Reprinted as Harper paperback in two volumes, Nos. TB508, 509.

Abraham Wolf, *A History of Science, Technology, and Philosophy in the Eighteenth Century* (2nd ed., London: Allen & Unwin, 1952; 814 pp., 345 illustrations). Reprinted, TB539, 540. See also previous work.

D. Histories of Technology Published before 1900

The beginning of the historiography of technology is placed around 1350 in **Lynn White, Jr.,** *Medieval Technology and Social Change* (p. 11), p. 129, note 3, which cites also a review of histories of technology through the sixteenth century. The following three titles, which are bibliographies of histories of technology and invention, are comprehensive. A few representative titles of books are then listed in chronological order. In a few cases information not in Feldhaus and Klinckowstroem's article is supplied. See also **Sarton,** *Guide* (p. 17), 167–168.

GENERAL WORKS

F. M. Feldhaus and Carl Graf von Klinckowstroem, "Bibliographie der Erfindungsgeschichtlichen Literatur," *Geschichtsblätter für Technik und Industrie*, 10 (1923), 1–21. Nearly 250 titles of general histories of invention and technology, in several languages, from Lilius (1496) and Polydore (1499) to Pfeiffer (1920). Only 25 titles are later than 1900. Biographies, dictionaries, and encyclopaedias are excluded. An impressively complete list; the several works and editions of various authors are indentified, such as those of Beckmann, Feldhaus, Figuier, Poppe, and others. Indispensable to any serious work in history of histories.

Carl Graf von Klinckowstroem, "Die älteren deutschsprachigen Bücher über die Geschichte der Erfindungen," *Börsenblatt für den Deutschen Buchhandel* (Frankfurter Ausgabe), 16 (July 1965), 1423–1427. Bibliographical data are given for seven early German works along with biographical information on authors. Authors are Maier (1519), Marperger (1704), Beckmann (1780–1805), Vollbeding (1795), Busch (1790–1798), Donndorff (1817–1821), and Poppe (1837).

John Ferguson, *Bibliographical Notes on Histories of Inventions and Books of Secrets* (2 vols., London: Holland, 1959). Brought together are Ferguson's papers printed in the *Transactions* of the Archaeological Society of Glasgow between 1882 and 1911. A compendium of substantial information on the contents of books of the sixteenth through nineteenth centuries. Bate, Cardano, Porta, Theophilus, Schott, and the like are here, as are Beckmann, Goguet, and other relatively modern histories of invention and technology. John Ferguson (not related by blood or tradition to the present compiler) demonstrates the virtues of knowing what is beyond title pages.

Polydore Vergil, *De Inventoribus Rerum* (Venice, 1499; English abridgment, London, 1546). A description of the book, a biography of the author, and a discussion of editions is in John Ferguson, "Notes on the Work of Polydore Vergil 'De Inventoribus Rerum'" (*Isis, 17* [1932], 71–93).

Guido Panciroli, *Rerum Memorabilium sive Deperditarum* (2 vols., Frankfurt, 1629–1631). Translated as *The History of Memorable Things Lost . . . and an Account of Many Excellent Things Found, Now in Use among the Moderns*, done into English . . . (n.p., 1715). First published 1599–1602. See **Feldhaus** (above).

Dennis de Coetlogon, *An Universal History of Arts and Sciences . . . All Arts, Liberal and Mechanical* (London, 1745). Listed in Ralph R. Shaw, *Engineering Books in America prior to 1830* (New York, 1933).

Johann Beckmann (1739–1811), *A History of Inventions, Discoveries, and Origins* (4th ed., 2 vols., London, 1846). Translated from the German, which first appeared in five volumes (1780–1805). The author made critical use of earlier sources, which are cited in footnotes. For other editions, see **Feldhaus** (p. 15); briefly noted in **Klinckowstroem** (p. 15).

Johann H. M. von Poppe (1776–1854), *Geschichte der Technologie seit der Wiederherstellung* (3 vols., Göttingen, 1807–1811). Bibliographic footnotes. Also by the same author, *Geschichte der Erfindungen in den Künsten und Wissenschaften, seit der altesten bis auf die neueste Zeit* (4 vols., Dresden, 1828–1829). See British Museum printed catalogue for other titles by Poppe, as also **Feldhaus** (p. 15) and **Klinckowstroem** (p. 15).

J. F. L. Williams, *An Historical Account of Inventions and Discoveries* (2 vols., London, 1820). Slight.

Francis S. White, *A History of Inventions and Discoveries* (London, 1827). Alphabetically arranged. Slight.

Das neue Buch der Erfindungen, Gewerbe und Industrien (6 vols., Leipzig, 1863–1868). Illustrated. For other editions, 1854–1907, see **Feldhaus** (p. 15).

Karl Karmarsch, *Geschichte der Technologie seit der Mitte des XVIII. Jahrhunderts* (Munich, 1872, 939 pp.). Recently reprinted. A volume in the Royal Academy of Bavaria's series, *Geschichte der Wissenschaft in Deutschland*. For other titles, see **Sarton**, *Guide* (p. 17), 124–125.

II. GENERAL BIBLIOGRAPHIES AND LIBRARY LISTS
(More than one subject. For single subjects, see Part XIV).

"No matter how diligent a scholar may be there are always some ancient tools which he managed to overlook."
—George Sarton
A Guide to the History of Science (New York, 1952), p. 74.

A. General Bibliographies

The works of Sarton and Russo, which in my opinion should be in the personal working collection of any scholar who expects to do serious work in the history of technology, are listed first. The rest of this section is arranged alphabetically.

George Sarton, *A Guide to the History of Science* (New York: Ronald, 1952, 316 pp.). This book, by the founder of the discipline of the history of science and founder and editor of *Isis* and *Osiris,* is one fortunate result of a lifetime of intense and (what is particularly impressive to me) systematic bibliographical interest. Because Sarton saw the history of technology as a part of his discipline, his bibliography contains much that is of interest to the historian of technology. Thorough familiarity with and frequent use of Sarton's *Guide* will not only save a scholar's time, it will also enrich and deepen his sensitivity to available source and reference materials. A series of introductory essays (pp. 3–66), particularly recommended, are based upon lectures given by Sarton at the University of London in 1948.

François Russo, *Histoire des Sciences et des Techniques: Bibliographie* (Actualités Scientifiques et Industrielles 1204, Paris: Hermann, 1954, 186 pp.). I have found this a very useful adjunct to Sarton's *Guide*. The author has, along with the history of science, "systematically treated the history of technology, too long neglected and too often artificially separated from the history of the sciences with which, especially after the eighteenth century, it is closely allied." Russo's work is not heavy with technology, however, and I think that his idea of a definitive combined bibliography is more a desideratum than a real possibility. Nevertheless, one excellent feature of this book is a list of 290 scientific and technical authors of the sixteenth to eighteenth centuries (pp. 58–91) which includes many of the pioneering works on technology: Agricola, Barba, Belidor, Besson, Bion, Biringuccio, Böckler, Camus, Caus, and so forth, and a list of over 200 authors of the nineteenth and twentieth centuries (pp. 96–116). I have seen a mimeographed *Supplement* (1955, about 30 pp.).

American Historical Association. A series of "Pamphlets," each containing a concise summary of interpretations and a bibliography of pertinent works in a particular historical field, has been published by the Service Center for Teachers of History, 400 A Street, S.E., Washington, D.C. 20003. A full list of (approximately) sixty-five titles will be sent upon request. Each pamphlet, 75 cents. For example, No. 4, *Industrial Revolution*, by Eric Lampard, thirty-five page text and four-page bibliography; No. 13, *History of Science* (2nd ed.), by Marie Boas Hall, seven-page text and eighteen-page bibliography.

The American Historical Association's Guide to Historical Literature (New York: Macmillan, 1961, 962 pp.). When background information is required to place technical history in context, this guide will provide an entry to adjacent fields. The section on "General Reference Resources," by Constance M. Winchell and Shepard B. Clough (253 titles), should be part of a scholar's basic store of information. The whole book contains over 15,000 briefly annotated titles.

Theodore Besterman, *A World Bibliography of Bibliographies* (4th ed., 5 vols., Lausanne: Societias Bibliographica, 1965–1966). This audacious work has 117,187 bibliographies in some 45 languages under more than 15,000 heads and subheads. Vol. 5 is a detailed name and subject index. I found perhaps 50 useful bibliographies, new to me, on the history of technology. The printed records of the British Patent Office are set out fully and are carefully indexed. Bibliographies that were designed to include current literature in rapidly growing technical fields may be helpful to the historian, particularly because many of those listed by Besterman are now 30 years old or older. A random sampling reveals bibliographies in the following subjects — and this is a tiny sample (the numbers in parentheses are Besterman's, indicating the number of items listed in each bibliography): explosives (3,000); condensers (700); bagasse (400); belting (71); buttons (41); joints, mechanical (250); gas turbines (1,665). There are 20 columns of bibliographies listed under "electricity and magnetism," compared to 12 in the third edition (1955–1956).

Arthur Birembaut, "Quelques réflexions sur les problèmes posés par la conservation et la consultation des Archives Techniques Françaises," *Archives internationales d'histoire des sciences,* 19 (Jan.–June 1966), 21–102. A report on the state of source materials for the history of technology in public institutional archives and museums in France.

The author, knowledgeable in the history of technology, has what is apparently a firsthand acquaintance with the sources he cites. The discussion of original and published sources on French patents, for example, occupies nearly twenty pages. A scholar intending to use archival materials in France should certainly be thoroughly familiar with this report. For any student of technology, however, many helpful titles and suggestions are given throughout the article.

Hans Bohatta, *Internationale Bibliographie der Bibliographie* (Frankfurt a.M.: Klostermann, 1939–1950; 4to, 652 pp., 8 parts, paged continuously). "Technik," pp. 505–523, lists about 700 titles of journals, bibliographies, and library catalogues.

R. A. Buchanan, "Bibliographical Review . . . The History of Technology," *The Technologist,* 2, No. 2 (London, Spring 1965), 199–207. A short essay on the literature of the field in general from "a predominantly British point of view." Critical comments are to my taste a bit bland.

Bulletin Signalétique, Section 22, Histoire des Sciences et des Techniques (Centre de Documentation du Centre National de la Recherche Scientifique, 15, quai Anatole-France, Paris 7e). Of particular interest since 1958, when present section was established. Quarterly. The list of (2,500) journals examined is given in Vol. 18, No. 1 (1964), pp. i–xlix; a typical volume (19) contains 3,510 annotated entries under six broad subject headings with various subdivisions; an index was published separately. The section on technology is divided into general history of technology; tools, mechanisms, machines; communications, transport, and civil engineering; building construction and buildings; textiles and clothing; paper and printing; military technology. Many references to technology are in the various science sections.

Alan Bullock and A. J. P. Taylor, *A Select List of Books on European History 1815–1914* (2nd ed., Oxford: Clarendon, 1957; 79 pp.). Books dealing with British history are excluded; a section on "Economic and Social History," pp. 15–18, is supplemented by lists on the same subject under individual countries. Marginal for the historian of technology, but I should hesitate to call it negligible.

[Centre de Documentation]. *Documents pour l'Histoire des Techniques.* Cahier No. 1 (1961, 63 pp.); Cahier No. 2 (Oct. 1962, 90 pp.). Published by Le Centre de Documentation d'Histoire des Techniques (Conservatoire National des Arts et Métiers, 292, Rue Saint-Martin, Paris 3e). The Cahiers of this newly organized research center contain bibliographic and informative articles (e.g., in No. 2, a bibliographic study of Beau de Rochas, pioneer in theory of the internal combustion engine); annotated current bibliography of articles; book reviews; and short descriptions of organizations that are concerned with the history of technology (e.g., Verein Deutscher Ingenieure, Italian Institute for History of Technology). Very good. Reviewed in *Technology and Culture, 4* (Winter 1963) 51. Cahiers Nos. 3 ff. appear annually as one quarterly number of *Revue d'Histoire des Sciences et de leurs Applications* (1963 ff.).

Robert L. Collison, *Bibliographies Subject and National; A Guide* (2nd ed., New York: Stechert-Hafner; London: Lockwood, 1962, 185 pp.). A wide-ranging but compact annotated list of standard bibliographies and many that are little known. Collison's comment, "There is something very satisfying in handling a well-constructed bibliography," describes my reaction to his own work.

[Cooper Union]. *A Guide to the Literature on the History of Engineering Available in The Cooper Union Library* (New York: The Cooper Union, 1946 [Engineering and Science Ser. Bull. 28]). Reprint with additions (1947). A selection of 665 titles, including standard and special bibliographies, summary histories, and a few articles. Many entries are annotated. Classified and indexed. The authors of this guide made the best of a difficult situation, in which satisfactory monographs and summary histories did not exist in many fields of technology. If minor titles are listed, it is because nothing better was available. I recommend the work highly, even though it was compiled more than twenty years ago. Single copies of this guide are available from Librarian, The Cooper Union Library, Cooper Square, New York 10003.

K. A. C. Creswell, *A Bibliography of the Architecture, Arts and Crafts of Islam to 1st Jan. 1960* (Cairo: American University at Cairo Press, 1961; fo., 1,330 cols.). Distributed by Oxford University Press, London. Arranged by subjects; geographic division within subjects. Arts and crafts include arms and armour, astrolabes, bookbinding, ceramics, enamel, lacquer, glass, ivory, jewelry, paper, metalwork, stone and marble, textiles, and woodwork. Descriptive annotations. This work, begun in 1912, was carried on for three months of each of thirty-nine years, during the author's "annual visit to London" from Cairo. Total number of books and articles is 12,300. A nicely made book, in the grand manner.

Lorna M. Daniells, *Studies in Enterprise: A Selected Bibliography of American and Canadian Company Histories and Biographies of Businessmen* (Boston: Baker Library, 1957). A classified list of more than 2,000 titles of books, articles, and pamphlets of over 20 pages. Additionally, a subject index, author index, and index of periodicals.

Continued annually in *Business History Review* through 1964; expected at less frequent intervals in the future.

Wilhelm Engelmann, *Bibliotheca Mechanico-Technologica . . . über alle Thiele der Mechanischen und Technischen Künste und Gewerbe* (Leipzig, 1844, 503 pp.). A general list of books current in 1843. I have not found it particularly helpful.

Robert J. Forbes, *Bibliographia antiqua: Philosophia naturalis* (Leiden: Nederlands Instituut voor het Nabije Oosten, 1940–1950). Comprises 10,751 entries in ten parts, issued in six volumes. Pt. 1, mining and geology; pt. 2, metallurgy (3,518 items, of which nearly two-fifths are on iron); pt. 3, building materials; pt. 4, pottery, faience, glass, glazes, beads; pt. 5, paints, pigments, inks; pt. 6, leather; pt. 7, fibres and textiles; pt. 8, paper, papyrus, etc.; pt. 9, agriculture, irrigation, animals, bread and baking, etc.; pt. 10, engineering, tools, instruments, metrology, commerce, coinage, vehicles, roads, ships, lighthouses, etc. Wide-ranging, without annotation, but ready to save any investigator an enormous amount of pick-and-shovel work. *Supplement I, 1940–1950* (1952), adds 2,500 entries and a list of periodical title abbreviations (11 pages); *Supplement II, 1950–1960*, adds another 2,000 entries. A short review of the work is in *Technology and Culture*, 5 (Fall 1964), 639–640.

H. E. Haferkorn and Paul Heise, Comps., *Handy List of Technical Literature. Reference Catalogue of Books Printed in English from 1880 to 1888 Inclusive; to which is Added a Select List of Books Printed before 1880 and Still Kept on Publishers' and Jobbers' Lists* (6 vols., Milwaukee, 1889–1893). Comprehensive coverage of British and American publishers. Pt. 1, useful arts in general, 1,596 titles; pt. 2, technology of boats, ships, and war, 2,001 titles; pt. 2a, electricity and magnetism, 557 titles; pt. 3, engineering and mechanics, 2,145 titles; pt. 4, mines and mining, 949 titles. Pts. 5–6, fine arts and architecture, I have not seen.

[Harvard Guide]. Oscar Handlin et al., *Harvard Guide to American History* (Cambridge, Mass.: Harvard University Press, 1954, 689 pp.). This guide not only answers questions about what materials are available and where but, having been written with the needs of graduate students in mind, also answers questions about the principles and mechanics of historical writing. An abridgment of Samuel E. Morison's "History as a Literary Art" is included. I recommend taking the trouble to get the whole essay (Old South Leaflet, Series 2, No. 1: Old South Assn., Old South Meeting House, Boston); also reprinted in Morison's *By Land and By Sea* (New York, 1953). The **Harvard Guide** has been issued in paperback (New York: Atheneum, 1966).

Brooke Hindle, *Technology in Early America: Needs and Opportunities for Study* (Chapel Hill: North Carolina University Press, 1966, 168 pp.). An interpretive essay and extensive bibliography, pp. 29–94, on American technology to *ca.* 1850. The book was written to guide a 1965 conference sponsored jointly by Institute of Early American History and Culture, Williamsburg, Va., and Eleutherian Mills-Hagley Foundation, Wilmington, Del. A directory of museum collections by Lucius F. Ellsworth is included, pp. 95–126.

The Historical Association, London, has issued more than sixty-seven pamphlets in a series entitled "Helps for Students of History." The one I have seen is *Medieval European History 395–1500,* by R. H. C. Davis, and consists of a thirty-five-page, closely packed, critical bibliographical essay. No. 52, "The Early History of Science" (1950), and No. 59, "Business History" (1960), are listed as recent numbers in the series. The Historical Association is at 59A Kennington Park Road, London, S.E. 11.

Isis, quarterly journal of the History of Science Society (Editorial office at Smithsonian Institution, Washington, D.C. 20560). The "Critical Bibliography of the History of Science and its Cultural Influences" now is published yearly in *Isis* but appeared more often in its earlier years. The "90th Critical Bibliography" appeared in Vol. 56 (Winter 1965). While the emphasis upon technology is slight, there are many works listed that are as important to the historian of technology as to the historian of science. A fifty-year index to *Isis* is being prepared. A table of contents and index of Vols. 1–20 were published in 21 (1934), 502–698.

Askel G. S. Josephson, Comp., *A List of Books on the History of Industry and Industrial Arts* (Chicago: John Crerar Library, 1915, 486 pp.). Only books in the John Crerar Library are included, but the collection was strong when the list was compiled. Classified by subject, with an adequate index. This is a companion volume to the earlier *List of Books on the History of Science* (1911).

Eugene Lacroix, *Bibliographie des Ingénieurs, des Architectes, des Chefs d'Usines Industrielles, des Elèves des Ecoles Polytechnique et Professionelles et des Agriculteurs* (3 vols., Paris, 1863–1867). Covers *ca.* 1800–1860, in alphabetical order by name. [This note by Nathan Reingold.]

David S. Landes, *Technological Change and Industrial Development in Western Europe, 1750–1914* (*The Cambridge Economic History of Europe,* 6 [1965]), 217–601 and 943–1,007. The bibliography, pp. 943–1,007, is a tour de force that provides (1) a knowledgeable guide to the primary sources and official publications of Great Britain, France, Belgium, Germany, Switzerland, Austria, Italy, and Sweden, and (2) a list of monographs and articles for each country, arranged within each country by subject. This work has been issued separately (1968) under the title *The Unbound Prometheus*.

Henrietta M. Larson, *Guide to Business History* (Cambridge, Mass.: Harvard University Press, 1948). This valuable work lists 4,904 items, books and articles; each entry is individually and carefully annotated. Histories of industries and companies occupy about 240 pages. A well-made index, pp. 1,037–1,181, was contributed by Elsie H. Bishop. See also **Daniells**, *Studies in Enterprise* (p. 20). Larson's *Guide* has been reprinted (Boston: Canner, 1964).

Louise-Noëlle Malclès, *Les Sources du Travail Bibliographique*, Vol. 3 (Geneva and Paris, 1958, 577 pp.), is subtitled *Bibliographies Spécialisées (Sciences Exactes et Techniques)*. Concerned mainly with current literature, but bibliographies and standard historical works of particular branches of science and technology are listed.

John Neu, "The History of Science," *Library Trends* [Urbana, Illinois], 15 (Apr. 1967), 776–792. An impressively systematic survey, by a librarian, of the bibliographies of books and articles on the history of science and technology. A large number of titles are sensibly analyzed. Recommended.

[New York Public Library]. Karl Brown, *A Guide to the Reference Collections of the New York Public Library* (New York: N.Y.P.L., 1941, 416 pp.). Reprinted from the N.Y.P.L. *Bulletin,* May 1935–Feb. 1941, this admirable guide keeps attention riveted on the collections in the Library, but in so doing it suggests exciting sources that have scarcely been looked at, much less exploited. The kind of information set forth is merely suggested by the following notes. The Science and Technology Division has an "Industrial Arts History" file, some 12,000 cards. There are manuscript materials on the building of Croton Reservoir, Grand Central Terminal, Holland Tunnel, and on railroad electrification, as well as much material on the Centennial and later exhibitions. Separate lists of books, published in the Library *Bulletin,* are referred to: American interoceanic canals (1916, 90 pp.); engineering books in America prior to 1830 (see **Shaw**, p. 24); electricity (1903); aeronautics (1908); illumination (1908, 49 pp.); typewriters (1913, 13 pp.); and many others. The value of extensive subject lists, which I am sorry to observe no longer appear in the *Bulletin,* may be restricted by the absence of annotations, but nothing has been put forward to take their place. I suspect that scholars and students spend countless hours that might be saved by referring to the lists.

Newcomen Society for the Study of the History of Engineering and Technology, *Transactions* (Headquarters of the society are in Science Museum, London, S.W. 7). Starting with Vol. 2 (1921–1922) and continuing through Vol. 25 (1946–1947), an "Analytical Bibliography of the History of Engineering" was compiled by members of the society who were connected with the Patent Office. This bibliography will repay careful reading. Topical indexes to these bibliographies are in decennial indexes of *Transactions,* Vols. 1–10 (1920–1929) and Vols. 11–20 (1930–1940) but not in the *General Index to Transactions, Volumes I-XXXII, 1920–1960* (1962). (See also p. 152.)

K. J. Rider, *The History of Science and Technology. A Select Bibliography for Students* [Special Subject List No. 48] (London: The Library Association, 1967, 60 pp.). 20s. An informatively annotated list of 250 titles "for students embarking on the subject and for the general reader." The history of science occupies half the book; the history of technology is treated separately in the other half. Included are bibliographies, general histories, periodicals, biographies, and subject histories. The selection, judicious and manageable in length, exhibits an enviable knowledge of available works.

The Science Library, South Kensington, London, S.W. 7, *Bibliographical Series* (No. 1, 1930, ff.). The individual numbers are generally short lists of current works in restricted fields, not intended to provide historical background, and thus are presently of limited help to the historian. An index to Nos. 1–555 was published in New York Public Library, *Bulletin,* 46 (Aug. 1942), 707–731. An index to Nos. 1–789 is in Science Museum, *List of Science Library Bibliographical Series* (London, 1965, 32 pp.), which is available from Science Museum.

Ralph R. Shaw, *Engineering Books in America prior to 1830* (New York: N.Y.P.L., 1933). Reprinted from the *Bulletin* of the N.Y.P.L. (37 [1933], 38–61, 157–160, 209–222, 301–310, 539–555). About 700 titles are listed; gives one or more present locations of copies. Very good, but **Hindle,** *Needs* (p. 22), p. 20, warns of the erroneous impression of French influence given by presence of French books.

Henry Sotheran & Co., *Bibliotheca Chemico-Mathematica: Catalogue of Works in Many Tongues on Exact and Applied Science* (2 vols., London, 1921). Compilation of a first-rate antiquarian bookseller's catalogues. Most items are annotated. Heavily technical, starting with the earliest printed works. There are 17,397 entries and many illustrations; very good subject index. *First Supplement* was issued in 1932 (7,198 titles, arranged under 13 general subjects, no index); *Second Supplement,* in 1937 (2 vols., 22,943 titles, refined subject headings, no index); and *Third Supplement,* in 1952 (5,951 titles). An extremely valuable work. Sotheran also issued two independent volumes entitled

Catalogue of Science and Technology (No. 1, 1918, 248 pp.; No. 2, 1919, 256 pp.). (See also p. 43.)

Verein Deutscher Ingenieure, Hauptgruppe Technikgeschichte, issues a mimeographed *Rundbrief* of news and bibliography two or three (?) times a year. A typical issue lists over 100 books and articles. Address, 4 Düsseldorf 1.

G. F. Westcott and H. P. Spratt, *Synopsis of Historical Events; Mechanical and Electrical Engineering* (London: H.M.S.O. for Science Museum, 1960, 44 pp.). Includes dates of happenings in energy conversion; atomic energy; pumping, blowing, and compressing machinery; and explosives and ordnance. Before 1550, pp. 1–4; 1556–1800, pp. 5–9; 1801–1900, pp. 10–21; 1901–1950, pp. 21–31; 1950–1958, pp. 32–40. This listing is valuable because it refers to sources of further information. Nearly all the sources are secondary works, but the authors have avoided the usual error of making a list of snippets of unsupported and frequently disjointed information.

Norman B. Wilkinson, *Bibliography of Pennsylvania History* (Harrisburg: Pennsylvania Historical and Museum Commission, 1957, 826 pp.). Contains 9,121 titles, unannotated, on all aspects and periods of Pennsylvania history. Due weight is given to technical matters, and the early arts and industries of Pennsylvania are, of course, of more than parochial interest. A revised edition is being prepared.

Judith B. Williams, *A Guide to the Printed Materials for English Social and Economic History 1750–1850* (2 vols., New York: Columbia University Press, 1926). This is the only guide to economic history that I know of. Within its declared limits it is a careful, valuable work, which includes titles on roads, vehicles (including pamphlets on Gurney's and Hancock's steam carriages and the Ackermann steering linkage), rivers and canals, railways (including titles on an overhead monorail of 1823 and the question of track gauges), shipping (including docks and ship propulsion machinery), electric telegraph, steam engine (all the standard and many lesser books in English), mining and metal working, textiles, and ceramics. Reprinted, 1966.

See also
Sarton, "Charles Fremont" (p. 255).

B. Library Lists

The following lists have been selected (from library lists that I have run across) for their emphasis upon technology. It should be noted, however, that I have made no systematic search for printed catalogues of libraries whose collections are heavily historical or technical.

Within recent years, **G. K. Hall & Co.**, 70 Lincoln St., Boston, Mass. 02111, has published dozens of catalogues, photoprinted from the original catalogue cards, of collections in the United States and Western Europe. Although expensive ($20 to $70 a volume; many catalogues run to several volumes), these catalogues are vastly more useful to a scholar and vastly less costly than proposed computer-controlled systems. Only a few of these catalogues are listed in this section and under the appropriate subject-matter rubrics. A list of all titles may be obtained from the publisher.

American Society of Civil Engineers, *Catalogue of the Library* (2 vols., New York, 1900–1902). Important for listing of collections of handbooks, works on exhibitions and education, and reports, specifications, and pamphlets on civil engineering works. Contains a convenient list of British *Abridgment of Specifications,* issued around the 1860's, covering patents from 1617 onward. Vol. 2 is a supplement. I have not checked this catalogue against the **Engineering Societies Library Catalog,** following.

John Crerar Library, Chicago, *Author-Title Catalog* (35 vols.); *Classified Subject Catalog* (42 vols., including Subject Index). These catalogues are available from G. K. Hall & Co. (See above.) The John Crerar Library, which has recently been moved to the Illinois Institute of Technology, is particularly strong in technical and scientific works, both books and serials. See also **Josephson,** *A List of Books* (p. 22).

Engineering Societies Library, New York, *Classed* [sic] *Subject Catalog* (13 vols., Boston: G. K. Hall, 1963). First supplement (1 vol., 1964); second supplement (1 vol., 1965). The library has some 185,000 volumes, including many reports, pamphlets, and similar documents not readily found elsewhere. A "history" class has nearly 5,000 entries (12, 287–524). Many subject bibliographies (1, 19–667) are old enough to be of historical interest. To locate histories and contemporary works on particular subjects, see the index (Vol. 13).

Catalogue of the Books Belonging to the Library of the Franklin Institute (Philadelphia, 1847, 117 pp.). Among other entries are listed many unbound reports of railroad and canal companies, promotional societies, and mechanics' institutes.

Catalogue of the Library of the Franklin Institute of the State of Pennsylvania, for the Promotion of the Mechanic Arts (Philadelphia: Franklin Institute, 1876, 472 pp.).

Germany, Reichspatentamt, *Katalog der Bibliothek* (3 vols., Berlin, 1923). Subject listing in Vol. 1 (1,983 pp.); Vols. 2–3 (1,487, 1,512 pp.) have author-title entries.

[Goss Library]. William M. Hepburn, *A Manual of the William Freeman Myrick Goss Library of the History of Engineering and Associated Collections* (Lafayette, Ind.: Purdue University, Engineering Experiment Sta., 1947, 218 pp.). A classified list of titles of a very respectable collection whose strength lies mostly in the nineteenth century, although it contains a sprinkling of works earlier than 1800. A *Supplement* to the manual was published in 1953. There is a total of 2,200 titles in the two lists. The professional collection of Frank Gilbreth, innovator in scientific management, was given to the Goss Library in 1939; its contents are listed on pp. 179–206; the Bitting Collection on glass and on expositions is listed on pp. 157–175.

Great Britain, Patent Office, *Catalogue of the Library of the Patent Office* (2 vols., London, 1881–1883). Vol. 1, authors; Vol. 2, subjects.

Great Britain, Patent Office, *Catalogue of the Library of the Patent Office* (London, 1898). Vol. 1, authors; the intended second volume, on subjects, was never published. However, a *Supplement* to the first volume was published in 1909.

Great Britain, Patent Office Library. A series of eighteen *Guides to the Patent Office Library* was issued, commencing in 1899. Two are keys to classification systems of foreign patents; one is a guide to the search department; one is a subject list on patent law; and one is an index of periodicals in the library (298 pp.). Others are book lists:

2. Fine and Graphic Arts (including Photography), etc. (2nd ed., 1904, 374 pp.)
6. Chemistry and Chemical Technology (1901, 106 pp.)
7. Certain chemical industries (1901, 100 pp.)
9. Domestic Economy, etc. (1902, 136 pp.). On foods, beverages, etc.
10. Textile Industries . . . including the culture and chemical technology of textile fibres (1902, 128 pp.)
11. General Science, Physics, Sound, Music, Light, Microscopy, and Philosophical Instruments (1903, 184 pp.)
12. Architecture and Building Construction (1903, 164 pp.)
13. Mineral Industries (1903, 302 pp.)
14. Electricity, Magnetism, and Electro-technics (1904, 286 pp.)
15. Agriculture, etc. (1905, 424 pp.)
16. Heat and Heat Engines (excluding Marine Engineering) (1905, 200 pp.)

17. Aerial Navigation and Meteorology (1905, 64 pp.)
18. Military and Naval Arts, including Marine Engineering (1907, 304 pp.).

A "new series" of fifteen parts was published in 1908–1919, under the general title *Subject Lists. New Series:*

AA-BE. Biography, Bibliography, the auxiliary Historical Sciences, etc. (1908, 336 pp.)
BF-BL. Law of Industrial Property (Patents, Designs and Trade Marks) and Copyright (1909, 84 pp.)
BM-BZ. Fine and Graphic Arts (excluding Photo-mechanical Printing and Photography) (1914, 228 pp.)
CD-CK. Silicate Industries (Ceramics and Glass) (1914, 88 pp.)
CK_{15}-CO_{17}. Enamelling, Art Metalwork, Furniture, Costume and Hair dressing and working (1914, 70 pp.)
CO_{20}-CZ. Textile Industries and Wearing Apparel, including the culture and chemical technology of Textile Fibres (1919, 334 pp.)
FO-FR. Horology (1912, 60 pp.). There are about 600 titles in this part.
FS-GF. General Physics (including Measuring, Calculating and Mathematical instruments, and Meteorology) (1914, 196 pp.)
GG-GP. Sound and Light (including Music, Musical instruments, and general Optical instruments) (1914, 140 pp.)
WN-XN_{39}. Mineral Industries. Part 1. Geological sciences, coal mining (1912, 300 pp.)
XN_{40}-XR. Mineral Industries. Part 2. Iron manufacture, alloys and metallography (1912, 74 pp.)
XS-YH. Mineral Industries. Part 3. Metallurgy, non-ferrous and general, assaying, and fuel combustion (1912, 138 pp.)
YK-YM. Peat, Destructive Distillation, Artificial Lighting, Mineral Oils and Waxes, Gas Lighting and Acetylene (1911, 108 pp.)
YN-ZB. Chemical Technology (including Oils, Fats, Soaps, Candles, and Perfumery; Paints, Varnishes, Gum, Resins, India-rubber; Paper and Leather industries) (1911, 176 pp.)
ZC-ZQ. Chemistry (including Alchemy, Electrochemistry and Radio-activity) (1911, 218 pp.).

The former Patent Office Library is now "National Reference Library of Science and Invention (A Part of the British Museum), Holborn Division," located at 25 Southampton Buildings, London, W.C.2. M. W. Hill, Deputy Librarian, has very kindly informed me that "Our main interest is in modern science and technology; therefore, most of our holdings of works published before 1910 have been sent to other libraries. However, we have retained in the Library all patent, trademark and design publications, all bibliographies and a few other old works of special value or merit. Many of the items we have disposed of have, in fact, been sent to the British Museum unless that organisation already had a copy. With the exception of a gap from 1888 to 1907, when records of the disposals of works were not kept, we can state to which library an item was sent."

Institution of Civil Engineers, *Catalogue of the Library* (London, 1851, 228 pp.). A list of about 4,000 titles.

[Kress Library]. Harvard University, Graduate School of Business Administration, Baker Library, *Kress Library of Business and Economics Catalogue* (4 vols., Boston: Baker Library, 1940–1964): Vol. 1, through 1776, 414 pp., 7,279 titles; Vol. 2, 1777–1817, 397 pp., 7,085 titles; Vol. 3, 1818–1848, 397 pp., 7,642 titles; Supplement, through 1776, 175 pp., 2,569 titles. Arranged chronologically and alphabetically within each year. Annotated throughout; all volumes indexed. A second supplementary volume is being prepared.

Paris, Ecole Nationale des Ponts et Chaussées, *Catalogue des Livres Composant la Bibliothèque de l'Ecole* . . . (Paris, 1872, 8vo, 626 pp.). About 6,000 titles. Index, classified by subject, of a good collection of books. Strongest in civil engineering, but many titles on mechanical engineering and industrial chemistry.

C. Miscellaneous

Some of the following titles are of marginal interest in the history of technology. I have chosen to list the marginal as well as the indispensable, however, in order that the reader may judge whether his personal inspection is warranted.

[American Historical Association]. A. P. C. Griffin, *Bibliography of American Historical Societies* (2nd ed., in *Annual Report of the American Historical Association for the Year 1905*, Vol. 2, 1907; 1,374 pp., 7,537 entries). Arranged by societies in the United States and Canada. Tables of contents of periodicals and collected works are given.

[American Historical Association]. *Writings on American History* (Washington: A.H.A., annually since 1902). *An Index 1902–1940* (1956) makes the earlier issues readily usable; however, American history writings have been infrequently concerned with technology.

Lee Ash, *Subject Collections: A Guide to Special Book Collections* (2nd ed., New York: Bowker, 1961, 20,000 entries). There are eight entries under Technology, history; two under Engineering history; and a few under particular branches. However, by gauging the size and ages of collections not labeled "history," the reader may find helpful clues. A third edition was published in 1967.

Henry P. Beers, *Bibliographies in American History: Guide to Materials for Research* (Paterson, N.J.: Pageant, 1959, 487 pp.). A reprint, without change, of the 1942 edition. Few titles on history of technology; no annotations. However, there are sections on transportation, mining, industry, and so forth, and standard general works are listed.

Bibliographic Index (New York: H. W. Wilson, since 1937). Cumulated in four-year volumes. About 1,500 periodicals are represented. Very thin on history of technology.

Harry J. Carman and Arthur W. Thompson, *A Guide to the Principal Sources for American Civilization, 1800–1900, in the City of New York: Printed Materials* (New York: Columbia University Press, 1962, 630 pp.). A list of books, arranged by subject. The general field is too big for the volume, so the lists cannot be comprehensive. Most of the books are available in other cities, so it is not needed as a locating device. The book has little on technology. But see the companion volume for manuscripts, by the same authors (p. 117), and the predecessor work, Evarts B. Greene and Richard B. Morris, *A Guide to the Principal Sources for Early American History (1600–1800) in the City of New York* (2nd ed., New York: Columbia University Press, 1953).

Dissertation Abstracts: Abstracts of Dissertations and Monographs in Microform (Ann Arbor: University Microfilms). Monthly, indexed annually. A continuation of a series of annual indexes of dissertations which originated in 1934. The number of dissertation subjects that have a bearing on the history of technology has increased sharply since about 1950. All except two of the twenty-odd theses that I listed in the serial publication of this entry (*Technology and Culture, 4* [1963], 328) are to be found in **Kuehl,** *Dissertations* (p. 31). The missing entries apparently were not written in history departments: D. F. Baker, *Study of the Evolution of Industrial Engineering* (Ohio State, 1957–1958), and T. H. Robinson, *Amalgamated Society of Engineers, 1851–1892* (Chicago, 1936).

The American Historical Association publishes also a triennial *List of Doctoral Dissertations in History in Progress* (Washington, 1961, 1964, 1967 ———), and each *Annual Report* of the Association carries a list of completed doctoral dissertations.

Richard W. Hale, Jr., *Guide to Photocopied Historical Materials in the United States and Canada* (Ithaca: Cornell University Press, for A.H.A., 1961, 4to, 241 pp.). Locates about 11,000 items of U.S. and foreign origin. Of marginal utility in history of technology. Some business papers are listed.

Hester R. Hoffman, *The Reader's Adviser* (10th ed., New York: Bowker, 1964, 1,292 pp.). "An Annotated Guide to the Best in Print" over a wide spectrum of classes and subjects. Of immediate interest are chapters on tools of the book trade, bibliography, encyclopaedias, dictionaries, reference books, biography, travel, science, and history. Added reasons for owning the book are to be found in the sections on essays, drama, fiction, poetry, and philosophy. Excellent for finding quickly a "bit" of information; even better for browsing.

Norma O. Ireland, Ed., *Local Indexes in American Libraries: A Union List of Unpublished Indexes* (Boston: Faxon, 1947, 221 pp.). A list of such indexes concerned principally with the history of technology would be useful; the information in this book could provide a point of departure in compiling such a list. The following card catalogues are mentioned in this book: Boston Public Library, scientific and technical books before 1800 (3,960 cards), pictorial materials on science and technology (15,840 cards); New York Public Library, technology-biography (70,000 cards by name, file active); technology-history (12,000 cards, inactive. Same as "industrial arts history," mentioned under **N.Y.P.L.** *Guide* [p. 23].). A science-technology subject index of 360,000 cards, 1858–1929, said to be in United Gas Improvements Company Library, Philadelphia, turned out to be a ghost; nobody I could find at U.G.I. or Franklin Institute in 1961 had ever heard of it. A card file on "history" in the Science and Technology Division of the Cleveland Public Library, called to my attention by an anonymous reader, was not listed by Miss Ireland.

Frances B. Jenkins, *Science Reference Sources* (3rd ed., Champaign, Ill.: Illini Union Bookstore, 1962, 135 pp.). A list of references for a course in a library school. Emphasis is, as it should be, upon current catalogues and reference services; however, sections on history, biography, organizations, patents, research reports, and serials are provided for the several branches of science and technology. Of particular interest is "Engineering Sciences," pp. 102–116, which covers engineering and not the engineering sciences properly speaking. **Winchell** (p. 33) is of course a much more comprehensive work.

Warren F. Kuehl, *Dissertations in History, An Index to Dissertations Completed in History Departments of United States and Canadian Universities 1873–1960* (Lexington: University of Kentucky Press,

1965, 249 pp.). About 7,600 titles are listed, few in technological history. An annual list of dissertations completed (see p. 30) should be consulted to supplement Kuehl's index.

Library of Congress, *A Guide to the Study of The United States of America* (Washington: G.P.O., 1960, 4to, 1,193 pp.). This guide is valuable mainly for collateral works. Annotations are unusually full and informative. It is weak on technology.

Robert W. Murphey, *How and Where to Look It Up. A Guide to Standard Sources of Information* (New York: McGraw-Hill, 1958, 721 pp.). Despite its title, this is a serious work that will repay a session of browsing.

R. A. Peddie, *Subject Index of Books Published before 1880* (London, 1933; 2nd series, 1935; 3rd series, 1939; new series, 1948). A general index of about 150,000 titles in all; many early works. An appreciable number of listings will be found under such rubrics as bridges, clocks, rockets, sewage, obelisks, weights and measures, telegraph, steam road carriages, and others.

Dorothea D. Reeves, Comp., *Resources for the Study of Economic History: A Preliminary Guide to Pre-Twentieth Century Printed Material in Collections Located in Certain American and British Libraries* (Boston: Baker Library, 1961, 62 pp.). A report on about thirty-five libraries and special collections pertaining to this field, whose source materials are often of interest to technological historians. Hours, regulations, loan and reproduction policies, a short statement on the nature of collections, and notice of guides and publications are given for nearly every institution. Very good. A revised edition is being prepared.

Technical Book Review Index (Pittsburgh: Carnegie Library for Special Libraries Assn., monthly). Between 100 and 150 books are intelligently reviewed each month, but only one or two are history titles. For example, Vol. 30 (1964) contains reviews of only twelve books on technical history. Except for annual author index, there is no means of discovery except page-by-page search.

Tekniska Museet, Stockholm. An index in the museum's library of 250,000 cards on the history of industry and technology is described briefly in *Daedalus* (Stockholm, 1949), pp. 116–119.

John L. Thornton and R. I. J. Tully, *Scientific Books Libraries and Collectors; a Study of Bibliography and the Book Trade in Relation to Science* (2nd ed., London: The Library Assn., 1962, 406 pp.). All sorts of interesting information on science is to be found here, but technology was excluded.

GENERAL BIBLIOGRAPHIES AND LIBRARY LISTS

A. J. Walford, Ed., *Guide to Reference Material*, Vol. 1, Science and Technology (London: The Library Assn., 1966, 483 pp.). Three thousand main entries arranged by subjects, of encyclopaedias, bibliographies, manuals, directories, etc. This guide is intended to lead to sources of current information; thus, there is little attention given to history. Vol. 2, Social and Historical Sciences, forthcoming.

Constance M. Winchell, *Guide to Reference Books* (8th ed., Chicago: American Library Association, 1967, 741 pp.). A standard work devoted to helping readers find bibliographies and basic works on many subjects. The coverage is comprehensive and remarkably complete. Sections on bibliographies and library catalogues, dictionaries, periodicals, newspapers, government publications, encyclopaedias, and biography are followed by subject-matter divisions which include "History and Area Studies," pp. 463–524, and "Pure and Applied Sciences," pp. 525–610. The history of science and technology will be found in the sciences division. The annotations, broadly informative, are models of clarity. This work describes in detail the bewildering series of printed catalogues of the Library of Congress, British Museum, and Bibliothèque Nationale. Supplements followed the seventh edition at three-year intervals. Perhaps the same policy will be followed in keeping the eighth edition current.

III. DIRECTORIES OF TECHNICAL, ACADEMIC, AND BUSINESS ORGANIZATIONS

(Histories of societies will be found in Part XII.)

[American Association for State and Local History]. *Directory [of] Historical Societies and Agencies in the United States and Canada 1965–1966* (Nashville: The Assn., 1965; 140 pp., index). Gives name, location, and director; indicates existence of museum, library, and publications. Published biennially. Available from the association at 132 Ninth Avenue, North, Nashville, Tenn. 37203.

Deutsche Technisch-Wissenschaftliche Forschungsstätten. Teil I, Die Technisch-Wissenschaftlichen Vereine (Berlin: VDI-Verlag, 1930, 135 pp.); *Teil II, Die Technisch-Wissenschaftlichen Forschungsanstalten* (1931, 445 pp.). Pt. 1 is a directory of technical, industrial, and scientific societies, giving founding dates but only general information on publications; pt. 2 lists institutes of research.

Deutscher Verband Technisch-Wissenschaftlicher Vereine, *Handbuch* (Düsseldorf, 1960, 100 pp.). A summary of organization and activities of West German engineering and scientific societies. Listed in *Directories in Science* (p. 36), No. 178.

Encyclopedia of Associations (4th ed., 3 vols., Cleveland: Gale Research Co., 1964). Vol. 1, "National Organizations," has sections for trade, business, government, scientific, engineering, technical, and educational associations; Vol. 2, "Geographic and Executive Index"; Vol. 3, "New Associations," a "bi-monthly updating service." An expensive but comprehensive directory.

Winifred Gregory, Ed., *International Congresses and Conferences 1840–1937. A Union List of Their Publications Available in Libraries of the United States and Canada* (New York: H. W. Wilson, 1938; fo., 229 pp.).

Sara E. Harcup, Comp., *Historical, Archaeological and Kindred Societies in the British Isles* (London: University of London, Institute of Historical Research, 1965, 53 pp.). About 900 societies are listed alphabetically; a geographical section lists societies by county; a short (3 pp.) subject list gives some indication of special subjects of inquiry of selected societies. This appears to be the first systematic listing of this kind in the United Kingdom.

Raymond Irwin and Ronald Staveley, *The Libraries of London* (2nd ed., London: The Library Assn., 1961, 332 pp.). Read before visiting London.

Etna M. Kelley, *Business Founding Date Dictionary* (Scarsdale, N.Y., 1954). Nine thousand entries: alphabetical and chronological listings. This directory "was designed primarily for those who wish to cultivate the anniversary market." Slight.

Anthony T. Kruzas, Ed., *Directory of Special Libraries and Information Centers* (Detroit: Gale Research Co., 1963, 767 pp.). Lists and describes very briefly about 10,000 collections in the United States and Canada. Subject index, pp. 737–767.

[Library of Congress]. *Library Reference Facilities in the Area of the District of Columbia* (7th ed., Washington: Library of Congress, Reference Dept., Loan Division, 1966). Descriptive information is included on more than 200 libraries. The scholar who expects to work in Washington can hardly avoid having his horizons extended if he will but peruse this guide.

Library of Congress, International Organizations Section, *World List of Future International Meetings*, Pt. I [science, technology, agriculture, medicine] (monthly). Subject, sponsor, and geographic indexes are in each issue but were not cumulated in the bound volumes I have seen.

A list of "periodic international congresses" is given in **World List of Scientific Periodicals** (p. 142), Vol. 3, pp. 1,789–1,824.

A chapter on international congresses is in **Sarton, Guide** (p. 17), pp. 290–302.

Library of Congress, Science and Technology Division, Reference Dept., *Directories in Science and Technology. A Provisional Checklist* (Washington: Library of Congress, 1963, 65 pp.). An annotated list, arranged geographically, of 331 titles of directories in many countries throughout the world; the directories are generally of institutions or individuals who are involved in science or engineering; some industrial directories.

DIRECTORIES   37

Katherine O. Murra, *International Scientific Organizations: a Guide to their Library, Documentation, and Information Services* (Washington: Library of Congress, 1962, 808 pp.). Reports on 449 organizations, giving description of facilities and services, publications 1959-date, and a few data on history, membership, management, etc.

National Academy of Sciences–National Research Council, *Scientific and Technical Societies of the United States and Canada* (7th ed., Washington, 1961 [Pub. No. 900]). Lists current information on 1,597 U.S. and 239 Canadian societies. Although some older societies are difficult to locate unless the current name is definitely known, this guide will tell when a society was founded, its purpose, membership, and current publications. The index is less than satisfactory.

Eloise G. ReQua and Jane Statham, *The Developing Nations: A Guide to Information Sources Concerning Their Economic, Political, Technical and Social Problems* (Detroit: Gale Research Co., 1965 [Management Information Guide: 5], 339 pp.). Recent (1950–present) literature in English can be located through some 500 annotated entries. Sections on Bibliographies on Underdeveloped Areas, pp. 223–230; Directories, pp. 233–235; Periodicals, pp. 239–248; Agencies and Institutions Administering Development, pp. 251–267.

Scientific and Learned Societies of Great Britain. A Handbook Compiled from Official Sources (60th ed., London: Allen & Unwin, 1962). Formerly *Yearbook of Scientific and* . . . . Comparable to but not as comprehensive as the N.A.S.–N.R.C. handbook in America. See particularly pt. VII, Engineering and Architecture, pp. 109–126.

Seeger and Guernsey, *Cyclopaedia of the Manufacturers and Products of the United States* (1892; 902 pp. of listings + 250 pp. advertising, illustrated). Useful for establishing which companies were active in 1892. This is a precursor of the *Thomas Register* and similar current analytical directories of manufacturers. There were others during the nineteenth century: *Commercial Directory* (Philadelphia, 1823, 283 pp.), was perhaps the earliest; J. M. Bradstreet, *Gazeteer of the Manufacturers and Manufacturing Towns of the United States* (New York, 1866, 172 pp.), is too general to be very helpful. A good discussion of business directories is in **Larson**, *Guide to Business History* (p. 23), pp. 845–854.

Union des Associations Internationales, *Les Congrés internationaux de 1681 à 1899. Liste complète* (Brussels: U.A.I., 1960, 76 pp.). A list of 1,400 congresses. The first, in 1681, was a medical congress in Rome; the next was in 1815, the next, 1838; 23 before 1851; 240 before 1876.

Union des Associations Internationales, *Les 1798 organisations internationales fondées depuis le Congrès de Vienne* (Brussels: U.A.I., 1957, 232 pp.). The Congress of Vienna was in 1815. An English version was also published. See also *World List of Future International Meetings* (p. 36).

Yearbook of International Organizations (9th ed., Brussels: Union of International Associations, 1962, 1,562 pp.). In English. According to *Directories in Science* (p. 36), this yearbook supplies reasonably detailed descriptions of 1,722 governmental and nongovernmental organizations. Indexed by subjects, abbreviations of organizations' names, and geographical locations. The subject index is given also in French. A French edition is issued in alternate years.

Willing's Press Guide (London: Willing's Press Service, annual). List of newspapers and other periodicals of Great Britain and many other countries, including the United States, giving year of establishment, when published, price, and publisher's name and address. The eighty-seventh edition appeared in 1961.

World of Learning (18th ed., London: Europa Publications, 1967–1968, 1,690 pp.). Addresses and founding dates of academies, (7,500) learned societies and research institutes, (5,000) libraries and museums, and (6,000) universities throughout the world. Names of responsible officials are given for nearly every entry. Issued annually.

IV. EARLY SOURCE BOOKS AND MANUSCRIPTS
(before 1750)

A. Bibliographies, Guides, and Descriptions of Books

There is an impressive body of important scholarly work having to do with Renaissance and early modern technical books and manuscripts already in existence. Several leading scholars in the history of technology are interested principally in the works of this period; the field is an active one, and it holds great promise for scholars who are prepared to read at least three languages. Nevertheless, there are many ideas and insights that can be gained from these works through translations and intelligent summaries in the reader's language. Unfortunately, there is no guide in English that addresses itself to the problem of providing an entry to the field for one who is neither a linguist nor thoroughly versed in the tools and procedures of scholarship.

Keller's *A Theatre of Machines* (p. 41) provides an illustrated overview of the early printed books. The books of **Gille** (p. 10) and **Parsons** (p. 11) will provide a context for them, and the strikingly original and fresh work being done by **Reti** (p. 42) will furnish suggestions enough to engage the best efforts of a generation of scholars. Beck's pioneering work of 1899 (p. 40), which describes and interprets many mechanical and hydraulic devices, generally devotes a chapter to each of the books described.

The series of metallurgical books that have been annotated and translated jointly by Smith and the classicists Gnudi, Sisco, and Hawthorne are of great and lasting value.

Abell, Leggat, and Ogden, *A Bibliography of the Art of Turning* (p. 278), lists only sixteen titles published before 1750, but bibliographic and descriptive information is very full for these.

R. C. Anderson, "Early Books on Shipbuilding and Rigging," *Mariner's Mirror, 10* (Jan. 1924), 53–64. Critical essay on European books from 1536 to 1720.

R. C. Anderson, "Eighteenth Century Books on Shipbuilding, Rigging, and Seamanship," *Mariner's Mirror, 33* (1947), 218–225.

Theodor Beck, *Beiträge zur Geschichte des Maschinenbaues* (Berlin, 1899; 559 pp., 806 figs.). Profusely illustrated summaries and commentaries on Hero, Pappus, Vitruvius, Frontinus, and Cato (87 pp.), Biringuccio, Agricola, Cardano, Besson, Ramelli, Lorini, Porta, Zonca, Turriano, Zeising, Fontana, Caus, and Verantius (260 pp.); a few manuscript sources (23 pp.); and three separate chapters on Leonardo (140 pp.). Parts of this book are explained in English in **Wolf**, *16th and 17th Centuries* (p. 14).

J. C. Brunet, *Manuel du libraire et de l'amateur de livres* (6 vols., Paris, 1860–1865; Supplement by P. Deschamps and G. Brunet, 2 vols., Paris, 1878–1880). A standard catalogue that is frequently useful for early works.

Maurice J. D. Cockle, *A Bibliography of English Military Books up to 1642 and of Contemporary Foreign Works* (London, 1900). Lists 166 English and about 450 foreign titles. For a checklist of holdings, see **Spaulding and Karpinski** (p. 43).

Horst de la Croix, "The Literature on Fortification in Renaissance Italy," *Technology and Culture, 4* (Winter 1963), 30–50. A critical bibliographical essay. Approximately 100 titles are listed of manuscript and printed works on military, architectural, and mechanical subjects. See also **Promis**, *Biografie* (p. 82).

E. Darmstaedter, *Berg-, Probier-, und Kunstbüchlein* (Munich, 1926, 111 pp.). According to **Besterman** (p. 18), this work describes eighty titles.

Bern Dibner, *Heralds of Science as Represented by Two Hundred Epochal Books and Pamphlets Selected from the Burndy Library* (Norwalk, Conn., 1955). Notes on early works in sections devoted to astronomy, chemistry, electricity, general science, metals, mathematics, medicine, physics, and technology.

John Ferguson, *Bibliotheca Chemica* (2 vols., Glasgow, 1906). Heavy with alchemical and pharmaceutical books, but the serious student of chemical technology before 1800 will wish to draw on the detailed annotations.

John Ferguson, "Some Early Treatises of Technological Chemistry," *Philosophical Society of Glasgow, Proceedings, 19* (1888), 126–159; *25* (1894), 224–235; *43* (1911), 232–258; *44* (1912), 149–189. This series supplements the author's *Bibliographical Notes* (p. 15).

Johann G. T. Grässe, *Trésor de livres rares et précieux; ou, Nouveau dictionnaire bibliographique* (7 vols. in 8, Berlin, 1922; orig. ed., 1859–1869). Over 100,000 books of all kinds are listed. A general reference work, not particularly recommended for browsing.

R. S. Hartenberg and J. Denavit, "Men and Machines, an Informal History," *Machine Design, 28* (1956), May 3, 74–82; June 14, 101–109; July 12, 84–93. Reprinted 1956. About thirty illustrations, some standard but many pertaining to the particular interests of the authors, who are important contributors to the modern theory of kinematics of mechanisms. Annotated bibliography, chiefly of modern works about early technology.

E. Wyndham Hulme, "Introduction to the Literature of Historical Engineering to the Year 1640," Newcomen Society, *Transactions, 1* (1920–1921), 7–15. This article surveys the field briefly. A list of books includes, among others, twenty-eight titles on surveying and mathematical instruments, 1531–1636, and a number of modern works. Reprinted in *The Engineer, 130* (Nov. 12, 1920), 483–485.

A. G. Keller, *A Theatre of Machines* (London: Chapman & Hall, 1964; 115 pp., including 52 pls.). Reproduced is a selection of plates (1570–1630) from Ramelli, Besson, Strada, Branca, Zonca, Scappi, and Verantius. A short (10 pp.) introduction and a caption (100–400 words) for each plate, intended for the general reader. No bibliography; no scholarly apparatus; but the author is one of the best-informed scholars of this subject. Reviewed in *Technology and Culture, 7* (Fall 1966), 527–528.

Edmond R. Kiely, *Surveying Instruments: Their History and Classroom Use* [National Council of Teachers of Mathematics, Nineteenth Yearbook] (New York: Columbia University, Teachers College, 1947, 411 pp.). History of instruments and surveying techniques from Egyptian times through the seventeenth century, pp. 1–237; bibliography of 557 items, including well over 100 books published during the sixteenth and seventeenth centuries.

Arnold C. Klebs, "Incunabula Scientifica et Medica," *Osiris, 4* (1938), 1–359. Except for Polydore Vergil, *De Inventoribus Rerum* (Venice, 1499), it is difficult to point to any nonmedical technological incunabula. This short-title list, therefore, is of limited immediate concern to most students of the history of technology.

M. Koch, *Geschichte und Entwicklung des bergmännischen Schrifttums* (Goslar: Herman Hübener Verlag, 1963, 176 pp.). A basic work, calling for translation. One chapter is devoted to medieval works on mining and metals and one to each century, sixteenth through the nineteenth. The bibliography of 501 titles of Quellenschriften lists over 200 earlier than 1750.

Jacob Leupold, *Prodromus Bibliothecae Metallicae*, corrected, continued, and augmented by Franz Bruckmann (Wolfenbüttel, 1732). According to **Besterman** (p. 18), this book gives some 1,750 titles of earlier works. John Ferguson, in his *Bibliotheca Chemica*, notes that it is "a good guide to the literature at that time." I have not seen a copy.

[Parsons Collection]. Karl Brown, "Catalogue of the William Barclay Parsons Collection," N.Y.P.L., *Bulletin*, 45 (1941), 95–108, 585–658. Reprinted (1941, 108 pp.). **William B. Parsons** (1859–1932), an American engineer, was author of *Engineers and Engineering in the Renaissance* (p. 11). His collection consisted of early technical works from 1485 onward, railroad and canal items (many obscure but valuable pamphlets), and manuscripts of Robert Fulton. Listed under "general engineering," the early technical works range from mathematics through machines to architecture. Three titles in the fifteenth century; thirty-three in the sixteenth; sixteen in the seventeenth. Full bibliographical information, but no annotations.

J. W. Perry, *Cogs in the Wheel: One Hundred Books from the Llewellyn Powell Collection of Early Technology* (Durban: University of Natal Library, 1961). A majority of the titles are classics of engineering and technology. Thirty-five were published before 1750. Short annotations for most titles. Many illustrations, poorly reproduced. Intended as a "library handlist." Reviewed in *Technology and Culture* (Winter 1964), pp. 133–134.

Ladislao Reti, "Francesco di Giorgio Martini's [1439–1501] Treatise on Engineering and Its Plagiarists," *Technology and Culture*, 4 (Summer 1963), 287–298 + 13 pp. illustrations. An impressive and significant pioneering study of manuscript precedents for certain machine illustrations published in the sixteenth to nineteenth centuries. Cites a 1475 manuscript of Martini that has marginal comments added by Leonardo da Vinci. The drawings of Martini, published by Strada (1617) and Zonca (1607), appear in books of Zeising (1612), Böckler (1661), and in Leupold (1724), Switzer (1729), and Borgnis (1818).

Gottfried Rosenthal, *Literatur der Technologie* (Berlin, 1795, 420 pp.). Appears at end of Vol. 8 of Johann K. G. Jacobsson, *Technolo-*

gisches Wörterbuch, Vols. 1–4 (Berlin, 1781–1784); Supplement Vols. 5–8 (1793–1795). The supplementary volumes were compiled by Rosenthal. The *Literatur der Technologie* lists early bibliographies and works on mining, machinery, raw materials, etc. This title was located through the patience and kindness of H. Drubba of Hannover, Germany. Reprinted, with foreword by Carl Graf von Klinckowstroem (listed in Dawson [Farnham] reprint catalogue 172).

George Sarton, *Introduction to the History of Science* (3 vols. in 5, Baltimore, 1927–1948). Vol. 1, From Homer to Omar Khayyam; Vol. 2, From Rabbi Ben Ezra to Roger Bacon; Vol. 3, Science and Learning in the Fourteenth Century. Occasional notes on technology, always with bibliographic information.

Science Museum, *Historic Books on Machines* (Science Museum [London] Book Exhibition No. 2, 1953, 28 pp.). About forty titles before 1750, briefly annotated. Reprinted, 1963.

Science Museum, *Historic Books on Mining and Kindred Subjects* (Science Museum [London] Book Exhibition No. 4, 1960). Sixty-two titles from Agricola (1556) to 1784, annotated.

Christian Ehrenfr. Seyfferts, *Bibliotheca metallica, oder bergmannischer Büchervorrath* (Leipzig: Zuckel, 1728). Cited in Rosenthal (p. 42). Listed also in Bestermann (p. 18) under Seiffert.

Cyril Stanley Smith, "Some Important Books in the History of Metallurgy . . . The Eisenman Memorial Collection," *Metals Review* [A.S.M.], 36 (Sept. 1963), 10–12, 14. Annotated list of thirty-three books displayed in connection with Sorby Centennial celebration in Cleveland in Oct. 1963. Fourteen of the books were printed before 1750.

Sotheran, *Bibliotheca Chemico-Mathematica* (p. 24). A more comprehensive and informative bibliography of early works than any other that I know of. The *Second Supplement* (2 vols., 1937), arranged by subjects, is the best place to start. The lists noted here generally include works printed before the end of the eighteenth century. See: Mining and Metallurgy, [title nos.] 13,960–14,277; General Engineering, 15,472–16,359; Early Machinery, 18,758–18,851; Navigation, Naval Architecture and Engineering, 19,873–20,090; Military Science, 20,918–21,202. Fully annotated. See also 127 plates from early works reproduced in Vols. 1–2 of the original work (1921).

Thomas M. Spaulding and L. C. Karpinski, *Early Military Books in the University of Michigan Libraries* (Ann Arbor, 1941; 45 pp. + plates of more than 120 title pages). See also Cockle (p. 40).

Margaret B. Stillwell, *Incunabula and Americana, 1450–1800. A Key to Bibliographical Study* (New York: Columbia University Press, 1931, 483 pp.).

Lynn Thorndike, *History of Magical and Experimental Science* (8 vols., New York: Columbia University Press, 1923–1958). Describes books of technology from the time of Vitruvius and Hero forward, to the end of the seventeenth century. Vol. 5, Chapter 16, is devoted to Cardano; Chapter 17, "Three Technologists: Taisnier, Besson, and Palissy"; Vol. 7, Chapter 21, "Artificial Magic and Technology," provides a context for the seventeenth-century books.

Usher, *A History of Mechanical Inventions* (p. 7). In the second edition, a brief but good short-title list is given: "Treatises on Pure and Applied Mechanics to 1740," pp. 435–437.

Wellcome Historical Medical Library, *A Catalogue of Printed Books in the Wellcome Historical Medical Library. I. Books Printed before 1641* (London, 1962, 407 pp.); *II. Books Printed from 1641 to 1850 A-E* (London, 1966, 540 pp.). Further volumes may be expected.

See also
Russo, *Bibliographie* (p. 17).
Wolf, *16th and 17th Centuries* (p. 14).
Klemm, *Technik* (p. 5).

B. Translations and Editions of Books

This list is highly selective. Perhaps each reader will extract his own circle of titles from the several bibliographies listed in the preceding section.

The distinction between books and books made from manuscripts is not clear-cut. Generally, manuscript works that have appeared in printed form over a period of a century or so I have listed as books.

R. d'Acres's *The Art of Water Drawing* [1659], Introduction by Rhys Jenkins (Cambridge: Heffer, 1930). Issued as Newcomen Society Extra Publication No. 2. An early version of the steam pump.

Georgius Agricola, *De Re Metallica* [1556], Herbert C. and Lou Henry Hoover, Trans. (London, 1912; reprinted New York: Dover, 1950). The best-known early work on mining and metallurgy. Exhaustively annotated. Numerous bibliographical suggestions.

Johann Comenius, *Orbis Sensualium Pictus* (3rd ed., London, 1672, 376 pp.). A schoolbook, in English, in which trades are depicted and from which ten pages are reproduced in J. Paul Hudson, *A Pictorial Booklet on Early Jamestown Commodities and Industries* (Williamsburg, Va.: 350th Anniversary Celebration Corp., 1957.) A facsimile edition may be expected in 1968.

John T. Desaguliers, *A Course of Experimental Philosophy* (2 vols., London, 1734–1744; numerous plates). A rich compendium of mechanical knowledge, colored but not vitiated by the author's opinions and Weems-like fables. Vol. 1 has plates of the first flanged railroad wheel; Vol. 2 has plates of the Marly Machine and steam engines.

Bern Dibner, *Moving the Obelisks* (Norwalk, Conn.: The Burndy Library, 1950). A handsome monograph based on Domenico Fontana, *Della Transportatione dell' Obelisco Vaticano* (Rome, 1590), from which several plates are reproduced.

Frontinus, *The Two Books on the Water Supply of the City of Rome* [ca. 100 A.D.], Clemens Herschel, Trans. (2d ed., New York, 1913).

Otto von Guericke, *Experimenta nova* (1672). Facsimile ed. (Aalen, Germany: O. Zellar, 1962). The Magdeburg hemispheres, now in Deutsches Museum, were illustrated in this book.

[Hero of Alexandria] Erone Allessandrino, *Degli Automati ovvero Macchine Semoventi* (Florence: Olschki, 1962). Facsimile reproduction of a work translated by Bernardino Baldi and published in Venice in 1589; a new introduction by Renato Teani.

Jacob Leupold, *Theatrum Machinarum* (10 vols., Leipzig, 1724–1739). This well-known work, which James Watt is said to have learned German in order to read, can be confusing if the entire set is not present or if a definitive listing such as Brunet's is not available. I have never seen an entire set, and I have been unable to determine the various reprintings; nevertheless, I have inspected the first six volumes, Mrs. Robert Multhauf has kindly supplied details on the last four volumes, and I have seen two volumes that were reprinted in 1774 from the original plates. Leupold (1674–1727) died before the series had been completed; Joachim Ernst Scheffler and others were responsible for the last two or three volumes. The entire work contains about 525 good engravings. Variously bound, the series can be divided into ten parts:

    I. Theatrum Machinarum Generale: Schauplatz des Grundes Mechanischer Wissenschaften (1724; 264 pp., 71 plates)
    II. Theatrum Machinarum Hydrotechnicarum: Schauplatz der Wasserbaukunst (1724; 184 pp., 51 plates)

III. Theatrum Machinarum Hydraulicarum: Schauplatz der Wasserkünste, Tomus I, Erster Theil (1724; 172 pp., 53 plates)
IV. Theatrum Machinarum Hydraulicarum: Schauplatz der Wasserkünste, Tomus II, Anderer Theil (1725; 165 pp., 54 plates)
V. Theatrum Machinarium [sic]: Schauplatz der Heb-Zeuge . . . Waltzenwerk, Wagen, Heb-Laden, Haspel, Erd-Winden, Kraniche, Flaschenzügen, Rader- und Schraubenwerk (1725; 164 pp., 56 plates)
VI. Theatrum Staticum Universale [4 parts, 1726]: Schauplatz der Gewicht-Kunst und Waagen (92 pp., 19 plates); Schauplatz der Wissenschaft und Instrumenten zum Wasser-Wägen (44 pp., 7 plates); Schauplatz der Machinen zu Abwiegung und Beobachtung aller vornehmsten Eigenschafften der Lufft (72 pp., 23 plates); Schauplatz von Wasser- oder Horizontal-Waagen (24 pp., 8 plates)
VII. Theatrum Pontificiale, oder Schauplatz der Brücken und Brücken-Baues . . . (1726; 153 pp., 57 plates)
VIII. Theatrum Arithmetico-Geometricum: Schauplatz der Rechen- und Mess-Kunst (1762 [sic]; 200 pp., 43 plates). Brunet dates this volume 1727; the title page mentions 45 plates.
IX. Theatri Machinarum supplementum: Zusatz zum Schauplatz der Machinen und Instrumentum (1739; 100 pp., 40 plates); followed by:
Allgemeines Register der vornehmsten Sachen . . . (a general index of the whole series). A bibliography completes this volume.
X. Theatrum Machinarum Molarium, oder Schauplatz der Mühlen-Bau-Kunst . . . von Johann Matthias Beyern und Consorten (Leipzig und Rudelstadt, 1735). Teil 1, 125 pp., followed by: Real Register, oder compendieuses Mühlen Lexicon, über die Kunst-Wörter, so an vorstehenden Machinen vorkommen (a glossary of this volume only), followed by:
Theil 2, Kern des Mühlen-Rechts (206, 49 pp., 43 plates).

Loeb Classical Library (Cambridge, Mass.: Harvard University Press), comprises about 450 volumes of English translations from the Greek and Latin. The utility of these works as technological source material is demonstrated by—among others, of course—R. J. Forbes in his writings on ancient technology.

Venturus Mandey and Joseph Moxon, *Mechanick-Powers: or the Mistery of Nature and Art Unvail'd* (London, 1696). Machines and devices, many drawn from French publications.

Joseph Moxon, *Mechanick Exercises, or the Doctrine of Handy-Works* (3rd ed., London, 1703). On smithing, joinery, turning, bricklaying, and so forth, issued earlier in parts. Tools of the trades are illustrated.

Joseph Moxon's *Whole Art of Printing* (1683–1684) has been reprinted recently with excellent annotations: Herbert Davis and Harry Carter, Eds. (Oxford University Press, 1958).

Giambattista della Porta, *Natural Magick* (1589; facsimile of English translation, 1658), preface by Derek J. de S. Price (New York, 1957).

Hans Sachs, *Das Ständebuch, 114 Holzschnitte von Jost Amman mit Reimens von Hans Sachs* (Leipzig, 1934; 134 pp., 12mo), Insel-Bucherei Nr. 133. Excellent reproductions of Jost Amman's illustrations of trades of about 1570. For a short description of the work, see *Technology and Culture*, 6 (Winter 1965), 110–111. Theodore K. Rabb is preparing an annotated edition for publication by The M.I.T. Press.

Cyril Stanley Smith, in collaboration with classicists, has published English translations of the following works, chiefly but by no means exclusively concerned with mining and metallurgy: Vannoccio Biringuccio's *Pirotechnia* of 1540 (with Martha Teach Gnudi, 1942, 1959); *Bergwerk- und Probierbüchlein* of 1524 (with Anneliese G. Sisco, 1949); Lazarus Ercker's *Treatise on Ores and Assaying* of 1580 (with A. G. Sisco, 1951); R. A. F. de Réaumur's *Memoirs on Steel and Iron* of 1723 (with A. G. Sisco, 1956); Theophilus's *On Divers Arts* of the twelfth century (with John G. Hawthorne, 1963). The footnotes and bibliographies in these volumes are full of leads to works in this and a number of allied fields.

Jacob de Strada, *Künstlicher Abriss allerhand Wasser, Wind, Ross, und Hand-Mühlen* (Frankfurt, 1617–1618). Alex Keller has kindly reduced my confusion by identifying this work as the publication, by a grandson, of a book of machines purported to be the work of his grandfather. The grandfather, born in Mantua, lived in Vienna and Prague. Keller also pointed out that other useful biographical and bibliographical information on both van der Straet and Strada can be found in Thieme and Becker, *Allgemeine Lexikon der Bildenden Kunstler* (37 vols., Leipzig, 1907–1950), which title is described in entry BE71 of **Winchell,** *Guide to Reference Books* (p. 33).

[Johannes Stradanus], *"New Discoveries," the Sciences, Inventions and Discoveries of the Middle Ages and the Renaissance as represented in 24 engravings issued in the early 1580s by Stradanus* (Norwalk, Conn.: Burndy Library, 1953). Loose plates, nicely reproduced, of clockmaker, foundry, print shop, distillery, silk culture, horse harness, oil mills and presses, sugar boiling, oil colormaking, spectacle-maker, armor polishing, copper engraving, windmill, water mill, and so forth. Stradanus was the latinized name of Jan van der Straet, a Flemish engraver who spent most of his life in Florence. He was also known in Italy as Giovanni della Strada.

Sung Ying-Hsing, *T'ien-Kung K'ai-Wa. Chinese Technology in the Seventeenth Century,* E-Tu Zen Sun and Shiou-Chuan Sun, Trans. (University Park: Penn. State University Press, 1966; 372 pp., illustrated). In the words of a review in *The Times Literary Supplement,* Sept. 8, 1966, this is a translation, by a mineralogist and his scholarly wife, of a 1637 book "which was a full-dress study, charmingly written, of the industrial technology of his culture." The reviewer warns of errors in the translation of terms outside the field of mineralogy but is duly grateful for a book which, "noble as it is, is a first approximation." A bibliography of Chinese sources, pp. 311–320, is followed by one on works in English, mostly modern and classified by subject, pp. 321–338. Books giving present-day understanding and books and articles on historical themes are listed. Reviewed by K. Yabauti, editor of Japanese translation of the same work, in *Technology and Culture,* 8 (Jan. 1967), 92–94.

Emanuel Swedenborg, *Regnum Subterraneum* (Stockholm, 1734). An abridged translation, entitled *Traité du Fer,* forms a section of Bouchu's *Art des Forges* (Paris, 1761) in the Académie des Sciences, *Descriptions des Arts et Métiers.* Cf. Herbert Dingle, "Scientific Work of Emanuel Swedenborg" (*Endeavour,* 17 [July 1958], 127–132).

Nicoló Tartaglia, *Quesiti et Inventioni Diverse* (Facsimile of the 2nd ed. of 1554, A. Masiotti, Ed.) (Ateneo di Brescia, 1959). Reviewed by A. G. Keller in *History of Science,* 2 (1963), 155–159.

Marten Triewald's Short Description of the Atmospheric Steam Engine, Published at Stockholm, 1734, Carl Sahlin, Trans. (Cambridge: Heffer, 1928). Issued as Newcomen Society Extra Publication No. 1.

Faustus Veranzio, *Machinae novae (ca.* 1595) (Munich: Heinz Moos Verlag, 1965). A facsimile edition. A copy of the original work was sold in 1964 or 1965 for $7000.

Vitruvius, *The Ten Books on Architecture* [*ca.* 1 A.D.], Morris H. Morgan and A. A. Howard, Trans. (Cambridge, Mass., 1914; New York, 1960). For editions, see Bodo Ebhardt, *Die Zehn Bücher der Architektur des Vitruv und ihre Herausgeber* (Berlin, 1918; Ossining, N.Y., 1961).

Wang Cheng, *Illustrated Unusual Western Machines* (3 vols., 1627). In Chinese; described in Lin Hsien-Chou, "Wang Cheng and the First Book on Mechanical Engineering in China" (*International Journal of Mechanical Science* [Oxford], 2 [Oct. 1960], 30–39, 10 illustrations).

Christoff Weigel, *Abbildung der gemein-nützlichen Haupte-Stände* (1670). A book of trades and crafts.

C. Descriptions and Editions of Early Manuscripts

W. H. Bond, Ed., *Supplement to Census of Medieval and Renaissance Manuscripts in the United States and Canada* (New York: Kraus Reprint, 1961). See **Ricci** p. 51).

C. E. Bosworth, "A Pioneer Arabic Encyclopedia of the Sciences: al Khwārizmī's Key of the Sciences," *Isis*, *54* (Mar. 1963), 97-111. Description of a work written around 980 A.D. Chapters on arithmetic, geometry, music, and mechanical contrivances have been translated into German by Eilhard Wiedemann (d. 1928).

On **Filippo Brunelleschi** (1379-1446), see Frank D. Prager, "Brunelleschi's Inventions and the 'Renewal of Roman Masonry Work'" (*Osiris*, 9 [1950], 457-554); and Giustina Scaglia, "Drawings of Brunelleschi's Mechanical Inventions for the Construction of the Cupola" (*Marsyas*, *10* [1960-1961], 45-68). The latter is based upon study of the notebooks of Buonaccorso Ghiberti (1451-1516).

A. G. Drachmann, *Ktesibios, Philon, and Heron* (Copenhagen: Munksgaard, 1948). A descriptive and critical work on these Greek technologists.

A. G. Drachmann, *Mechanical Technology of Greek and Roman Antiquity* (Copenhagen: Munskgaard, 1963). A systematic commentary on the works of Hero (*ca.* 50 A.D.) and others, based upon manuscript sources in Greek, Arabic, and Latin.

B. Gille, *Renaissance Engineers* (p. 10), includes an extensive and important catalogue of manuscript works. See principal entry.

Das Hausbuch. Bilder aus dem deutschen Mittelalter von einem unbekannten Meister (Leipzig: Insel-Verlag). Insel-Bucherei Nr. 452. Introduction by Richard Graul. Title listed on another Insel-Bucherei; no further information.

Villard de Honnecourt (*ca.* 1260), Robert Willis, Trans. and Ed., *Facsimile of the Sketch-Book of Wilars de Honnecourt* (London, 1859).
A useful inexpensive edition is Villard de Honnecourt, *Sketchbook* (Theodore Bowie, Ed.) (New York: Geo. Wittenborn, 1962; 80 pp., 64 pls.).

Leonardo da Vinci. A definitive listing of editions of facsimile reproductions of the Leonardo manuscripts is in **Elmer Belt and Kate Trauman Steinitz**, *Manuscripts of Leonardo da Vinci* (Los Angeles: The Elmer Belt Library of Vinciana, 1948, 69 pp.). Sumptuous facsimile editions are still being issued. I should hope that a selection of Leonardo's mechanical devices can be produced for a price low enough to interest working libraries. A comprehensive collection of editions is available for study in the Elmer Belt Library of Vinciana, which Dr. Belt recently gave to University of California, Los Angeles. See note in *Technology and Culture*, 2 (Summer 1961), 303.

Facsimile editions of Leonardo's works are listed also in **Russo**, *Bibliographie* (p. 17), pp. 51–52.

Another listing of editions is in **Leonardo da Vinci**, *I Libri di Meccanica* nella Ricostruzione Ordinata di Arturo Uccelli (Milan: Hoepli, 1940, 673 pp.), pp. 561–564. Uccelli, in this book, systematically presents Leonardo's writings and drawings on mechanics. Profusely illustrated. Uccelli has also attacked the aeronautical works, in **Leonardo da Vinci**, *I Libri del Volo* nella Ricostruzione Critica di Arturo Uccelli con la Collaborazione di Carlo Zammattio (Milan: Hoepli, 1952, 233 pp.).

The holdings of the Lieb Memorial Library of Vinciana, at Stevens Institute of Technology, Hoboken, N.J., are listed in **Maureen C. Mabbott**, Comp., *Catalogue of the Lieb Memorial Collection of Vinciana* (Hoboken, 1936, 103 pp.). This collection, assembled by John W. Lieb (1860–1929), is presently inactive.

Leonardo da Vinci, *The Notebooks of Leonardo da Vinci*, Edward McCurdy, Trans. (2 vols., New York, 1938). Large parts, but not all, of the texts have been systematically arranged.

Ivor Hart, *The World of Leonardo da Vinci* (New York: Viking, 1962, 374 pp.). This grew out of the same author's *Mechanical Investigations of Leonardo da Vinci* (Chicago, 1925), but the later work reflects a thoroughgoing reconsideration of the man and his times. A paperback "revised and corrected" reprint of the 1925 work, to which have been added eight new pages of plates and an introduction by Ernest A. Moody, has been published (1963) by University of California, Berkeley.

Ladislao Reti, "Leonardo da Vinci nella Storia della Macchina a Vapore," *Rivista di Ingegneria* (Milan, 1956–1957), pp. 3–31. A connecting thread is shown between Leonardo's work and that of later investigators. The author has considerately prepared an English résumé, mimeographed.

Ladislao Reti, "The Leonardo da Vinci Codices in the Biblioteca Nacional of Madrid," *Technology and Culture*, 8 (Oct. 1967), 437–445, 4 plates. The leading scholar of Leonardo's technical works describes the two codices (about 350 folios) that he, in 1964, had searched for and that were found by chance in 1967. Further information and additional plates are in Reti's "Die wiedergefundenen

Leonardo-Manuskripte der Biblioteca Nacional in Madrid," *Technikgeschichte, 34,* No. 3 (1967), 193–225, 23 plates. Reti also will edit the definitive facsimile edition.

Das Mittelalterliche Hausbuch, Waldburg–Wolfegg–Waldsee, H. T. Bossert and W. F. Stork, Eds. (Leipzig, 1912, fo.). A facsimile edition of this much-used manuscript work of *ca.* 1480. Johannes Graf Walburg-Wolfegg, *Das Mittelalterliche Hausbuch* (Munich: Prestel-Verlag, 1957) is an inexpensive abridgment in which forty-eight plates are reproduced, some in color.

S. A. J. Moorat, *A Catalogue of Western Manuscripts on Medicine and Science in the Wellcome Historical Medical Library. I. MSS. Written before 1650 A.D.* (London, 1962, 650 pp.).

Ladislao Reti, "The Codex of Juanelo Turriano (1500–1585)," *Technology and Culture, 8* (Jan. 1967), 53–66. This important but neglected work has been studied by Reti and Alex Keller, who are preparing a critical translation for publication by The M.I.T. Press.

Seymour de Ricci, *Census of Medieval and Renaissance Manuscripts in the United States and Canada* (3 vols., New York: H. W. Wilson, 1935–1940). See also **Bond** (p. 49).

John R. Spencer, "Filarete's Description of a Fifteenth Century Italian Iron Smelter at Ferriere," *Technology and Culture, 4* (Spring 1963), 201–206. Critical comment and an excerpt (1,400 words) from Filarete's manuscript treatise on architecture. Entire treatise to be published by Yale University Press. See also important comments on the subject in *Technology and Culture, 5* (Summer 1964), 386–407; *6* (Summer 1965), 428–441.

Daniel V. Thompson, Jr., "Trial Index to Some Unpublished Sources for the History of Medieval Craftsmanship," *Speculum, 10* (1935), 410–431. List of manuscripts earlier than 1500, intended to deal with the materials of painting. Subject index includes such subjects as gilding, stained and painted glass, metals, alloys, solders, cements, textile cleaning and dyeing, parchment, leather, varnish, bone, horn, ivory, etc. White's *Speculum* article (p. 11) mentions Thompson's "delightful *Materials of Medieval Painting* (London, 1936)."

Lynn Thorndike, "Marianus Jacobus Taccola," *Archives Internationales d'Histoire des Sciences, 8* (1955), 7–26. Report on a fifteenth-century manuscript work on machines.

Wilhelm Treue et al., *Das Hausbuch der Mendelschen Zwölfbrüderstiftung zu Nürnberg; Deutsche Handwerkbilder des 15. und 16.*

Jahrhunderts (2 vols., Munich: Bruckmann, 1965, large 4to; 275, 156 pp.). The Bildband contains 335 handsome illustrations, 64 in color; the Textband (which includes another 45 illustrations) has a series of articles on the details and setting of the *Hausbuch* by K. Goldmann, W. V. Stromer, W. Treue, R. Kellermann, H. Zirnbauer, K. Schneider, F. Klemm, and A. Wissner. There are about 85 crafts represented in the drawings, most in more than one version. Represented are craftsmen in textiles, leather, metal, wood, and stone; transport workmen; traders and scribes; farmers, gardeners, hunters; and preparers of food. The reproductions are from the first of three manuscript volumes, which in all contain over 800 pictures. **Cyril Stanley Smith,** in his notes on *Biringuccio* (p. 47), p. xii, mentions the illustrations of the Nürnberg Hausbuch . . . made for the merchant house of Mendel, published as *Deutsches Handwerk im Mittelalter* (Leipzig: Insel-Verlag, n.d.), Insel-Bücherei Nr. 477. I have not seen this last title.

Three illustrated articles in *Beiträge zur Geschichte der Technik und Industrie* describe manuscripts:

H. T. Horwitz, "Technische Darstellungen in Bilderhandschiften des 13. bis 17. Jahrhunderts," *11* (1921), 179–184, 7 illustrations from Nationalbibliothek zu Wien.

H. T. Horwitz, "Giuliano da San Gallo (1445–1516)," *16* (1926), 200–216, 29 illustrations of ms. notebooks in Vatican Library, State Library of Siena, and others scattered.

E. Marx, "Bericht über ein Dokument Mittelalterlicher Technik," *16* (1926), 317–322, 13 illustrations from ms. of 325 pp. in Staatsbibliothek zu Weimar.

V. ENCYCLOPAEDIAS, COMPENDIA, AND DICTIONARIES OF TECHNOLOGY

A. Encyclopaedias

While current encyclopaedias are convenient secondary sources, older encyclopaedias are useful for their delineation of contemporary practice and their reflection of contemporary knowledge and attitudes. It should be noted that some of the older encyclopaedias have required twenty years or more to complete and that the first volumes have thus appeared nearly a generation before the last. Thus the actual date of publication of a particular volume may be of interest in assessing its information or point of view. Furthermore, it is unrealistic to expect the editor of as complex a work as an encyclopaedia to have found enough contributors of superior and precise knowledge to make all articles of equal value. Therefore, critical use of even the best encyclopaedias is advisable.

The discussion of encyclopaedias in **Sarton, Guide** (p. 17), pp. 78–83, is quite full, and the reader is referred to that work for information about the French *Grand Dictionnaire, Grande Encyclopédie, Larousse,* and the German *Konversations-Lexikons* of Brockhaus and others, all relatively recent. A comprehensive history of encyclopaedias is in the article "Encyclopaedia," in the eleventh edition of the *Encyclopaedia Britannica*; however, Rees's *Cyclopaedia* and the important **Descriptions des Arts et Métiers** (p. 58) are not mentioned in the *Britannica* article. The most comprehensive reference work on encyclopaedias is that by **Collison** (p. 54).

As pointed out by Sarton, old encyclopaedias on closed shelves or in storage are nearly useless, while those on open or accessible reserve shelves are likely to be consulted frequently when their value is recognized. The older series (before 1850), being of rag paper, will yet survive several centuries of handling.

1. *Bibliographies of Encyclopaedias*

Robert Collison, *Encyclopaedias: Their History Throughout the Ages* (New York and London: Hafner, 1966, 334 pp.). A remarkable tour de force of bibliographical scholarship, in which encyclopaedias in more than a score of languages, dating from 370 B.C. (Speusippos) to 1965 A.D., are described in more or less detail. A full chapter is devoted to Diderot, one to the **Britannica**, and one to Brockhaus. Chronology, pp. xiii–xvi; bibliography of articles about encyclopaedias, pp. 296–297; list of encyclopaedias not mentioned in text, pp. 298–313. Like the **Britannica** Eleventh (p. 57), Collison's book often tells me more than I want to know, but it is nevertheless a valuable and welcome addition to my reference shelf.

Gert A. Zischka, *Index lexicorum: Bibliographie der lexikalischen Nachschlagewerke* (New York: Hafner, 1959, 290 pp.). A closely packed and impressive bibliography of many classes of systematic printed compilations of information. A list of sources upon which the author drew is at the head of each chapter. The following chapters are pertinent for the history of technology:

I. Enzyklopädien und Konversationslexika, pp. 1–16. A list, briefly annotated, of encyclopaedias from *ca.* 1500 to *ca.* 1957, in many languages.

VII. Biographie, pp. 79–91. Over 250 titles of biographical dictionaries, Who's Whos, and other biographical compilations.

XIX. Technik und Mathematik, pp. 223–244. Includes technical compendia such as those listed in Part V. B. 1; glossaries of technical terms; bi- and multilingual technical dictionaries; dictionaries of inventions; and handbooks and glossaries under the following subject headings: mining (twelve titles before 1850), clockmaking, textiles, dyes (from 1782), arts and manufactures (four titles of the eighteenth century), electrical technology, railroads, autos, aircraft, machine tools, steam engine. Many multivolume works are listed. This chapter in particular requires careful study, so rich and varied are the titles, from (e.g.) William Hooson's *The Miner's Dictionary* (Wrexham, 1747) to *Dictionary of Guided Missile Terms* (1950).

2. *Encyclopaedias in English*

John Harris, *Lexicon technicum: or, an Universal English Dictionary of Arts and Sciences Explaining Not Only the Terms of Art, but the Arts Themselves* (2 vols., London, 1704–1710, fo.). This is the first general technical encyclopaedia in any language. The first volume, which treated 8,200 terms, appeared in 1704; a supplementary volume of 1,900 terms was published in 1710. The fifth and last edition (2 vols., 1736; 12,000 items) was followed by a supplement (1744; 4,000 items). Harris died in 1719. See Douglas McKie, "John Harris and his *Lexicon technicum*" (*Endeavour,* 4 [1945], 53–57).

ENCYCLOPAEDIAS, COMPENDIA, AND DICTIONARIES   55

Ephraim Chambers, *Cyclopaedia, or an Universal Dictionary of Arts and Sciences* (2 vols., London, 1728, fo.). According to **Collison** (p. 54), p. 104, "The influence of Chambers's encyclopaedia has been incalculable." Diderot's *Encyclopédie*, Rees's *Cyclopaedia*, *Encyclopaedia Britannica*, and thus nearly all subsequent encyclopaedias were affected by Chambers's work. His first edition has a plate of Savery's steam engine (1698), but there is no mention of Newcomen's (1712).

Abraham Rees, Ed., *The Cyclopaedia; or, Universal Dictionary of the Arts, Sciences, and Literature* (39 vols. text, 6 vols. pls., London, 1819) is the richest work in English for technology of the period, with detailed text and handsome, carefully prepared plates. Contributors are identified, not always clearly, in Vol. 1, pp. iv–v, and on the "covers of the several parts of the works," which covers I have not seen (the work was issued in paper-bound parts, generally two parts to a volume). The English edition was issued at intervals from 1802 to 1820, although all title pages of bound volumes show 1819, and of plates, 1820. For actual dates, see Benjamin D. Jackson, *An Attempt to Ascertain the Actual Dates of Publication of the Various Parts of Rees's Cyclopaedia* (London, 1895), a pamphlet of seven pages (copy in Harvard University Herbarium).

A Philadelphia edition of Rees was completed in 1822 (Scharf and Westcott, *History of Philadelphia*, Philadelphia, 1884, p. 605). A few American plates were added: Burr's and Wernwag's bridges, Perkins's ship pump, and the Columbian printing press; I have not discovered textual revisions.

Rees's *Cyclopaedia* has neither index nor table of contents, nor are any pages numbered. Thus time and patience are required to exploit fully the contents of the work. The following table is indicative rather than comprehensive and betrays as well my predilections. However, articles on industrial chemical processes can be found under the name of the substance; articles on metallurgical processes are likewise under the name of the metal. Scientific and astronomical instruments are treated separately, under barometer, microscope, micrometer, etc. The military arts are slighted in my list, as are the theoretical articles on mathematics, mechanics, perspective drawing, and engraving. The approximate number of pages of each article is given in parentheses.

Aerostation; balloons (12)   Escapement (25)
Canal (136)   Glass (20)
Cannon (30)   Iron (25)
Chronometer (50)   Machine (12)
Clocks and Clock Making (55)   Machinery, Portsmouth Block (18)
Coal (15)   Manufacture of Cotton (28)
Cotton (20)   Mill-work (8)
Electrical (40)   Mining (12)

Paper (25)
Plough (10)
Printing (20)
Quarrying (10)
Rocket Artillery (6)
Rope Making (20)

Shipbuilding (80)
Steam (10)
Steam Engine (90)
Threshing Machine (5)
Wool (15)
Woolen Manufacture (35)

Edinburgh Encyclopaedia, conducted by David Brewster . . . First American edition, corrected and improved (18 vols., Philadelphia, 1832). Fairly extensive revisions are evident; some articles were rewritten (e.g., several articles on steam engines and boats were rewritten by James Mease, and American geography and biography were added); I found two added plates, one of a Sellers and Pennock fire engine, one of the Trenton bridge, Fitch's steamboat, and Evans's Columbian steam engine. The detailed perspective drawings of the Portsmouth block machinery I had not seen elsewhere.

Several other encyclopaedias in English were published before 1850, but all are less attentive to technology than Rees and Brewster, just described. It may be helpful, however, to list those that I have seen.

Pantologia, A New Cyclopaedia . . . of Human Genius, Learning, and Industry . . . John Mason Good, Olinthus Gregory, Newton Bosworth (12 vols., London, 1813, 8vo). Several drawings by John Farey, Jr., not duplicated in **Gregory's** *Mechanics* (p. 61).

William Nicholson, *American Edition of the British Encyclopaedia* (3rd. ed., 12 vols., Philadelphia, 1819, 8vo); and *Penny Cyclopaedia* (29 vols., London, 1833-1846, 4to) are marginal on technical matters.

Encyclopaedia Americana, Francis Lieber, Ed. (13 vols., Philadelphia: Carey & Lea, 1829-1833). Supplementary volume, 1846. Based on the seventh edition of the German Brockhaus, *Conversations-Lexicon* (1827-1829). I found little under gunpowder, iron, steel, paper, steam, or steamboat that was not available in better form elsewhere.

Encyclopaedia Metropolitana (29 vols., London, 1845). One volume—"Mixed and Applied Sciences, Vol. 6"—deals with technology. Eighty-seven plates.

The two following works, translated from the German, are notable for the quality of their illustrations.

Spencer F. Baird, Trans. and Ed., *Iconographic Encyclopaedia of Science, Literature, and Art,* systematically arranged by J. G. Heck

(4 vols. text, 2 vols. pls., New York, 1851–1852). Unusually sharp and attractive plates, with considerable information on American practice. See Vol. 4, "Technology" (163 pp.), and thirty-five plates in second volume of plates.

Iconographic Encyclopaedia of the Arts and Sciences, translated from the German of the Bilder-Atlas (Iconographische Encyclopaedie) revised and enlarged by eminent American specialists (published by arrangement with F. A. Brockhaus, 7 vols., Philadelphia, 1886–1890). See Vol. 5, "Building and Engineering" (62 plates comprising 950 figures) and Vol. 6, "Applied Mechanics" (128 plates comprising 1,100 figures). Frontispiece of Vol. 6 has a fine tinted lithograph of the Centennial Corliss engine. Good selection of pictures; many superb steel engravings.

Encyclopaedia Britannica (11th ed., 29 vols., 1910–1911) is a generally reliable, scholarly, and amazingly comprehensive work that is unlikely to be superseded by later editions. The "Handy Volume Issue" (8vo, on India paper), compact enough for an apartment, often turns up in old-book stores for $30 to $40. The ninth edition (1875–1889) is noted for the eminence of its contributors; the tenth edition consists of the ninth plus eleven supplementary volumes; the twelfth and thirteenth editions were similarly composed of additions to the eleventh. The current fourteenth edition (1929) is partially revised each time it is reprinted. The publication dates of the first eleven editions are listed on the flyleaves of the eleventh edition. See also the chapter on the *Britannica* in **Collison** (p. 54), pp. 138–155.

3. *Encyclopaedias in French*

The output of French encyclopaedic works in the eighteenth century is important, handsome, and utterly confusing unless all the various series can be seen and compared. The following listing, in order of first appearance, attempts to identify each major series. **Collison,** *Encyclopaedias* (p. 54), being primarily concerned with general works, does not include a full discussion of the *Académie des Sciences* series, which follows.

Académie des Sciences, Paris, *Machines et inventions approuvées par l'Académie royale des sciences, depuis son établissement jusqu'à présent* (7 vols., Paris, 1735–1777, 4to). Some 500 plates of pumps, machines, timekeepers, structures, ferry boats, and so forth. In the absence of a patent office, this 42-year series served as a sort of patent digest. It was not planned as an encyclopaedia. It is mentioned here because its title can easily be confused with the Académie's **Descriptions des Arts et Métiers** (p. 58).

Encyclopédie, ou dictionnaire raisonné des sciences, des arts, et des métiers (17 vols. text, Paris, 1751–1765, fo.); *Recueil des planches* (11 vols., Paris, 1763–1772); *Supplément à l'Encyclopédie* (4 vols. text, Amsterdam, 1776–1777); *Suite du recueil des planches* (1 vol., Paris, 1777); *Table analytique et raisonnée des matières* (2 vols., Paris, 1780).

This is the great encyclopaedia of Diderot and d'Alembert, whose influence in preparing France for the Revolution has given rise to a vast literature about the work. A guide to the study of the *Encyclopédie* is in the introductory essay of Charles C. Gillispie, Ed., *A Diderot Pictorial Encyclopedia of Trades and Industry* (2 vols., New York: Dover, 1959). Some 485 (of 3,100) plates are nicely reproduced in Gillispie, whose work serves as a ready reference but an uneven condensation of, and in no sense a substitute for, the original series. Another selection of (63) plates from the *Encyclopédie* is in Jürgen Dahl, *Jugend der Maschinen* (Ebenhausen bei München: Langewiesche-Brandt KG, 1965). Good reproduction of plates, many folding; perfunctory text. A facsimile edition of the entire *Encyclopédie* was begun in 1966 by Friedrich Fromman Verlag, Stuttgart.

Called to my attention by a bookseller's catalogue, the Yverdon "Encyclopédie Suisse," which was a 56-volume quarto edition of the Encyclopédie, largely revised and rewritten, is described in George B. Watts, "The Swiss Editions of the *Encyclopédie*," *Harvard Library Bulletin*, 9 (Spring 1955), 213–233. Like Watts's other studies of French encyclopaedias, this article is interesting, precise, and highly informative.

The great *Encyclopédie* is separate and distinct from the *Descriptions des Arts et Métiers*, which are here described.

Descriptions des Arts et Métiers, faites ou approuvées par messieurs de l'Académie royale des sciences (45 vols. [Smithsonian copy], Paris, 1761–1788, fo.).

In the Smithsonian Institution Library copy, one of two complete sets in the United States (the other is in the Metropolitan Museum, New York), there are some 12,000 pages of text and 1,700 plates. The work is accurately described by its title as a series of detailed, painstaking descriptions of existing trades and industries. The complete set consists of perhaps 80 cahiers, or parts (some of fewer than 20 pages, some of several volumes), and many libraries have some of the cahiers. Each part treats a single trade or industry. Some of the bulkier cahiers describe the arts of coal mining, ironmaking, joinery, coopering, turning, shipbuilding, tanning, dyeing, the making of cutlery, pipe organs, textiles, harness, shoes, locks, chinaware and paper, and the vocation of fishing. Roubo, *L'art du Menuisier*, which has appeared in many later editions, is a part of this series. The entire work is ably described and the parts are collated in **Arthur H. Cole and George B. Watts**, *The Handicrafts of France as recorded in the De-*

scriptions des Arts et Métiers (Boston, 1952, 43 pp.), which also lists translations and other derivative works, pp. 18–20. A convenient listing of parts appears also in *The British Museum Catalogue of Printed Books 1881–1900* (58 vols., Ann Arbor, 1946; Vol. 1, cols. 755–759).

Seven manuscript volumes "prepared for but not published in" the *Descriptions*, by Réaumur, Duhamel du Monceau, and Fougeroux de Bondaroy, held by Harvard's Houghton Library, are listed in **Hamer,** *Guide* (p. 113), p. 252.

J. E. Bertrand, *Descriptions des Arts et Métiers, faites ou approuvées par messieurs de l'Académie royale des sciences de Paris* (19 vols., Neuchâtel, 1771–1783, sm. 4to) is a shrunken but very attractive version of the work just listed, issued at a lower price so artisans might afford it. Collated also in **Cole and Watts,** (p. 58), pp. 36–37.

Encyclopédie méthodique, ou par ordre des matières (Paris, from 1782, 4to). An outgrowth of and a frequent borrower from Diderot's *Encyclopédie,* this work is often confused with Diderot's; it is, however, another series. Pushed onward by succeeding editors for 50 years, the work was never entirely completed. By 1832, 166½ volumes of text and 51 of plates (6,439 individual plates) had been issued; however, only the parts of immediate interest will be mentioned here. Three series of text volumes: *Arts et Métiers Mécaniques* (8 vols., 1782–1791); *Marine* (3 vols., 1783–1787); and *Manufactures, Arts et Métiers* (3 vols., 1785–1790), all refer to one series of plates: *Recueil des Planches* (8 vols., 1783–1790). Total number of plates is 2,550. All but part of Vol. 7 of *Planches* are in the Smithsonian Institution Library. There are other volumes on architecture (3 vols.), art militaire (4 vols.), chimie et métallurgie (6 vols. text and 1 vol. plates). Because of various schemes of binding, the reader may well be forced to do his own collating. In this task he may be helped by **Russo,** Bibliographie (p. 17), pp. 94–95; **Brunet,** *Manuel du Libraire* (p. 40); **Collison,** *Encyclopaedias* (p. 54), pp. 110–112; and Dawson's of Pall Mall, *Catalogue No. 162* (1966), pp. 24–25, who offers "a complete set" of 197 volumes (for £6,800). A definitive list of the work is in George B. Watts, "The Encyclopédie méthodique" (P.L.M.A. [Publications of the Modern Language Association], 73 [Sept. 1958] 348–366).

Le Dictionnaire de l'Industrie (3 vols., Paris, 1776, 8vo). Comparatively negligible, this work is mentioned here merely to acknowledge its existence. Reprinted nearly without change in six thin volumes in 1795, it was revised extensively for the third edition (6 vols., 1801). No plates. The work is described in Jean Dautry, "Une oeuvre inspirée de l'*Encyclopédie*: le *Dictionnaire de l'Industrie* de 1776"

(*Revue d'Histoire des Sciences et de Leurs Applications*, 5, No. 1 [Jan.-Mar. 1952], 64-72).

4. Encyclopaedias in German

The significance of the following encyclopaedias was suggested to me by Robert P. Malthauf, who has used them with profit in his studies of industrial chemistry. Later, the useful articles and plates on lathes and turning in Krünitz were pointed out to me by Warren G. Ogden, Jr.

Deutsche Encyklopädie; oder, Allgemeines Real-Wörterbuch aller Künste und Wissenschaften (23 vols. and one vol. of pls., Frankfurt a.M., 1778–1807). This work was never completed, reaching only the letter K. See **Collison**, *Encyclopaedias* (p. 54), p. 110.

Johann S. Ersch and J. G. Gruber, Allgemeine Encyclopädie der Wissenschaften und Künste (167 vols., Leipzig: Brockhaus, 1818–1889). This work was not completed; the published portions were A-Ligature and O-Phyxios. See **Collison**, *Encyclopaedias* (p. 54), pp. 179, 182.

Johann G. Krünitz, *Oekonomisch-technologische Enzyklopädie, oder allgemeines System der Staats-, Stadt, Haus- und Landwirtschaft* (242 vols., Berlin: J. Pauli, 1773–1858). **Collison**, *Encyclopaedias* (p. 54), pp. 108–109, says "The short articles and the handy size of the octavo volumes made them of use for leisure reading in a way few encyclopaedias ever achieve."

Johann H. Zedler, *Grosses vollständiges Universal-Lexicon aller Wissenschaften und Künste* (64 vols., Halle and Leipzig, 1731–1750). Four supplementary volumes, A-Caq, were issued 1751–1754. See **Collison**, *Encyclopaedias* (p. 54), 104–105.

B. Handbooks and Compendia of Technology

A welcome procession of mechanical "dictionaries"—more accurately compendia—marched through the nineteenth century, gathering recruits to its ranks in each decade. Exhibiting hundreds of woodcuts and steel engravings and encompassing the whole of industrial technology, the dictionaries were popular until nearly 1900. For detailed contemporary information on machines, structures, and other material objects, nearly all of the old works are constantly useful.

In the first half of the nineteenth century the line between dictionaries, handbooks, and other works such as millwrights' guides and

ENCYCLOPAEDIAS, COMPENDIA, AND DICTIONARIES 61

"engineers' assistants" was not distinct enough to make further subdivision of the following lists practicable; the works were few enough to make it unnecessary. Therefore, all works except multilingual dictionaries and handbooks of essentially modern form and content are grouped together under *Compendia*. See my notes on Chapter XIX of **Zischka**, *Index lexicorum* (p. 54). The following lists are arranged chronologically.

1. Compendia (mechanical "dictionaries" and "guides")

Oliver Evans, *The Young Mill-wright and Miller's Guide* (1st ed., Philadelphia, 1795) was a remarkably early work, which incorporated an unequivocal plate illustrating the first fully automatic flour mill. The fifth edition (1826) was revised by Thomas P. Jones, long-time editor of the Franklin Institute *Journal*; the fifteenth and last edition was in 1860. Editions are collated in **Bathe**, *Oliver Evans* (p. 170), p. 344.

Oliver Evans, *Abortion of the Young Steam Engineer's Guide* (Philadelphia, 1805) was, as the title indicates, something less than what the author hoped it would be; but it is an equally remarkable book, considering the meager literature that Evans had to draw upon.

John Banks, *Treatise on Mills* (London, 1795). There are sections on moving bodies, the velocity of effluent water, experiments in circular motion, and so forth. There is no practical *do thus, do so* instruction, but the way is shown to "make the most of a given stream" (preface).

Olinthus Gregory, *A Treatise of Mechanics* (3rd ed., 3 vols., London, 1815). This edition was probably the first of dozens of English publications to copy Hachette's synoptic chart of mechanical movements. Gregory's first edition (2 vols., 1806) has the earliest account I have seen of Maudslay's lathe slide rest.

James Smith, *The Mechanic, or Compendium of Practical Inventions* (2 vols., Liverpool, 1818, 8vo). One hundred and six engraved plates.

Giuseppe A. Borgnis, *Traité Complet de Mécanique Appliquée aux Arts* (9 vols., Paris, 1818–1821). A descriptive and analytical work; 250 plates.

Gerard Joseph Christian, *Traité de Mécanique Industrielle* (3 vols. text, 8vo, 1 vol. pls., 4to, Paris, 1822–1825). Christian was director of the Conservatoire des Arts et Métiers. Sixty double-spread plates.

Dictionnaire Technologique, ou nouveau dictionnaire universel des arts et métiers (22 vols. text, sm. 8to, 2 vols. pls., 4to, Paris, 1822–

1835). Wide variety of subjects treated; several hundred attractive plates. Copy in Smithsonian Institution Library.

Robertson Buchanan, *Practical Essays on Mill Work and other Machinery* (2nd ed., 2 vols., London, 1823). The third edition, of 1841 (1 vol. text., 8vo, and 1 vol. pls., fo.) was importantly enlarged; essays by Nasmyth on machine tools and Willis on gears were added. The seventy plates, bound together in folio size, bear the title George Rennie, *Illustrations of Mill Work and other Machinery* . . . (London, 1841).

John Nicholson, *The Operative Mechanic, and British Machinist* (1st American from 2nd London ed., 2 vols., Philadelphia, 1826). Of primary interest.

Alexander Jamieson, *Dictionary of Mechanical Science* (7th [sic] ed., London, 1832; 4to, 1,066 pp. First ed., 1827).

Zachariah Allen, *Science of Mechanics* (Providence, 1829). One of several books on the borderline between handbook and dictionary. Speeds and feeds for machining cast iron are on pp. 356–357. An informative "Comparative View of England, France and the United States, as Manufacturing Nations," pointed out to me by Nathan Rosenberg, is on pp. 346–356.

Jacob Bigelow, *Elements of Technology, taken chiefly from a course of lectures delivered at Cambridge, on the application of the sciences to the useful arts* (2nd ed., Boston, 1831). About twenty plates. This widely known work yields upon inspection a pretty thin potion, compared to Evans, for example. Bigelow, a botanist and physician, compiled these Rumford lectures from other books, which sources he fully acknowledged. Although sometimes given credit for it, Bigelow did not invent the word technology, nor did he claim to have done so.

Joh. Jos. Prechtl, *Technologische Encyklopaedie* (20 vols. text, 11 vols. pls., Stuttgart, 1830–1855); Supplement (5 vols. text, 4 vols. pls., 1857–1869). About 700 fine double-spread plates, all volumes 8vo.

Luke Hebert, *The Engineer's and Mechanic's Encyclopaedia* (2 vols., London, 1836–1837, 2,000 illustrations). Also later editions.

Karl Karmarsch, *Handbuch der Mechanischen Technologie* (3rd ed., 2 vols., Hannover, 1857–1858). First edition, which I have not seen, was in 1837–1841. The third edition has no illustrations but does have English and French indexes, and French and English words are given for many technical terms in the text.

ENCYCLOPAEDIAS, COMPENDIA, AND DICTIONARIES 63

Andrew Ure, *Dictionary of Arts, Manufactures, and Mines* (1 vol., London, 1839, 1,240 illustrations; 4th ed., 2 vols., 1853; 3 vols., 1860, 2,000 illustrations). By a champion of machine civilization. Useful for points of view as well as information. Ure was the author also of *The Philosophy of Manufactures* (London, 1835; 3rd ed., 1861, 480 pp.).

George William Francis, *The Dictionary of the Arts, Sciences, and Manufactures, illustrated with 1000 engravings* (London, 1846). Inferior to other similar works.

Edward Cresy, *Encyclopaedia of Civil Engineering, Historical, Theoretical, and Practical* (2 vols., London, 1847). Similar in approach to Ure, preceding, and **Appleton's**, following.

Appleton's Dictionary of Machines, Mechanics, Engine-Work, and Engineering (2 vols., New York, 1850–1851; 2nd ed., 1867). Added in the second edition were twenty-eight engraved portraits; most text pages remained unchanged from first edition, but editorial changes were numerous enough to require checking with first edition if dating is involved.

Peter Barlow, *Encyclopaedia of Arts, Manufactures, and Machinery* (London, 1851; 4to, 834 pp., 87 pls.). Charles Babbage's "Introductory View of the Principles of Manufactures" occupies pp. 1–84.

Charles Tomlinson, Ed., *Cyclopaedia of Useful Arts and Manufactures* (2 vols., London, 1852–1854; Appendix [Vol. 3], 1866). Well illustrated; introductory essay (160 pp.) on the Great Exhibition of 1851.

Oliver Byrne, *The American Engineer, Draftsman, and Machinists' Assistant* (Philadelphia, 1853, sm. fo.). Two hundred wood engravings, fourteen large lithographed plates. See also **Byrne**, *Practical Model Calculator* (p. 66).

Frederick Moné, *Treatise on American Engineering* (New York, 1854, fo.), with twenty-five double-folio plates of steam engines, machine tools, pumps, locomotives, and a cotton factory building. Text explains plates.

Karl Karmarsch and Friedrich Heeren, *Technisches Wörterbuch oder Handbuch der Gewerbekunde* (3 vols., Prague, 1854–1857). Much enlarged, to eleven volumes, for the third edition (1876–1892).

David Scott, *The Engineer and Machinist's Assistant* (New and improved ed., 2 vols., Glasgow, Edinburgh, and London, 1856, fo.). The first volume has treatises on the steam engine, "mill geering," hand

and machine tools, water wheels, and a description of plates. There are 133 plates in Vol. 2.

William Johnson, *The Imperial Cyclopaedia of Machinery* (Glasgow, Edinburgh, London, and New York, ca. 1857, sm. fo.). Over 100 double pages of plates. History of screw propulsion; history of railways of Great Britain; machines from the Great Exhibition.

Gustavus Weissenborn, *American Engineering . . . Stationary, Marine, River Boat, Screw Propellor, Locomotive, Pumping and Steam Fire Engines, Rolling and Sugar Mills, Tools, and Iron Bridges* (1 vol. text, 4to, 1 vol. pls., fo., New York, 1861; 212 pp. text, 52 pls.). See also his *American Locomotive Engineering* (1871).

Eighty Years' Progress of the United States (2 vols., New York and Worcester: L. Stebbins, 1861; 457, 455 pp., 200 illustrations). "American genius," says one of the contributors to this compilation, "is the engineer of this locomotive, 'Progress'; his hand is on the throttle-lever, which he opens wider each day" (Vol. 2, p. 268). Extensive and detailed treatment of agricultural implements, printing, metals industries, coal and oil, transportation, and manufactures of many kinds. Valuable for points of view and frequently for details of practice.

Charles P. F. de Laboulaye, *Dictionnaire des Arts et Manufactures* (3 vols., Paris, 1861). Relatively few illustrations, but text has proved useful.

Julien François Turgan, *Les Grandes Usines, Etudes Industrielles en France et à l'Etranger* (14 vols., Paris, 1861–1882). Well but not profusely illustrated; the plates that do appear are exceptionally fine.

Max Becker, *Handbuch der Ingenieur-Wissenschaft* (5 vols. text, 8vo, 5 vols. pls., sm. fo., Stuttgart, 1863–1873). Finely detailed mechanical drawings and perspective plates.

Spon's Dictionary of Engineering, Civil, Mechanical, Military, and Naval (8 vols., London, 1871–1874). Very good, and profusely illustrated.

Horace Greeley et al., *The Great Industries of the United States* (Hartford: Burr & Hyde, 1872; 1,304 pp., 509 illustrations). Enthusiastic descriptions, in more or less detail, of a very large number of industries and devices. The chapter entitled "Education: Economical and Efficient" is on the "education business." A few of the subjects treated are axes and plows, brass, brushes, clothing, cutlery, glass, glue and sandpaper, horseshoe nails, iron, petroleum, printing presses, safes, sewing machines, soap, silver mining, and stereotyping.

Edward H. Knight, *Knight's American Mechanical Dictionary* (3 vols., New York, 1874–1876). Profusely illustrated. Cross-referencing is particularly helpful. Many later editions. A London edition (Cassell, n.d. [1883?]) has three volumes totaling 2,831 pages with 7,395 illustrations. Vol. 4, a supplement, which includes index references to technical journals, 1876–1880, was published in 1884; 960 pages, 2,549 illustrations. In all, an underexploited source of contemporary practice.

Encyclopaedia of Chemistry . . . as applied to the Arts and Manufactures (2 vols., Philadelphia: Lippincott, 1877–1879). Illustrated.

Park Benjamin, Ed., *Appleton's Cyclopaedia of Applied Mechanics* (2 vols., New York: Appleton, 1878–1880; 960, 959 pp., 4,326 illustrations); Supplement (titled *Modern Mechanism*, 1892, 924 pp.). Consists of short, well-illustrated articles giving basic technical descriptions of machines, machine elements, and devices, such as sewing machines, ore stamps, shingle machinery, cotton gin, riveting machines, railroad cars, pumps, etc. This is a new work, not merely a reissue of *Appleton's Dictionary* (p. 63).

2. Handbooks

The historian of technology must of course determine the state of theoretical knowledge at any particular time as it is reflected in learned journals; but he must also know how much of that knowledge was available to the engineer or mechanician in the field who was charged with getting things built. Handbooks show the state of knowledge at the practical level. It is generally safe to assume that the practical man used little reference material beside the handbooks, at least until very late in the nineteenth century.

This list of handbooks is confined to civil, mechanical, and electrical engineering. Even in these fields the list is not exhaustive. I hope that enough titles have been given to suggest the wide variety of handbooks and their numberless revisions (which attest to their popularity and usefulness) and to point up the need for a critical work on the development of handbooks.

In the **Reichspatentamt,** *Katalog* (p. 27), Vol. 2, under "Handbuch," there is a list ten columns long with twenty-one cross references.

Henry Adcock, *Adcock's Engineers' Pocket-Book, for the year 1838* (London, 1838). First edition. Mathematical tables, strength and weight of materials, hydraulics, and so forth.

Charles H. Haswell, *Engineers' and Mechanics' Pocket Book* (New York: Harper, 1844; 264 pp., 12mo). The author, sometime Chief

Engineer of the U.S. Navy, acknowledged his debt to Adcock, Grier, Gregory, *Library of Useful Knowledge*, the *Ordnance Manual*, and to the officers of West Point Foundry, particularly B. H. Bartol, engineer.

Oliver Byrne, *The Practical Model Calculator, for the Engineer, Mechanic, Machinist* [etc.] (Philadelphia, 1851). Published in twelve parts. Well organized, modern in format.

John W. Nystrom, *Pocket-Book of Mechanics and Engineering* (Philadelphia, 1855). Nystrom recommended that every engineer carry a notebook in which to record new facts for his professional repertory. John Trautwine, for one, had been doing so for nearly thirty years when he read Nystrom; Trautwine's *Pocket-Book* was not ready, however, until 1872.

Akademischer Verein Hütte, Berlin. *"Hütte," des Ingenieurs Taschenbuch* (3rd ed., Berlin, 1860). Many later editions.

William Templeton, *Engineer's, Millwright's and Machinist's Practical Assistant* (London, 1862).

William J. M. Rankine, *Useful Rules and Tables* (4th ed., London, 1873). First edition was in 1868.

Thomas Dixon, *Millwright's and Engineer's Ready Reckoner* (2nd ed., Philadelphia, 1868).

Scribner's Engineers' and Mechanics' Companion (Hartford, 1868).

John C. Trautwine, *Civil Engineer's Pocket-Book* (1st ed., Philadelphia, 1872). A real perennial. The twenty-first edition was published in 1937.

Utica Steam Engine Company, *The Engineers' and Mechanics' Hand-Book* (Utica, 1872, 32 pp.). Actually an advertising pamphlet, this title is included here to remind the reader that handbook material and much other valuable information is in older as well as in later catalogues.

William V. Shelton, *The Mechanic's Guide . . . for the use of Engineers, Mechanics, Artizans, &c.* (London, 1875).

Daniel K. Clark, *A Manual of Rules, Tables, and Data for Mechanical Engineers* (1st ed., London, 1877).

Stephen Roper, *Engineer's Handy Book* (Philadelphia, 1881).

E. Hospitalier, *Formulaire Pratique de l'Electricien* (Paris, 1883). The English edition of this work appeared a year later: Gordon Wigan, Trans., *The Electrician's Pocket Book* (London, 1884).

M. Krieg, *Taschenbuch der Elektrizität* (Leipzig, 1888).

John Roebling's Sons Co., *Hand-Book of Tables for Electrical Engineers* (Trenton, 1892). This 119-page book is the earliest American electrical handbook that I have seen.

William Kent, *Mechanical Engineer's Pocket-Book* (1st ed., New York, 1895). Kent heeded Nystrom's dictum that every engineer should keep his own notebook. Kent's latest edition is currently one of the standard handbooks.

Henry H. Suplee, *The Mechanical Engineer's Reference Book* (1st ed., Philadelphia, 1903).

Harold Pender and William A. Del Mar, *Handbook for Electrical Engineers* (1st ed., New York, 1914). Currently, in a later edition, one of the favorites.

Lionel S. Marks, *Mechanical Engineers' Handbook* (1st ed., New York, 1916). This comparative latecomer now ranks with Kent's as a modern authority.

3. *Trade Catalogues*

Trade catalogues are valuable for the pictures and drawings they contain, for specifications, prices, dates, and sometimes for patent numbers with which to identify an inventor. Catalogues may also furnish significant information on the styles, fashions, and designs of domestic equipment such as plumbing and heating fixtures and supplies, rain spouting, stoves, laundry equipment, and so forth.

There are a few notable collections of trade catalogues in large libraries, and I hope others will be formed as the value of the information in catalogues becomes more generally understood.

The Columbia Collection of some 200,000 catalogues, which was assembled by Miss Granville Meixell primarily as a library resource in the teaching of engineering, is now in the Museum of History and Technology Library, Smithsonian Institution, Washington. An article by Jack Goodwin describing the collection is listed on p. 68.

The Library of the University of Southern Illinois at Carbondale has a collection of over 10,000 trade catalogues assembled by a Nevada mining engineer. The earliest catalogue is dated 1830, and most are of the 1890–1915 period.

The Franklin Institute Library also has many thousands of trade catalogues. A sampling of the card catalogue indicates that most are of the 1920–1940 period. There are many earlier ones, however. I noted, for example, a Thomson-Houston catalogue of 1883, scientific instrument catalogues of *ca.* 1870, and a Glasgow, Scotland, machinery catalogue of 1878.

The F. Hal Higgins Collection of agricultural materials at the University of California, Davis, and the Bella C. Landauer Collection in the New-York Historical Society of nineteenth-century advertising are other important collections described in the articles listed here.

Lawrence B. Romaine, *A Guide to American Trade Catalogs 1744–1900* (New York: Bowker, 1960, 422 pp.). A pioneering work by a bookseller. Adequate listing, with locations, for 10,000 trade catalogues (probably three-fourths 1880–1900) in some 150 repositories. Perhaps its chief value now is the encouragement that it will give to librarians and others who may have had misgivings about the saving of old catalogues. A revised edition or supplementary volume probably will be undertaken by Jack Goodwin, of Smithsonian Institution. Romaine died in 1967.

Jack Goodwin, "The Trade Literature Collection of the Smithsonian Library," *Special Libraries,* 57 (Oct. 1966), 581–583. The Museum of History and Technology Library now has a collection of about 240,000 catalogues, consisting of the Columbia University collection, substantial transfers from Baker Library at Harvard and University of California Library at Davis, and many thousands already in the library. Other collections have been added since 1966.

Wayne D. Rasmussen, "The F. Hal Higgins Library of Agricultural Technology," *Technology and Culture,* 5 (Fall 1964), 575–577. A short description of the collection in the University of California at Davis. See also **Pursell and Rogers** (p. 214).

James J. Heslin, "Bella C. Landauer," in *Keepers of the Past,* Clifford Lord, Ed. (Chapel Hill: University of North Carolina Press, 1965), pp. 180–189. While this article is for the general reader, the outlines of the Bella C. Landauer Collection of trade cards are made reasonably clear. Mrs. Landauer is quoted as saying "All collectors gather good items, but . . . whose judgment is infallible? What may be rejected as inconsequential trash, I gather and preserve as a nucleus for historic reconstruction." The vital necessity of having single-minded collectors, recognizing what will be valuable and collecting it long before scholars are even aware of their needs, is demonstrated by Mrs. Landauer's collection. The collection, comprising over 35,000 items, chiefly nineteenth-century United States, is described in Business History Society, *Bulletin,* 5, No. 3 (Apr. 1931), 1–6; 8, No. 1 (Jan. 1934), 2; 9, No. 3 (May 1935), 33–38.

Detroit Public Library, *Automotive History* (p. 235) lists some 55,000 catalogues, brochures, and manuals in addition to the extensive manuscript and picture collections.

A few notable individual catalogues follow.

Appleby's Illustrated Handbook of Machinery (5 vols. (?), London, 1877–1895). Appleby, Ltd., apparently a jobber, seems to have handled nearly everything in a mechanical line. The handbooks are of Appleby products; the lists of items required to outfit, for example, a machine shop are minutely detailed, for shipment I suppose to British colonies. Publication was intended to be in eight parts or two volumes. I have seen five parts, and suspect that the project was not completed.

Section 1. Prime Movers (1880, 74 pp.)
Section 2. Hoisting Machinery (1887, 158 pp.)
Section 3. Pumping Machinery (1880, 172 pp.)
Section 4. Machine and Hand Tools (1897, 166 pp.)
Section 5. Contractors' Plant and Railway Materials (published?)
Section 6. Colonial and Manufacturing Machinery, Part A, Mining Machinery (1895, 62 pp.; other parts?)
Section 7. Miscellaneous Ironwork (published?)
Section 8. Useful Tables and Memoranda (published?)

First edition in one volume, 1863; second edition also in a single volume, 1869.

Manning, Maxwell, and Moore, *Illustrated Catalogue of Railway and Machinists' Tools and Supplies* (New York, Sept. 1884, 4to, 660 pp.), has hundreds of clear wood-engraved illustrations. A copy is in Smithsonian Museum of History and Technology Library.

Arkell and Douglas, *Illustrated Polyglot Catalogue of American Manufactures* (New York, 1886, fo., 378 pp.), is profusely and attractively illustrated. Items for the export trade include tools, hardware, chairs, appliances, coaches, and so forth. Copy in Library of Congress.

C. Glossaries

In the Library of Congress classification scheme, glossaries are to be found in each discipline under "9." That is, glossaries of engineering in general, T9; civil engineering, TA9; building, TH9; mechanical engineering, TJ9; electrical engineering, TK9; and so forth.

London. Science Library, *Glossaries of Technical Definitions (Exclusive of Botany and Zoology)* [Bibliographical Series, No. 707] (5th ed., London: Science Library, 1952; 82 pp.). Supplement No. 1

(1953, 16 pp.). The list and supplement together identify some 1,500 titles.

Bernice Simpson, *A List of Glossaries in the Fields of Science and Technology* [Reference List no. 36] (Chicago: The John Crerar Library, 1936; 40 pp.). A list of over 400 glossaries of special subjects, such as acoustics, bookbinding, brewing, cement, foundries, paper, tires, wire, and wool. Most of the glossaries are those that appear as appendices or short sections in reference books.

Eugen Wüster, *Bibliography of Monolingual Scientific and Technical Glossaries* (2 vols., Paris: UNESCO, 1955–1959). Vol. 1: National Standards, lists about 1,600 works (in two dozen languages) that define "correct" or "standard" technical terminology. The authority may be a national standards association, a government bureau, or other agency. Vol. 2: Miscellaneous Sources, contains 1,046 titles in 26 languages. Bibliographies and reference works used by the author are given on pp. 16–18 of Vol. 2.

Zischka, *Index lexicorum* (p. 54), Chap. XIX, lists a large number of glossaries in many subjects and several languages. It should be recognized also that compendia of technology (Part V.B.1) are often, in effect, expanded glossaries.

D. Bi- and Multilingual Dictionaries

Robert L. Collison, *Dictionaries of Foreign Languages. A Bibliographical Guide to the General and Technical Dictionaries of the Chief Foreign Languages, with Historical and Explanatory Notes and References* (New York: Hafner, 1955, 210 pp.). See particularly "Technical Dictionaries," pp. 163–190. A second edition is being prepared.

Alfred Schlomann, *Illustrated Technical Dictionary in Six Languages* (17 vols., New York: McGraw-Hill; London: Constable, 1906–1932). Catalogued variously as Schlomann-Oldenbourg (the latter is the original publisher in Munich) and *Illustrierte Technische Wörterbücher*. The languages are German, English, French, Russian, Italian, and Spanish. Arranged by subjects; full indexes. Thousands of small line drawings aid in identification. While any specialist can find errors of nomenclature in his field, this work is the best I know of for identifying details of devices and processes. Definitive listings are in Zischka (p. 54) and **UNESCO** (p. 71). New editions and reprintings have occurred, but only scattered numbers are in print.

Vol. 1:   Machine elements, metal- and woodworking tools, 2,000 entries, 823 illustrations; index, pp. 257–403

Vol. 2: Electrical engineering, 3,965 illustrations
Vol. 3: Steam boilers, engines, turbines, pumps, auxiliaries
Vol. 4: Internal combustion engine
Vol. 5: Railway construction and operation
Vol. 6: Railway rolling stock
Vol. 7: Hoisting and conveying machinery
Vol. 8: Reinforced concrete construction
Vol. 9: Machine tools for metal and wood, 4,000 entries, 2,400 illustrations; index, pp. 507–706
Vol. 10: Motor vehicles and aircraft (1910)
Vol. 11: Iron and steel
Vol. 12: Hydraulics, pneumatics, refrigeration
Vol. 13: Construction above and below ground
Vol. 14: Textile raw materials
Vol. 15: Spinning and spun fabrics
Vol. 16: Weaving and fabrics
Vol. 17: Aeronautics (1932; reprinted 1957)

UNESCO, *Bibliography of Interlingual Scientific and Technical Dictionaries* (4th ed., Paris, 1961, 236 pp.). For sale in the United States by Columbia University Press. A list of more than 2,000 dictionaries and glossaries, representing about 80 languages. Arranged by subjects; name and subject indexes. Reviewed in *Technology and Culture,* 4 (Winter 1963), 125–126. Supplement (1965, 83 pp.), in which another 450 titles are listed.

A. J. Walford, Ed., *A Guide to Foreign Language Grammars and Dictionaries* (London: The Library Assn., 1964, 132 pp.). An annotated list describing strengths and weaknesses of various dictionaries. Specific recommendations are given. Spanish, Portuguese, and Russian chapters were written by the editor; French, Italian, German, and Scandinavian chapters by other specialists.

Wolfram Zaunmüller, *Bibliographisches Handbuch der Sprachwörterbücher. Ein internationales Verzeichnis von 5600 Wörterbüchern der Jahre 1460–1958 für mehr als 500 Sprachen und Dialekte. A Critical Bibliography of Language Dictionaries* (Stuttgart: Anton Hiersemann Verlag; New York/London: Hafner, 1958, 496 cols.).

Zischka, *Index lexicorum* (p. 54). Chapter XIX, pp. 223–244, lists many bilingual and multilingual dictionaries in various disciplines.

VI. BIOGRAPHY

Although the search for biographical information about people whose names are not household words must often be lengthy and time-consuming, there is a surprisingly large number of guides and indexes that can at least make the quest more systematic and, in some cases, sharply reduce the time required.

Here general finding aids are listed first; standard biographical dictionaries come next, followed by works of the *Who's Who* variety which enable one to locate people of the recent past. These relatively straightforward groups are followed by a somewhat amorphous body of titles and suggestions that may be useful to readers endowed with diligence and an optimistic nature. No attempt has been made to explore the world of genealogy.

A. Biographical Dictionaries

1. *Published Finding Aids*

Albert M. Hyamson, *A Dictionary of Universal Biography* (2nd ed., London, 1951). A listing of over 55,000 names with references to the principal biographical dictionaries of United States, England, France, Germany, Italy, Canada, Australia, Ireland, and several other countries. The British *Annual Register* (1850–1949) and the *Encyclopaedia Britannica,* eleventh edition, are included also. By a single dictionary reference in this book, one is able to do the equivalent of searching some two dozen separate works, many of which are usually not readily at hand. While this is merely a start in the direction of assembling the names of engineers and mechanicians, Hyamson's work is a monumental catalogue that deserves emulation and refinement in more limited fields.

Robert B. Slocum, *Biographical Dictionaries and Related Works* (Detroit: Gale Research Co., 1967, 1,056 pp.). A major new guide to biographical dictionaries, Who's Whos, and other biographical compilations in 100 countries. Author, title, and subject indexes.

2. *Dictionaries of Biography*

Webster's Biographical Dictionary (Springfield, Mass., 1943, 1,697 pp.), is a handy volume for a ready-reference shelf. Some minor corrections and changes have been made in successive printings, but the 1963 printing still carries "First Edition" on its title page.

Dictionary of Scientific Biography (New York: Scribner). This multi-volume work, whose first volume probably will appear in 1968, is intended to include only those technologists who have contributed significantly to science. Instrument makers will be fairly numerous, but other kinds of mechanicians will be excluded. This work will not eliminate the need for a dictionary of biography of technologists.

Dictionary of American Biography, Allen Johnson and Dumas Malone, Eds. (20 vols., New York: Scribner, 1927–1936). Index (Vols. 1–20, 1937), is a useful adjunct; supplementary volumes in 1944 and 1958. The entire work, except the index, is in print at this writing, twenty-two volumes in eleven. Engineers and mechanicians are well represented. Articles are authoritative; a list of sources for each article facilitates further work. A condensed version of this work is listed immediately following.

Joseph G. E. Hopkins, Ed., *Concise Dictionary of American Biography* (New York: Scribner, 1964). A single-volume one-to-fourteen condensation of *DAB* (preceding), including the two supplemental volumes. Nearly 15,000 articles.

National Cyclopaedia of American Biography (13 vols., New York: J. T. White, 1898–1906). *Conspectus and Index* (1 vol.); 2nd ed. of *Conspectus and Index* (1937). The original work, which has small woodcut portraits of many obscure individuals, is useful; and the series is still alive. Vol. 15 (1916)–52(1958)+ and *Current Volumes* A(1930)–I(1960)+ of living Americans, are indexed in loose-leaf *Indexes* (4th ed., 1959).

Appleton's Cyclopaedia of American Biography, James G. Wilson, Ed. (6 vols., New York, 1886–1889). Supplementary volume in 1900. Also still useful.

Dictionary of Canadian Biography, George W. Brown, Marcel Trudel, André Vachon, Eds. (Toronto: University of Toronto Press,

1966———). Vol. 1, 1000–1700, contains 594 biographies. Succeeding volumes will cover a specific number of years; each volume will be self-contained, with index and bibliography. French-language editions are published by les Presses de l'université Laval, Québec. This will presumably supersede *Macmillan Dictionary of Canadian Biography* (3rd ed., London: Macmillan, 1963).

Dictionary of National Biography, Leslie Stephen and Sidney Lee, Eds. (63 vols., London, 1885–1900). Supplement (3 vols., 1901); *Epitome and Index* (2nd ed., 1903, 1,456 pp.); *Errata* (1904, 209 pp.); three supplementary volumes (1912); three more (1927); then decennial volumes (1930———). Later supplements have cumulative indexes from 1901. Perhaps the best dictionary of biography in existence, this British work was used as a pattern for the *Dictionary of American Biography*. The practice of giving the location of a subject's portrait, not followed in DAB, is a good one. See the **Concise DNB**, following.

Dictionary of National Biography. The Concise Dictionary. Part I, From the Beginnings to 1900 (2nd ed., Oxford, 1906). Tenth reprinting (1965, 1,503 pp.); Part II, 1900–1950 (Oxford, 1961, 528 pp.). Errata are keyed to main entries.

Allgemeine Deutsche Biographie (56 vols., Leipzig, 1875–1912). Signed articles giving lives and works, with notes on other biographical notices of subjects. Vols. 46–55 repeat alphabet after 1899. Vol. 56, a general index, is helpful.

Die neue Deutsche Biographie [published by the Historical Commission of the Bavarian Academy of Science] (Berlin: Duncker & Humblet, 1953———). Seven volumes (partly through H) had been published by 1965. Adequate treatment of scientists and technologists is expected.

Johann Christian Poggendorf, *Biographisch-literarisches Handwörterbuch zur Geschichte der exakten Wissenschaften* (6 vols., Leipzig, 1863–1936). Reprint (10 vols., Ann Arbor, Mich., 1945): Vols. 1–2 to 1858; Vol. 3: 1858–1883; Vol. 4: 1883–1904; Vol. 5: 1904–1922; Vol. 6: 1923–1931. Standard and indispensable work. Concise biographical information and lists of subjects' works.

Biographie Universelle (Michaud) Ancienne et Moderne . . . (2nd ed., 45 vols., Paris: Chez Madame Desplaces, 1854–1865). The existence of two *Michauds*, in forty-five and eighty-five volumes, with different publishers and overlapping dates, calls for a few words of explanation. This second edition is a "nouvelle édition" of Joseph and Louis Gabriel Michaud's *Biographie Universelle, Ancienne et Moderne*

(85 vols., Paris, 1811–1862). The earlier work completed the alphabet in fifty-two volumes (1811–1828); mythological names are in Vols. 53–55 (1832–1834); supplementary Vols. 56–85 (1834–1862) consist of a new alphabetical listing that extended through Vil. The second edition (1854–1865) was commenced by the younger Michaud and completed by Desplaces. It was superior to the first edition, which was often bitterly antirevolutionary, but both editions contain contributions of distinguished men in many fields. See also comments under *Nouvelle Biographie Générale,* which follows.

Nouvelle Biographie Générale, Ferdinand Hoefer, Ed. (46 vols., Paris: Didot, 1853–1866). Similar in scope and treatment to *Michaud.* Much was borrowed from the earlier Michaud. **Sarton,** in his *Guide* (p. 17), p. 84, preferred N.B.G., perhaps because of more systematic entries of vital data and bibliographies giving titles in original language. **Winchell** (p. 33) notes the better editing of Michaud, the fuller bibliographies, and the larger number of names in the second half of the alphabet, N–Z. Hoefer has more names, especially minor ones, in A–M. For description of litigation between Desplaces and Didot, see Anon., "Biographical Dictionaries" (*Quarterly Review,* 157 [1884], 187–230). The article, overlong for its content, is on these two works.

Dictionnaire de Biographie Française, Michel Prévost, et al., Eds. (Paris, 1933——), had advanced by 1965 through Desplagues (Vol. 10). At the present rate of appearance, most of this otherwise promising work will belong to a future generation.

Conrad Matschoss, *Männer der Technik* (Berlin, 1925, 4to, 306 pp.). Over 850 short biographies. Not to be confused with Matschoss' *Grosse Ingenieure* (Munich, 1937, 334 pp.), a rather slight work promptly translated into English as *Great Engineers* (London, 1939). This earlier book is a specialized biographical dictionary. Its appearance provoked a pamphlet war by Feldhaus and others, but the inaccuracies pointed out by the pamphleteers seem less than alarming. The small portraits (about 100) are uniformly poor, however.

Louis Leprince-Ringuet, Ed., *Les Inventeurs Célèbres, Sciences Physiques et Applications* (New ed., Paris: Mazenod, 1962, large 4to, 455 pp.). This attractive book, with many portraits nicely reproduced, describes innovators as well as inventors in the usual English sense. For example, Carnot, Mayer, Clausius, and Helmholtz occupy about two pages each. Men such as Vaucanson, Cugnot, Watt, Trevithick, Fourneyron, Lenoir, Daguerre, Siemens, Gramme are treated in some detail. A long section, pp. 364–439, contains very short sketches of many others, arranged under subject headings, which include electricity, hydrodynamics, instruments, navigation, aerostation, printing, photography, automobiles, materials of war, and others.

Académie des Sciences, Paris, *Index Biographique des Membres et Correspondants de l'Académie des Sciences . . . 1666 [–] 1954* (2nd ed., Paris, 1954) has very short notices without further references.

3. *Who's Whos*

The current biographical works have been in existence long enough to make their earlier issues, and their *Who Was Who* compilations, of value in historical studies.

There are a great many periodical and occasional *Who's Whos*, covering Canada, Germany, France, Italy, Denmark, U.S.S.R., Australia, Japan, Egypt and the Near East, Southern Africa, and many other countries. In addition to geographical compilations, there are numerous *Who's Whos* devoted to specific vocational fields and to other more or less distinctive modes of segregation. **Slocum** (p. 74) has an extensive list, and there are about fifty entries of possible interest to a historian under "Who's Who" in **Winchell**, *Guide* (p. 33).

The following list indicates the nature of works available.

American Men of Science, Physical and Biological Sciences, A Biographical Dictionary, Jacques Cattell, Ed. (10th ed., 4 vols., Tempe, Ariz., 1960–1961). Academician engineers are included. This is the latest edition of a work that first appeared in 1906 (1st ed., 364 pp.).

Who's Who (London: A. & C. Black), an annual work starting in 1849.

Who Was Who (London: A. & C. Black, Vol. 1, 1897–1915; Vol. 2, 1916–1928; Vol. 3, 1929–1940; Vol. 4, 1941–1950; Vol. 5, 1951–1960).

Who's Who in America (Chicago: A. N. Marquis Co., "published continuously since 1899," Vol. 34, 1966–1967).

Who Was Who in America (Chicago: A. N. Marquis Co., Vol. 1, 1897–1942; Vol. 2, 1943–1950; Vol. 3, 1951–1960).

Wer ist Wer? Das Deutsche Who's Who (14th ed., 2 vols., Berlin-Grunewald: Arani Verlags-GmbH., 1962; Vol. 2, 1965). The first volume has 22,000 names in Bundesrepublik and West Berlin; the second volume has 5,000 names in DDR.

International Who's Who (30th ed., London: Europa Publications, 1966).

B. Bibliographies and Indexes of Biography

Because **Hyamson** (p. 73) has the most extensive consolidated alphabetical listing, his work is in a class by itself so far as ease of reference is concerned. However, since many, if not most, of the people of particular interest to historians of technology do not appear in biographical dictionaries, more devious finding methods are required.

One is inclined merely to wag one's head sadly at the voluminous but largely unrecoverable biographical information in periodicals. Many trade publications, such as *Iron Age* (from 1873), *American Machinist* (from 1877), and *Machinery* (from 1894), had in their early years short sketches and reminiscences of leading technical men; but there are few general indexes, and only a systematic search can do more than intensify one's awareness of things unknown. *Engineering News, Railroad Gazette, Harper's New Monthly Magazine, The Engineer,* and *Engineering*—the last two of London—are serials, mentioned almost at random, in which biographical data have appeared. The cumulated index of *Cassier's Magazine* (p. 157) records more than 300 biographical sketches that appeared between 1891 and 1913.

It would be a large but not impossible task to construct an index of biographies of technical people. The materials for such an index are suggested here. Indeed, if this part wears the aspect of a set of notes for an indexer, it is only because it is intended as such. It will be gratifying some day to learn of a single sturdily bound index that will make this entire section quite superfluous.

In **Zischka,** *Index lexicorum* (p. 54), Chap. VII, pp. 79–91, more than 250 titles are listed of biographical dictionaries, Who's Whos, and other biographical compilations. A helpful section on biography and genealogy will be found in **Larson,** *Guide to Business History* (p. 23), pp. 1,007–1,015. A few biographical dictionaries are noticed critically in **Sarton,** *Guide* (p. 17), pp. 84–85.

In publications printed in the English language, Thomas James Higgins has reduced the undiscovered book-length biographies of engineers to a nearly irreducible minimum; and for electrical engineers and electrophysicists he has attacked the periodical literature.

Thomas James Higgins, "Book-Length Biographies of Engineers, Metallurgists, and Industrialists," *Bulletin of Bibliography,* 18, No. 9 (Jan.–Apr. 1946), 206–210; 18, No. 10 (May–Aug. 1946), 235–239; 19, No. 1 (Sept.–Dec. 1946), 10–12.

Thomas James Higgins, "A Biographical Bibliography of Electrical Engineers and Electrophysicists," *Technology and Culture,* 2, No. 1 (Winter 1961), 28–32, and No. 2 (Spring 1961), 146–165. A large variety of sources were systematically explored.

Other lists by Higgins, of mathematicians, chemists, physicists, and astronomers, are noted in **Sarton,** *Guide* (p. 17), p. 85.

E. Scott Barr has compiled lists of "biographical fragments" contained in the first hundred volumes of *Nature* (1869–1918) and the first eighty-seven volumes of *Scientific Monthly* (1872–1915). The lists are noted but not published in *Isis,* 56 (1965), 455.

Biography Index (New York: H. W. Wilson, 1946———). A cumulative index to biographical material in books and magazines. Within limits of heavy American and nontechnical bias, the *Index* lists all biographies—of all people, not only those just deceased—published during the years indexed. Vol. 1, 1946–1949; Vol. 2, 1949–1952; Vol. 3, 1952–1955; Vol. 4, 1955–1958; Vol. 5, 1958–1961; Vol. 6, 1961–1964; issued monthly and accumulated at three-year intervals. Analytical (subject) indexes.

Carnegie Library of Pittsburgh, *Men of Science and Industry; a Guide to the Biographies of Scientists, Engineers, Inventors, and Physicians, in the Carnegie Library of Pittsburgh* (Pittsburgh, 1915). This 189-page work indexes about 60 titles, including collected works such as Howe, Stuart, and Smiles, Smithsonian Institution *Annual Reports*, *Proceedings* of the Royal Society, American Philosophical Society, American Academy of Arts and Sciences, and the *Transactions* of the engineering societies: A.S.C.E., A.I.M.E., A.S.M.E., and A.I.E.E. About 600 engineers are listed in a subject index. The usefulness of this guide has been somewhat reduced but certainly not destroyed by the *Dictionary of American Biography*, which has better information on perhaps half of the engineers in the guide.

William Matthews, *American Diaries; an Annotated Bibliography of American Diaries Written prior to the Year 1861* (Berkeley, Calif., 1945; Boston, 1959, 383 pp.). A carefully constructed guide.

William Matthews, *British Autobiographies; an Annotated Bibliography of British Autobiographies Published or Written before 1915* (Berkeley, Calif., 1955, 376 pp.). A subject index makes this somewhat easier to use than the other Matthews' titles.

William Matthews, *British Diaries; an Annotated Bibliography of British Diaries Written Between 1442 and 1942* (Berkeley, Calif., 1950, 339 pp.).

William Matthews, *Canadian Diaries and Autobiographies* (Berkeley, Calif., 1950, 130 pp.). Entries are annotated.

Louis Kaplan, *A Bibliography of American Autobiographies* (Madison, Wis., 1961). Similar in scope to the works of Matthews, Kaplan's has subject and regional indexes, 6,377 entries. Engineers, artisans, inventors, and industrialists are represented.

Kirby and Laurson (p. 220), have provided very short "biographical outlines" of 114 early civil engineers and portraits of 28. Lists of sources for each sketch, and for the series, are given. Wide-ranging.

Pearl I. Young, *Octave Chanute 1832–1910: A Bibliography* (San Francisco: Edward L. Sterne, 1963, 28 pp.). Exhaustive listing of works by and about Chanute, designed to provide important help to anyone doing serious work on Chanute. Reviewed in *Technology and Culture,* 6 (Winter 1965), 147–148.

The various publications of the American Society of Civil Engineers (from 1867), American Institute of Mining Engineers—now Mining, Metallurgical, and Petroleum—(from 1871), American Society of Mechanical Engineers (from 1880), and American Institute of Electrical Engineers (from 1884) have obituary notices of members. This information was indexed until recent years, when economies of the present have further encouraged ignorance of the past. But see the following indexes: **A.S.C.E.** *Transactions* Index, Vols. 1–83 (1867–1920); **A.I.M.E.** *Transactions* Index, Vols. 1–35 (1871–1904); **A.S.M.E.** *Transactions* Index, Vols. 1–45 (1880–1923). This last index lists some 1,400 obituaries. Later indexes are named in **Titus,** *Union List of Serials* (p. 142).

It appears that there is not even a card file of obituary notices of the last thirty years in the four "Founder Society" (A.S.C.E., etc.) publications. The Engineering Societies Library suggested that an inquiry directed to an individual society might yield a date of death, which would reduce somewhat the searching otherwise required.

Contrasted with this is the attention still paid to people in **Chemical Abstracts** (Easton, Pa.: American Chemical Society, from 1907). There are now five decennial indexes and a sixth "collective" (five-year, 1957–1961) index. The *Fifth Decennial* [subject] *Index* for Vols. 41–50 (1947–1956), for example, lists some 600 names under "Obituaries" and 1,600 more under "Biographies."

The reader is also referred to **Titus,** *Union List of Serials* (p. 142) for the numerous indexes to the British society serials: Institution of Civil Engineers, *Minutes of Proceedings* (from 1837); Institution of Mechanical Engineers, *Proceedings* (from 1847); and Institution of Electrical Engineers, *Journal* (from 1872).

Biographical data for many British engineers are in the **Woodcroft Collection** in H.M. Patent Office Library, London, according to citations in the "Bibliography of Engineering and Applied Science," Newcomen Society *Transactions,* Vol. 2 (1921–1922), *et seq.* The collection was made in preparing a "Gallery of Inventors" in Woodcroft's Patent Office Museum. See *ibid., 14* (1932–1933), 9.

A "Biographical Index" of engineers and scientists, compiled since 1906 by the late **Edgar C. Smith,** historian of marine engineering, is now in Science Museum Library, London; noted in *The Newcomen Bulletin,* No. 80 (Mar. 1967). No notion of depth or breadth of the collection is suggested, however.

Several hundred biographical references are given in each of the three volumes (in four) of **Royal Society,** *Subject Index* (p. 146);

Vol. 1, Mathematics; Vol. 2, Mechanics; Vol. 3, pt. I, Heat, Light, Sound; Vol. 3, pt. II, Electricity and Magnetism.

The New York Public Library has an extensive card file on biography in the Science and Technology Division. A file of portraits is also maintained in this division.

An occasionally helpful last resort is *The Times* [London], *Palmer's Index to The Times Newspaper* (601 quarterly vols., 1790–1941)— see "Death Notices"; *Annual Index to The Times* (1906–1913); and its continuation *Official Index to The Times* (1914——).

The New York Times Index appeared first in 1913. Accumulated annually.

Finally, no search for biographical data is complete until the leads suggested in *Harvard Guide* (p. 21), pp. 188–206, have been exhausted.

C. Other Biographical Works

1. *Collected Biographies*

Phyllis M. Riches, *An Analytical Bibliography of Universal Collected Biography* (London, 1934, 4to, 709 pp.). The collected works, in English, touch many fields. The name index, pp. 1–541, is keyed to the list of collected biographies, pp. 577–632. A subject index, pp. 633–704, is followed by a subject index of collected biographies, pp. 705–709.

Most single-volume collections of short biographical sketches are valuable in direct proportion to the obscurity of their subjects, for they seldom contribute important verifiable information about well-known men.

The five following titles and many more (about seventy-five) are analyzed in **Higgins** (p. 78). At the end of each of his sections are listed the pertinent collected works, with brief summaries of contents.

Henry Howe, *Memoirs of the Most Eminent American Mechanics* (New York, 1844). Popular in style.

Charles B. Stuart, *Lives and Works of Civil and Military Engineers of America* (New York, 1871). Popular.

George Iles, *Leading American Inventors* (New York, 1912). Popular.

Samuel Smiles, *Industrial Biography: Iron-workers and Tool-makers* (London: John Murray, 1863, 410 pp.). Sketches of Yarranton, Huntsman, Cort, Roebuck, Mushet, Nielson, Bramah, Maudslay, Clement, Fox, Murray, Fairbairn and so forth. Generally valuable. Reprinted recently (Newton Abbot, Devon: David & Charles).

Samuel Smiles, *Lives of the Engineers.* Many editions, listed in **Hughes,** *Selections,* following. Full-length biographies of Watt, Boulton, Smeaton, George and Robert Stephenson, and others. Smiles was a romantic and didactic biographer, but in his books much factual information is preserved that would otherwise have perished.

Thomas P. Hughes, Ed., *Selections from Lives of the Engineers . . . by Samuel Smiles* (Cambridge, Mass.: The M.I.T. Press, 1966, 447 pp.). An introduction of twenty-nine pages is followed by the lives of Brindley, Rennie, and Telford, a section of editor's notes, pp. 418–429, and a bibliography of Smiles's books.

Samuel Smiles, *Men of Invention and Industry* (London: John Murray, 1884). Listed in **Hughes** (preceding); I have not seen this book. John Lombe, silk worker; John Harrison, chronometer builder; Friedrich Koenig, printing-press builder; and others are noticed.

Bishop, *American Manufactures* (p. 256), in Vol. 1, pp. 633–639, has short articles on over two dozen inventors and builders. Seth Boyden, Hotchkiss, Sawyer, LeVan, Esterly, Gray, Lamb, Knowles, and Pennock are among the names that were known in the 1860's and should be readily recoverable now. Other biographical information is in Vols. 2 and 3, *passim.*

R. S. Kirby, Ed., *Inventors and Engineers of Old New Haven* (New Haven, 1939). Six papers, including one on Yale engineers and one on the founding of Sheffield School.

J. D. Van Slyck, *New England Manufacturers and Manufactories* (2 vols., Boston, 1879) is one of those lavishly illustrated "prestige" books, paid for by a few wealthy subscribers, that make one thankful for the vanity of one's forebears. In addition to the nearly 250 steel-engraved portraits and views of works, the 750 pages of text contain an important quantity of biographical data on Corliss, Billings, Brown, Fairbanks, Scovill, Sturtevant, Wheeler, Wilson, Whitin, Benedict, Burnham, and many others.

A list of French collected works is in **Russo,** Bibliographie (p. 17), pp. 29–31.

F. P. H. Tarbé de St. Hardouin (catalogued under Tarbé), *Notices Biographiques sur les Ingénieurs des Ponts et Chaussées* (Paris, 1884, 276 pp.) was among the titles cited by **Kirby and Laurson** (p. 220).

Carlo Promis, *Biografie di Ingegneri Militari Italiani dal Secolo XIV alla Metà del XVIII* (Torino, 1874, 858 pp.). Short biographies and notices of printed and manuscript works of one hundred Italian military engineers, mostly of the fifteenth and sixteenth centuries. Brunelleschi, Ramelli, Lorini are included; Filarete, Ghiberti, Sangallo, Tartaglia, Vigevano, Zonca are not. Only eighteen of the men treated by Promis are also listed in **Horst de la Croix**, "Literature on Fortification" (p. 40).

The National Academy of Sciences, Washington, D.C., has published *Biographical Memoirs*, Vols. 1–38 (1877–1965), a continuing series now published by Columbia University Press. An index to the memoirs of Vols. 1–35 is in 36 (1962), 331–336. In *A History of the First Half-Century of the National Academy of Sciences 1863–1913* (Washington, 1913), there are biographical sketches of the founders, lists of members, living and dead, and contents of the first seven volumes of *Memoirs*. The general level of writing is quite good. One finds in N.A.S. such technical men as Dahlgren, Saxton, Eads, Meigs, Langley, William Sellers, and John G. Barnard.

Obituary notices of deceased Fellows of the Royal Society have been published at least since 1854. See **Royal Society of London**, *Obituaries of Deceased Fellows chiefly for the period 1898–1904, with a General Index to Previous Obituary Notices* (London, 1905 [*Proceedings*, Vol. 75]); from 1905 to 1932, notices appeared in the *Proceedings*. After 1932, see *Obituary Notices of Fellows* (9 vols., Nos. 1–23, London, Dec. 1932–Nov. 1954); and *Biographical Memoirs of Fellows*, Vol. 1 (London, 1955), the first of the current series.

2. *Biographies in Local Histories*

Works of local and regional coverage sometimes contain the only recorded data on locally prominent men. Many of these books of the 1880–1900 period are well printed on good paper and have steel-engraved portraits that are infinitely better than the halftone cuts that we tolerate today. More modest local histories and publications of local history societies are likely to have more extensive treatment of those people who are noticed.

The *Harvard Guide* (p. 21), pp. 217–237, has a full discussion of bibliographies and finding aids for the appropriate local histories, plus a selective list of state and local works. It is unnecessary, therefore, to repeat what appears there; it is only necessary to emphasize that the *Guide* can lead one directly to a fruitful source of fresh, often unique, information. Three examples follow.

J. A. Spalding, *Illustrated Popular Biography of Connecticut* (2 vols., Hartford, 1891–1901). The first volume contains capsule biographies and passable line portraits of about 20 (then living) engineers and mechanics, out of a total of perhaps 600 subjects. J. M. Allen, of Hartford Steam Boiler Inspection and Insurance Co., Ezra Bailey, of

Horton Co., Francis H. Richards, Pratt, Whitney, Billings, Gatling, H. R. Towne, Seth Barnes, and Christopher M. Spencer are among those represented. The second volume followed the same general pattern.

J. Thomas Scharf and Thompson Westcott, *History of Philadelphia 1609–1884* (3 vols., 1884) has a section of 120 pages on industrial history, including biographical data, and has also handsome full-page portraits of R. D. Wood, William Sellers, Disston, Harrison, Bement, Wheeler, and others.

Charles Nutt, *History of Worcester and Its People* (4 vols., New York, 1919). Biographies are in Vols. 3 and 4. An outline of industrial history, pp. 1,068–1,079, contains references to biographies. Some portraits: O. S. Walker (magnetic chuck) and Osgood Plummer (machine tools), for example, but no portraits of such important men as Washburn and Moen (wiremakers).

Thomas L. Bradford, *The Bibliographer's Manual of American History* (5 vols., Philadelphia: S. V. Henkels, 1907–1910). Some 6,000 titles of state, territory, county, and town histories. **Larson,** *Guide* (p. 23), describes this work as "useful though incomplete."

Clarence Stewart Peterson, *Consolidated Bibliography of County Histories* (Baltimore: the Author, 1961). About 3,000 titles; the author has tried to list all works of at least 100 pages.

Dorothea N. Spear, *Bibliography of American Directories through 1860* (Worcester: American Antiquarian Society, 1961). A checklist of 1,647 city and town directories, giving generally several locations of each title. A very substantial amount of biographical information can be developed through systematic use of city and town directories. Library of Congress has a good collection, but not on open shelves.

3. *Individual Autobiographies*

The following list is indicative rather than comprehensive or even representative. Such works can sometimes be located through biographical sketches in dictionaries and through the indexes listed in previous sections.

Harry Brearley, *Knotted String. Autobiography of a Steel-Maker* (London: Longmans, Green, 1941). Memorable account of Englishman who first (*ca.* 1912) saw possibilities in chromium-alloy stainless steel. Vivid descriptions of slums of Sheffield around 1880, from the inside looking out. Many penetrating comments on industrial facts and fancies by a maverick industrialist.

John Thomas Broderick (b. 1866), *Forty Years with General Electric* (Albany, N.Y.: Fort Orange Press, 1929; 218 pp., portrait).

John Brunton, *John Brunton's Book* (Cambridge, Mass., 1939; 164 pp., portrait). Autobiography of a British civil engineer in India and England.

Alexander G. Christie (1880–1964), *What Does an Engineer Do?* (New York: Vantage, 1963, 306 pp.). Autobiography of mechanical engineer, steam-turbine and heat-power consultant, teacher at Cornell, Wisconsin, and Johns Hopkins.

Mortimer E. Cooley (b. 1855), *Scientific Blacksmith* (Ann Arbor: University of Michigan Press, 1947, 290 pp.). The author was dean of engineering at Michigan.

Grenville M. Dodge (1831–1916), *How We Built the Union Pacific Railway and Other Railway Papers and Addresses* (Council Bluffs, Iowa, ca. 1911–1914; reprinted Denver: Alan Swallow, 1965, 171 pp.).

S. C. Ells, *Recollections of the Development of the Athabasca Oil Sands* (Ottawa: Canadian Dept. of Mines and Technical Surveys, July 1962 [Information Circular IC-139], 114 pp.). An account of exploration and exploitation of the oil sands, located 250 miles north of Edmonton, Alberta, for road-paving material and crude petroleum, 1912–1945. The author's manuscript memoirs, in two volumes, are in the Public Archives in Ottawa.

William L. Emmet (b. 1859), *The Autobiography of an Engineer* (2nd ed., New York: A.S.M.E., 1940; 233 pp., portrait). An American electrical engineer writes of the central-station electric-power industry.

Henry Ericsson (1861–?), *Sixty Years a Builder* (Chicago: A. Kroch, 1942; 388 pp., illustrated). Autobiography of a Chicago builder.

J. A. Fleming, *Fifty Years of Electricity. The Memories of an Electrical Engineer* (London, 1921; 384 pp., illustrated). Fleming (1849–1945) was a teacher and innovator; perhaps he is best known as inventor of the Fleming "valve."

Francis Fox (1844–1927), *Sixty-three Years of Engineering* (London, 1924, 338 pp.). Builder of railways, consultant on Simplon Tunnel, restorer of cathedrals and other venerable buildings.

John Fritz (1822–1913), *Autobiography of John Fritz* (New York: A.S.M.E., 1912; 326 pp., portrait). The author was one of the leading iron- and steelmakers in the United States.

John B. Jervis (1795–1885), "A Memoir of American Engineering," American Society of Civil Engineers, *Transactions*, 6 (1877), 39–67. Valuable in conveying the point of view of an early and prominent American engineer.

Dexter S. Kimball (1865–1952), *I Remember* (New York: McGraw-Hill, 1953, 259 pp.). From engineering in California in the 1880's, the author turned finally to engineering education.

Benjamin Garver Lamme (1864–1924), *An Autobiography* (New York: Putnam, 1926, 271 pp.). The author was an electrical engineer employed by Westinghouse.

Paul W. Litchfield (b. 1875), *Industrial Voyage* (New York: Doubleday, 1954, 347 pp.). Autobiography of the president of Goodyear Tire & Rubber Co., a curiously shallow man who had enthusiasm, energy, and a full measure of the team spirit. He was more concerned about labor than many managers; he came close to recognizing the atrophying influences of mass production, but he supplied sports and other diversions instead of attacking the problem directly.

Hiram P. Maxim, *A Genius in the Family* (New York: Harper, 1936). Reprint (New York: Dover, 1962). A son's recollections of Hiram S. Maxim (1840–1916), brother of Hudson.

Hiram P. Maxim (1869–1936), *Horseless Carriage Days* (New York: Harper, 1937). Reprint (New York: Dover, 1962). On the author's experiences in early automobile development.

Hudson Maxim, *Reminiscences and Comments as Reported by Clifton Johnson* (Garden City: Doubleday, 1924, 350 pp.). Author was responsible for smokeless powder.

Charles T. Porter (b. 1826), *Engineering Reminiscences Contributed to "Power" and "American Machinist"* (New York: Wiley, 1908; 335 pp., illustrated). Adventures in the United States and England of a designer and builder of steam engines.

George Escol Sellers, *Early Engineering Reminiscences, 1815–1840*, edited by Eugene S. Ferguson (Washington: Smithsonian Institution, 1965 [U.S.N.M., *Bulletin 238*], 203 pp., 84 illustrations). Detailed contemporary information on Philadelphia machine and fire-engine trade, U.S. mint, papermaking, and engraving. Author visited England in 1832, saw shops of Maudslay, Donkin, and others.

Werner von Siemens (1816–1892), *Inventor and Entrepreneur; Recollections of Werner von Siemens* (2nd English ed., London: Lund,

Humphries, 1966, 314 pp.). Reissue of the first English edition (1893), to which have been added many annotations and illustrations. Siemens was concerned with the electric telegraph, submarine cables, rotating electrical machinery, etc. An informative review by John J. Beer is in *Science, 153* (July 22, 1966), 405.

John F. Stevens (1853–1943), *An Engineer's Recollections* (New York, 1935; 70 pp., illustrated). On the Canadian Pacific and Great Northern railroads and the Panama Canal. Reprinted from *Engineering News-Record* between Mar. 21 and Nov. 21, 1935.

George Templeton Strong, *Diary*; edited by Allan Nevins and Milton H. Thomas (4 vols., New York: Macmillan, 1952). Valuable for frequent sharp comments on railroad and steamboat travel and other encounters with technology in the United States, 1835–1875.

Frank G. Tatnall, *Tatnall on Testing* (Metals Park, Ohio: American Society for Metals, 1966, 234 pp.). A lively and well-constructed narrative of the author's career in developing and promoting testing machines and devices, particularly the SR-4 strain gauge. Very good.

D. Portraits

There is one standard finding list for portraits: the **A.L.A. Portrait Index**, William C. Lane and Nina E. Browne, Eds. (Washington: Library of Congress, 1906, 1,600 pp.). While there are fewer technical people represented than one might wish, there are some, and reference is easy. In this *Index* are analyzed such titles as Royal Society of London *Proceedings* (1830–1904), *Appleton's Cyclopaedia of American Biography* (full-page plates only), *Gleason's, Ballou's, Century, Harper's New Monthly Magazine, Illustrated London News,* Van Slyck (p. 82), and some state histories and proceedings of historical societies. First rate.

The Library of Congress has a continuation of the *A.L.A. Portrait Index,* unpublished, on three-by-five cards, as it was prepared during the W.P.A. days (*ca.* 1940).

New York Public Library, Science and Technology Division, maintains a portrait file. The pictures range from good to mediocre, but in the present state of technological history it is often an accomplishment to locate any portraits at all of a good many men who are remembered.

In the **Smithsonian Institution, Division of Mechanical and Civil Engineering,** Robert M. Vogel has assembled and listed the nucleus of an important technical portrait file.

Library of Congress, Prints and Photographs Division, has an extensive but general portrait file. Perhaps two dozen American and English nineteenth-century tool builders—Maudslay, Roberts, Spencer, Pratt, Saxton, for example—were sought here recently, with negative results. It should be added, however, that they were sought here because they were not in the Smithsonian file at that time.

Paul Vanderbilt, *Guide to the Special Collections of Prints & Photographs in the Library of Congress* (Washington, 1955, 200 pp.) has an especially good analytical index. "Portraits," with about 180 entries, includes some technical people: M. C. Meigs and associates, Wright brothers and associates, Jones collection of aviators and inventors, and others not directly identifiable. This invaluable *Guide* is further described on p. 124.

The Frick Art Reference Library (10 East 71st St., New York City) has extensive photographic files of painted portraits—for example, most of the Charles Willson Peale portraits—gathered from many galleries and private collections. Because the staff is very small, it is necessary to visit the Library when inquiring about more than one or two names. Written permission of a portrait's owner is required before a photograph can be supplied.

Great Britain, Patent Museum, *Catalogue of the* [National] *Gallery of Portraits of Inventors, Discoverers, and Introducers of Useful Arts* (formed by Bennet Woodcroft). (Editions 1–5.) (London, 1855–1859.) This entry has been copied from a register of the Patent Office Library. I have not seen the title nor do I have further details.

British Museum, *Catalogue of Engraved British Portraits Preserved in the . . . British Museum* (6 vols., London, 1908–1925). Slightly better for British engineers than the *A.L.A. Portrait Index*.

David Piper, Comp., *Catalogue of Seventeenth Century Portraits in the National Portrait Gallery 1625–1714* (Cambridge: The University Press, 1963, 410 pp., illustrated). The first volume of a projected series to describe all portraits in the National Portrait Gallery in London. The next volume will describe earlier portraits; later volumes will treat those up to the present.

The Science Museum (London), "Photographs and Lantern Slides," (p. 125), records an available list of "Portraits" (List No. 60) which includes about 300 technologists and scientists. With few exceptions, these are the more prominent men in each field.

Deutsches Museum, Munich, has a very good portrait file as well as an outstanding collection of technical illustrations. Replies to

BIOGRAPHY 89

my inquiries for individual pictures have been uniformly prompt and helpful.

Paris, Bibliothèque Nationale, *Catalogue de la Collection des Portraits Français et Etrangers* (6 vols., Paris, 1896–1907). Completed only through "Louis Philippe." Of very limited value.

Collison, *Bibliographies* (p. 20), p. 79, lists four portrait indexes in German and notes that other indexes "less familiar in the libraries of the English-speaking world" are described in **Mary W. Chamberlin,** *Guide to Art Reference Books* (Chicago: American Library Association, 1959).

Portraits often appear in book-length biographies. Very occasionally a portrait will be found in British learned society proceedings. **Bishop,** *American Manufactures* (pp. 82, 256), has dozens of good, though small, portraits. Portraits in **Van Slyck** (p. 82) and in local histories have already been noted. *Appleton's Dictionary of Machines* (p. 63) has steel engravings of Steers, McKay, Daguerre, Senefelder, R. M. Hoe, Whistler, Perkins, Wells, Peter Nicholson, Evans, Slater, Goodyear, Hutton, James Reynolds, Bigelow, N. Wheeler, and a dozen others. Portraits of Evans and Perkins, perhaps others, should be approached critically. Schuselle's 1862 engraving of *American Scientists and Inventors* (p. 17) is attractive and probably accurate.

Because of the difficulties of locating portraits (and other pictorial material), the firms of The Bettmann Archive (136 E. 57th St., New York, N.Y. 10022), Culver Service (660 First Ave., New York, N.Y. 10016), and Brown Brothers (220 W. 42nd St., New York, N.Y. 10036) have been successful. A charge is made for service, and restrictions are placed upon reproduction.

See also Part IX, Illustrations.

VII. GOVERNMENT PUBLICATIONS AND RECORDS

A. Guides to the Records

The several well-known manuals of government publications and records were not written from the point of view of the historian of technology, and the scholar may come away from his study of such manuals with the impression that materials in his subject field are either unimportant or so buried among vast general collections as to be economically irretrievable. For example, he might miss completely what is probably the most important single body of manuscript material on American technology—patent records in the National Archives—because the significance of the records has not yet been evident to more than a handful of people; also, on the most superficial test of usefulness—namely, whether the records have been used—this whole great collection has thus far failed miserably.

In this section, I have called particular attention to the existence of some records and finding aids that appear to me to be significant. Generally, only records through the nineteenth century are listed, later ones being better categorized and indexed.

1. *United States*

An enormous amount of contemporary evidence and a surprising number of historical works on canals, railroads, bridges, steamboats, boiler explosions, international exhibitions, manufactures, mineral deposits, and so forth, are contained in the printed document series of the government.

It is easy to be baffled by the very quantity of paper involved, and there is no simple key to the series; but a moderate investment of time is likely to yield choice and important material, much of which will be fresh and often quite lively.

Two standard works that are useful to an understanding of the nature of the many series and of occasional works and that list available indexes and guides follow:

Anne M. Boyd and Rae E. Rips, *United States Government Publications* (3rd ed., New York: H. W. Wilson, 1949).

Laurence F. Schmeckebier and Roy B. Eastin, *Government Publications and Their Use* (3rd ed., Washington: Brookings Institution, 1960).

For a critical discussion of these and other guides and checklists, see the *Harvard Guide* (p. 21), pp. 112–115. See also in *ibid.*, pp. 119–125: "Federal Public Records."

One may reasonably wonder why a satisfactory general catalogue of government publications does not exist, at least on cards. Lacking such a catalogue, however, one can be grateful for the vision and industry of Clerk of the Senate Poore and Public Printer Ames, whose works are now indispensable.

Ben Perley Poore, Comp., *A Descriptive Catalogue of the Government Publications of the United States, September 5, 1774–March 4, 1881* (Washington, 1885, [Senate, Misc. Doc. 67, 48 Cong., 2 sess.], 4to, 1,392 pp.). Reprint (Ann Arbor: J. W. Edwards, 1953).

Poore's *Descriptive Catalogue* lists some 60,000 documents, varying in length from one page to several volumes. An index of over 50,000 entries is incomplete and exasperating but useful if the reader will supply patience and resourcefulness. Poore is castigated in most bibliographies, but there would be an alarming vacuum if this catalogue did not exist. Index entries are full, if not always complete, for such subjects as, for example, B. & O. R.R., C. & D. and C. & O. Canals, Cumberland Road, Mississippi River bridges, Potomac River bridges (see Washington city), Oliver Evans's patents, Rumsey's steamboats, Samuel Colt's lobbying, Fulton's claims and torpedo experiments, military academy instruction, armories and arms, international exhibitions (see city involved: there is no index entry "international exhibitions"), and many others. However, a line-by-line search of the catalogue, which can be made in a few evenings, will uncover many titles that are not indexed, some quite important ones. The following titles, not readily found in the index, will indicate the nature of the buried treasure. Subject and date can be translated into complete titles by reference to Poore, which is arranged chronologically.

Summary of western river navigation and bridges (1,354 pp.) June 7, 1878; history of steam marine of U.S. (119 pp.) Jan. 20, 1852; the bulkiest of over fifty reports on boiler and steamboat safety are (192 pp.) May 18, 1832, (95 pp.) Feb. 26, 1836, (472 pp.) Dec. 12,

1838, (94 pp.) Dec. 15, 1847, (184 pp., 10 pls.) Dec. 30, 1848; statistics of railroads (185 pp.) Nov. 27, 1856; report on southern railroads (1,057 pp.) Mar. 2, 1867; history of plans for railroads and canals between Atlantic and Pacific oceans (679 pp.) Feb. 20, 1849; heating and ventilating the Capitol (10 pp.) Feb. 19, 1817, (20 pp.) Jan. 8, 1844, (254 pp.) Jan. 26, 1860, (96 pp.) May 7, 1866, (212 pp.) Mar. 3, 1871; atmospheric telegraph for rapid transmission of packages (7 pp.) July 2, 1856; report on immigration, giving wages and costs of rent and groceries in industrial districts (231 pp.) Mar. 14, 1871; report on performance tests of American coals (607 pp.) June 6, 1844.

U.S., Department of the Interior, Division of Documents, *Comprehensive Index to the Publications of the United States Government, 1881–1893* (2 vols., Washington, 1905). This alphabetical listing of documents by subject was prepared by John G. Ames to fill the gap between Poore's and the first of the periodical catalogues of the Superintendent of Documents. For details of the latter series, see *Harvard Guide* (p. 21), p. 114.

U.S., Superintendent of Documents, *Checklist of United States Public Documents, 1789–1909* (3rd ed., Vol. 1, Washington, 1911, 1,707 pp.). Vol. 2, the index, never appeared. This work is useful in at least four ways. First, the serial numbers of the Congressional Documents series are listed. A powerful finding aid is thus provided, because many libraries shelve the Document series by serial numbers. This numbering scheme, originated by John G. Ames (1834–1910), begins with documents of the Fifteenth Congress and makes more manageable a vast body of government publications. Second, the preface, pp. vii–xxi, is an informative historical essay describing the many document catalogues and indexes. Third, this checklist is a definitive guide to a series of publications whose existence may be suggested but whose extent is not given by a particular library's catalogue. Finally, the grouping of publications by office of origin provides the searcher with a different and sometimes fruitful approach. I recommend careful perusal of this work, whose value only gradually became evident to me.

A finding list for printed minutes of congressional hearings is **Harold O. Thomen,** *Checklist of Hearings before Congressional Committees through the 67th Congress* (7 vols., Washington: Library of Congress, 1957–1959). There are relatively very few entries before 1900, but a reading will reveal among other things material on Sutro Tunnel (1872), Panama and other canals, and bridges (from 1869), "sweating system" (1892), "National multiroad highways" (1916), and "The Ladd Metric Bill, its fallacy and futility" (1922).

If you use **Thomen,** just cited, you should be aware also of **U.S. Congress, Senate Library,** *Index of Congressional Committee Hearings . . . prior to January 3, 1935 in the United States Senate Library* (Washington, 1935). Subject index, pp. 533–918.

Adelaide R. Hasse, *Index of Economic Materials in Documents of the States of the United States; Pennsylvania, 1789–1904* (1 vol. in 3, Washington: Carnegie Institution of Washington, 1919–1922, 4to., 1,711 pp.). A monumental compilation, by subjects, of printed reports of administrative officers, legislative committees, special commissions, and governors. See, for example, canals, ice, industries and manufactures, railroads, roads, water supply, and sewerage.

Volumes were issued also for California, Delaware, Illinois, Kentucky, Maine, Massachusetts, New Hampshire, New Jersey, New York, Ohio, Rhode Island, and Vermont. For detailed list, see entry CH8 in **Winchell,** *Guide to Reference Books* (p. 33).

2. *Other Countries*

In addition to treating British domestic engineering and industry, the voluminous **Parliamentary Papers of Great Britain** contain a moderate amount of useful information on American technology. Pertinent transcripts of minutes of hearings and reports of commissions provide rather full descriptions from an often fresh point of view of particular segments of American industry and engineering. For example, in the House of Commons series one finds enlightening reports and testimony on Artizans and Machinery (1824, Vol. 5); Exports of Tools and Machinery (1825, Vol. 5); American Cotton Mills (1833, Vol. 6); Exportation of Machinery (1841, Vol. 7); Reports on the New York Industrial Exhibition of 1853 (1854, Vol. 36); Report on the Committee on the Machinery of the United States (1854–1855, Vol. 50). The Royal Commission on Technical Instruction issued its report in six volumes, 1882–1884, which contained an account of a visit by a commissioner to American and other foreign technical colleges. Several indexes are listed below.

Starting in 1968, the Irish University Press, 77 Marlboro Street, Dublin 1, will issue a monumental series of reprints of nineteenth-century Parliamentary Papers. Some 700 volumes are planned. Microform reproduction of the same series will also be available from the Irish University Press.

Great Britain, Parliament, House of Commons, *Hansard's Catalogue and Breviate of Parliamentary Papers 1696–1834* (reprinted with an introduction by Percy and Grace Ford, Oxford, 1953, fo., 220 pp.).

Percy and Grace Ford, *Select List of Parliamentary Papers 1833–1899* (Oxford, 1953). Contains a subject index.

For one who expects to make extensive use of the papers, there is **Percy and Grace Ford**, *A Guide to Parliamentary Papers, What they are: How to find them: How to use them* (Oxford, 1955, 79 pp.).

Great Britain, Parliament, House of Commons, *General Index to the Accounts and Papers, Reports of Commissions, &c, &c . . . 1801–1852* (London, 1938, fo., 1,080 pp.).

Great Britain, Parliament, House of Lords, *General Index to the Sessional Papers . . . 1801–1859* (London, 1860). Reprinted, 1938.

David Landes, in the bibliography of his *Technological Change* (p. 22) provides a convenient key to public records of Great Britain, France, Belgium, Germany, Switzerland, Austria, Italy, and Sweden. Landes lists the available published guides and bibliographies.

The American Historical Association's Guide (p. 18) lists under individual countries finding aids for government records and publications.

Winifred Gregory, Ed., *List of the Serial Publications of Foreign Governments 1815–1931* (New York: H. W. Wilson, 1932, fo., 720 pp.). Incomplete but not yet superseded, this work serves to identify as well as locate older government serials.

New Serials Titles (p. 142) lists many recent government periodicals of the United States, Canada, Great Britain, Germany, and other countries.

B. Departmental and Other Publications (U.S.)

1. Census Records and Publications

Many of the original census schedules are in the **National Archives.** For information on holdings and guidance in their use, see Part D, on National Archives. See particularly **Fishbein**, "Censuses" (p. 111).

Of primary importance also are the publications of the Census Office, starting with the well-known **Tench Coxe**, *A Statement of Arts and Manufactures . . . for the Year 1810* (Washington, 1814) and extending to the present. Particularly valuable are the volumes resulting from the censuses of 1860 and 1880.

U.S., Census Office, *Manufactures of the United States in 1860* (Washington, 1865), while largely statistical, contains an introductory dissertation of 217 pages on the history and present state of manufactures.

U.S., Census Office, *Tenth Census* (22 vols., Washington, 1883–1888). Very lengthy prose reports were prepared by a number of able consultants. Vol. 2 contains a "Report on the Manufactures . . ." (476 pp.), "Statistics of Power Used . . ." (33 pp.), "Factory System in U.S." (78 pp.), "Manufacture of Interchangeable Mechanism" (88 pp.), and "Manufacture of Hardware, Cutlery, etc." (19 pp.). Vol. 4 is a report on transportation; Vol. 8 is on shipbuilding; Vol. 10 on petroleum, coke, and quarry industry; Vol. 15 on mining industries; Vols. 16 and 17 on water power; Vol. 22 on power and machinery employed in manufactures. Many good illustrations.

Henry J. Dubester, *Catalog of United States Census Publications 1790–1945* (Washington, 1950, 320 pp.) is a complete and convenient guide to Census publications. The first eleven censuses occupy fewer than thirty pages of the total.

Edward C. Lunt, "Key to the Publications of the United States Census, 1790–1887," *Publications of the American Statistical Association* (Boston, June, Sept., 1888 [N.S., Nos. 2,3]), pp. 63–125. A history of the U.S. Census, pp. 71–93, is followed by chronological and subject listings.

Carroll D. Wright and William C. Hunt, *The History and Growth of the United States Census* (Washington: G.P.O., 1900, 967 pp.). A history, pp. 7–130, is followed by schedules of inquiries used in the various censuses, pp. 131–910, publications following each of the first eleven censuses, pp. 911–914, and the census acts. This book is of particular value for the voluminous schedules of inquiries. From these may be learned the kinds of information that one may seek in the census return, in both published and unpublished form.

All census publications through 1890 have been put on forty-two rolls of microfilm, copies of which are available through the **National Archives.**

Not listed by Dubester or Lunt, because it is not a census document, is the "McLane Report," which was issued by the Treasury Department. This widely used statistical work is (**Louis McLane**) U.S., Treasury Dept., *Documents Relating to the Manufactures in the United States* (2 vols., Washington, 1833).

U.S., Bureau of the Census, *Historical Statistics of the United States, Colonial Times to 1957* (Washington, 1960, 789 pp.), is a carefully compiled series of tables, covering varying periods, on manufacturing, transportation, communications, power, population, productivity, technological development, and so forth. Sources for each table are discussed. A handy compendium of numerical information. A *Continuation to 1962, and Revision* was published in 1965.

2. Smithsonian Institution Publications

The searcher might summarily dismiss the publications of the Smithsonian Institution as being concerned mainly with bugs and fossils because there is a gap in the published indexes that is not readily evident. The several series of publications, some of which have changed in nature and emphasis over the years, are quite bewildering to contemplate, and I can point to no single, straightforward discussion of the scope of the various series. Suffice it to say, however, that there has been published by the Smithsonian Institution a large quantity of material that can be useful to the technological historian.

The only complete indexes of publications are those prepared by William J. Rhees, whose last list appeared in 1904. His 328-page exhaustive *Catalogue of Publications of the Smithsonian Institution, 1846-1882, with an alphabetical index of articles in the Smithsonian Contributions to Knowledge, Miscellaneous Collections, Annual Reports, Bulletins and Proceedings of the U.S. National Museum* . . . appeared in Vol. 27 of Smithsonian Miscellaneous Collections (Washington, 1882).

An expanded version of 383 pages, including later publications, was made a part of the 1886 *Annual Report of the Board of Regents of the Smithsonian Institution.*

Subsequently, Rhees published several checklists, the last being *List of Smithsonian Publications 1846–1903* (Washington, 1904 [Misc. Coll., Vol. 44]). This 100-page list, which gives short titles, is a valuable supplement to the earlier catalogues.

A List and Index of the Publications of the U.S. National Museum 1875–1946 (Washington, 1947 [U.S.N.M., *Bulletin 193*], 306 pp.). This publication is misleading to one not versed in the organizational scheme of the Smithsonian Institution because it does not mention the existence of the great number of titles issued by other branches of the Institution during the years covered. Pertinent works can be captured, however, by (1) reading the catalogues and lists named here, then (2) perusing the tables of contents of Miscellaneous Collections, 1903–present.

Ruth M. Stemple, *Author-Subject Index to Articles in Smithsonian Annual Reports 1849–1961* (Washington: Smithsonian Institution, 1963, [Publication 4503], 200 pp.). A very welcome key to the many first-rate articles appended to the *Annual Report of the Board of Regents.* See, for example, Joseph Henry on testing of building materials, with an account of the marble used in the extension of the U.S. Capitol (1856), several articles on the telegraph (1857 ff.), report on the state of knowledge of radiant heat (1859), account of roads and bridges (1860), history of petroleum (1861), history of the Paris

Académie des Sciences (1862), an article on the scientific education of mechanics and artisans (1872), memoir of Charles Babbage (1873), Francis Fox on great Alpine tunnels (1901), Charles Parsons on turbines (1907), Frank Sprague on electric locomotives (1907), review of forty years of technological chemistry (1908).

The *Annual Report of the Board of Regents* has always carried in an appendix a series of valuable articles, indexed as shown. From 1884 through 1904 the *Annual Report of the U.S. National Museum* contained an entirely separate series, including such classics as Hough's "Fire-making Apparatus in the U.S. National Museum" (1888), Watkins's "Development of American Rail and Track" (1889), and McGuire's "A Study of the Primitive Methods of Drilling" (1894). The *Annual Report of the U.S. National Museum* was published as Part II of the *Annual Report of the Board of Regents,* although "Part II" does not usually appear on title page.

An ambitious bibliographic undertaking, in conjunction with A.A.A.S., yielded such works as the following:

Alfred Tuckerman, *Index to the Literature of Thermodynamics* (1890 [Misc. Coll., Vol. 34], 243 pp.).

H. C. Bolton, *Select Bibliography of Chemistry, 1492–1892* (1893 [Misc. Coll., Vol. 36], 1,225 pp.), with massive supplements in 1899, 1901, and 1904.

H. C. Bolton, *A Catalogue of Scientific and Technical Periodicals, 1665–1895, together with Chronological Tables and a Library Checklist* (2nd ed., 1897 [Misc. Coll., Vol. 40], 8,600 titles).

There were also extensive bibliographies on metals such as platinum, manganese, columbium, uranium, thallium, and zirconium, citing literature from as early as 1815.

Smithsonian Annals of Flight. Published since 1964 by the National Air Museum, this series has included short (under 100 pages) monographs on the first U.S. coast-to-coast nonstop flight and the first airplane Diesel engine. I have see only two numbers thus far, both published in 1964.

George Brown Goode, Ed., *The Smithsonian Institution 1846–1896. The History of Its First Half Century* (Washington, 1897; 856 pp., illustrated). A useful compilation of much information.

The Library of the Museum of History and Technology has strong collections of the publications of technical and scientific museums, which publications are frequently difficult to identify or locate. In addition, a significant group of 5,000 volumes of technical monographic works (including a set of the **Académie des Sciences**, *Descriptions* [p. 58]) was transferred in 1964 from the Scientific Library of the Patent Office.

3. *Army, Corps of Engineers*

U.S. Army, Corps of Engineers, *Report of the Chief of Engineers* (1851/1852 +). There have been several indexes, as listed, but none for the years 1851–1865.

U.S. Army, Corps of Engineers, *Analytical and Topical Index to the Reports of the Chief of Engineers and the Officers of the Corps of Engineers . . . 1866–1900* (3 vols., Washington, 1902–1903). Vols. 1–2, river and harbor works; Vol. 3, fortifications, bridges, miscellaneous. According to a note in the preface, this index supersedes earlier volumes for 1866–1879, 1880–1887, and 1888–1892.

U.S. Army, Corps of Engineers, *Index to the Reports of the Chief of Engineers, U.S. Army (including the Reports of the Isthmian Canal Commissions, 1899–1914) 1866–1912* (2 vols., Washington, 1915–1916, 3,055 pp.). A detailed, annotated index which superseded all earlier indexes. A supplementary volume for 1913–1917 was published in 1921.

U.S. Army, Corps of Engineers, *Professional Papers* (34 vols., Washington, 1841–1940). A series of monographs on diverse subjects, such as bitumen, sea-walls, bridges, piling, forts, artillery, the arch, lime and cement, mortar and concrete, geodesy, rivers, meteorology, geology, North Sea Canal in Holland, wave action, railroads, and so forth. Although this series has ended, occasional monographs may continue to appear as, for example, the following:

U.S. Army, Corps of Engineers, *The Hopper Dredge; Its History, Development, and Operation* [Frederick C. Scheffauer, Editor-in-chief] (Washington, 1954; 399 pp., illustrated). Bibliography, pp. 375–376.

Printed Papers of the Essayons Club (Willet's Point, N.Y.: Battalion Press), Nos. 1–50, 1868–1882 (no more). For example, No. 11 (Apr. 13, 1868) is Henry L. Abbot, "Notes on the Practical Gauging of Rivers."

U.S. Army, Corps of Engineers, *Professional Memoirs* (11 vols., Washington, 1909–1919). This was continued as *Military Engineer*, journal of the Society of American Military Engineers.

Government surveys of western United States—in the 1850's for railroad routes and both earlier and later for mapping—can be found readily through **Harvard Guide** (p. 21), p. 73.

4. *Consular Reports*

U.S., *Consular Reports*. Issued in monthly numbers, half-numbers used for second issue in a month, bound in volumes. Nos. 1–273 (1880–1903) were published by Bureau of Foreign Commerce, Department of State; Nos. 274–297 (1903–1905) were published by Statistics Bureau, Department of Commerce and Labor; and Nos. 298–357 (1905–1910) were published by Manufactures Bureau, Department of Commerce and Labor. Discontinued with No. 357. This series contains news and notes of interest to traders and manufacturers. A great deal of technical information can be discovered through page-by-page search. I am unaware of any considerable use that has been made of these in the history of technology. A number of cumulative indexes, through 1900, were compiled and published; they are listed in *Checklist . . . 1789–1909* (p. 93), p. 939.

U.S., *Special Consular Reports* (53 vols., published by the three bureaus as listed just previously for *Consular Reports*, 1890–1912). In these reports are extended discussions of products of foreign countries, conditions and regulations in foreign countries, and markets for U.S. products. Subjects include cotton textiles, files, carpets, beer, refrigeration, roads, canals, irrigation, coal, rubber, gas, staves, lead, zinc, flour, lumber, drugs, screws, nuts and bolts, disposal of sewage and garbage, paper, wood pulp, gas and oil engines, creameries, briquettes, and so forth. A complete list of subjects may be found on the Library of Congress printed cards for this series.

U.S., Manufactures Bureau, Department of Commerce and Labor, *Special Agents Series* (Vols. 1–226, Washington, 1906–1924). Another series of reports on foreign trade and industry.

5. *Other Agencies*

U.S. Congress, *American State Papers* (38 vols., Washington, 1832–1861, fo.). A wide and in many ways rich collection of documents of the period 1789 through 1823 to 1838, the terminal dates varying among volumes. The work is divided into ten classes, as follows: Class 1, Foreign Relations (6 vols.); class 2, Indian Affairs (2 vols.); class 3, Finance (5 vols.); class 4, Commerce and Navigation (2 vols.); class 5, Military Affairs (7 vols.); class 6, Naval Affairs (4 vols.); class 7, Post Office (1 vol.); class 8, Public Lands (8 vols.); class 9, Claims (1 vol.); class 10, Miscellaneous (2 vols.). The series and its publishing history are described in *Checklist* (p. 93), pp. 3–4.

U.S., Department of Agriculture, *Index to the Annual Reports of the U.S. Department of Agriculture for the Years 1837 to 1893, Inclusive* (Washington, 1896, 252 pp.). The reports were issued as a part of every annual report of the Commissioner of Patents (1837–1848), as a separate volume of the same (1849–1861), and, after 1863, as the report of the Secretary of Agriculture. Among the many subjects listed in the index are export ice trade (1848, 1850, 1863), plan of an industrial university (1851), Haarlem Lake drainage (1855), steam cultivators, traction engines (1859+), milk preservation (1861), Vermont marble (1862), road construction (1863+), cloth and paper from corn (1863), barn construction (1867–1868), corn harvester, cotton harvester (1874), testing machine for binder twine (1891).

U.S., Department of Agriculture, *Yearbook* (Washington, 1894+). Indexes were issued as follows: 1894–1900, 1901–1905, 1906–1910, and 1911–1915.

U.S., Board for Testing Iron, Steel, and Other Metals, *Report of the U.S. Board Appointed to Test Iron, Steel, and Other Metals* (2 vols., Washington, 1881). See also the following entry.

U.S., Ordnance Bureau, *Report of the Tests of Metals and Other Materials for Industrial Purposes . . . at Watertown Arsenal* (38 vols., Washington, 1882–1919). This series is a continuation of the reports of the U.S. Board for Testing, just listed. Extensive mechanical tests and photomicrographs were made of iron and other metals: guns, shapes, fabricated assemblies, rails, springs, rotating shafts, and so forth. Timber and concrete were also tested. Also issued was one *Index to the Reports . . . from 1881 to 1912* (1913, 240 pp.).

U.S., Industrial Commission, *Reports of the Industrial Commission* (19 vols., Washington, 1900–1902). A series of thick octavo volumes in which hearings were recorded on many aspects of trusts and industrial combinations, including labor and management relations and the regulation of public utilities. In the preface to Vol. 19, readers are told that they will find in the reports "many revelations of business methods and complications of which they knew little before, and will have a liberal education in the economic problems of the day." Vol. 19 stands alone as a summary, but the testimony in other volumes is richer in detail and insight. In Vol. 19 also is a series of indexes: list and contents of reports, pp. 1,135–1,138; list of witnesses, pp. 1,139–1,157; general index, pp. 1,159–1,236.

6. *Miscellaneous*

A. Hunter Dupree, *Science in the Federal Government* (Cambridge, Mass.: Harvard University Press, 1957). The "Bibliographic Note," pp. 387–394, is valuable for its comprehensive view of the record, and the reference notes, pp. 395–441, are bibliographic in scope. The author makes evident the bearing of nongovernmental, as well as official, sources on technical work in government offices.

Graham Adams, Jr., *Age of Industrial Violence, 1910–15: The Activities and Findings of the United States Commission on Industrial Relations* (New York: Columbia University Press, 1966, 316 pp.) is an example of a work based upon the records of a government commission, supplemented by contemporary newspapers and a wide variety of other sources. The government publications used by Adams in this study are listed in his bibliography, pp. 279–305. See also p. 300.

Several volumes of the **Brookings Institution** series entitled "Service Monographs of the United States Government" (66 vols., 1918–1934) are pertinent to the history of technology. The following sampling of titles indicates the nature of the series.

William Stull Holt, *The Office of the Chief of Engineers of the Army* (Monogr. No. 27, Baltimore, 1923). History is outlined on pp. 1–61; see also bibliography, pp. 155–161.

Darrell H. Smith, *The Panama Canal* (Monogr. No. 44, Baltimore, 1927). History, pp. 1–144; bibliography, pp. 375–394.

Gustavus A. Weber, *The Bureau of Standards* (Monogr. No. 35, Baltimore, 1925). History, pp. 1–75; bibliography, pp. 271–282.

C. Patents

Perhaps the largest body of primary source material that remains nearly untouched by historians is that of official records of patents for inventions. Starting with printed patent No. 1 (issued on July 13, 1836, to Senator John Ruggles of Thomaston, Maine, who framed the bill for reorganization of the Patent Office) for coglike wheels to increase locomotive traction, a practically unbroken record of American invention extends to the present. Many, but by no means all, patents issued from 1790 to 1836 (that is, before the Patent Office fire of 1836, which destroyed all models and records) are retrievable in manuscript form or in contemporary publications.

The records of U.S. patents are in three groups: printed patents in Patent Office (1836–present); manuscript materials in the National Archives (1790–about 1905); and the patent models.

1. *Finding Aids (U.S.)*

To locate a patent, given the name of the inventor, the subject, or the date, the following indexes cover the years since the patent system was inaugurated:

Edmund Burke, Comp., *List of Patents for Inventions and Designs, Issued by the United States, from 1790 to 1847* (Washington, 1847). The patents are classified by subject, and there is a separate index of patentees. (There were other lists and reports issued before 1847: for example, *A List of Patents . . . 1790 to . . . 1836* [Washington, 1872; facsimile of earlier list] and *A Digest of Patents . . . 1790 to . . . 1839*[Washington, 1840]. All, however, are lists only, not digests or abstracts, and all were superseded by Burke's list of 1847.)

From 1848 until 1919, Vol. 1 of each *Annual Report of the Commissioner of Patents* lists patents by subject and indexes patentees. From 1853 to 1868, the second volume of each Annual Report gives pictorial abstracts of patents, six to ten to a page, from 1854 to 1858 by classes, and in other years numerically.

The weekly *Official Gazette of the United States Patent Office* (from 1872) illustrates and summarizes patents.

From 1920 to the present, an annual alphabetical list of patents and patentees is given in the *Index of Patents Issued from the United States Patent Office.*

United States Patent Office, *Subject-Matter Index of Patents for Inventions . . . 1790 to 1873, inclusive* (3 vols., Washington, 1874) is useful, though often exasperating because of strange and erratic nomenclature.

Merle Randall and Evelyn B. Watson, *Finding List for United States Patent . . . Numbers* (Berkeley: University of California Press, 1938, 31 pp.). A useful handbook containing (1) list of patents in Congressional Document series, 1836–1871, giving dates, patent numbers, and serial numbers of documents; (2) analysis of *Official Gazette,* 1872–1938, giving dates and patent numbers (through 2,128,889; also design, trademark, and reissue numbers); and (3) list of indexes to U.S. patents, 1790–1938.

All of these lists and indexes are available in the Patent Office, and portions of the series are to be found in many other libraries.

I have run across the following indexes of patents for specific subjects. Others undoubtedly exist.

James T. Allen, *Digest of United States Patents for Air, Caloric, Gas, and Oil Engines, 1789–1905* (5 vols., Washington, 1906; 2,058 pp. of pls. in 2 vols., 2 vols. of claims, 1 vol. index).

James T. Allen, *Digest of Cycles or Velocipedes . . . 1789 to 1892* (2 vols., Washington, 1892). The second volume contains some 4,000 illustrations.

James T. Allen, *List of United States Patents for Cycles and Velocipedes* . . . (Supplement [to preceding title], Washington, 1894).

James T. Allen, *Digest of United States Automobile Patents from 1789 to July 1, 1899* (Washington, 1900; 471 pp. of pls.).

United States Patent Office, *Index of Patents Relating to Electricity granted by the United States prior to July 1, 1881, with an Appendix . . . to June 30, 1882.*

United States Patent Office, *Index of Patents relating to Electricity, embracing patents issued by the United States Patent Office from July 1, 1882, to June 30, 1897* (Washington [issued as Appendixes II–XVI to preceding title], 1883–1898).

A useful guide to court records on patents involving litigation (which serves also as a rough guide to the economically more important patents) is **Shepard's Citations, Inc.,** *Shepard's Federal Reporter Citations,* Vol. 1, (1938), known in legal libraries as "Shepard's Digest." This volume lists patents involving litigation by name of patentee from 1790 to 1836, and numerically since 1836, pp. 2,319–2,716.

Files of printed patents, arranged numerically, are in the Public Libraries of Boston, Buffalo, Chicago, Cincinnati, Cleveland, Detroit, Los Angeles, Milwaukee, Newark, N.J., New York, Providence, St. Louis, and Toledo. Sets are also in the libraries of University of the State of N.Y., Albany; Ohio State University, Columbus; State Historical Society of Wisconsin, Madison; Franklin Institute, Philadelphia; Carnegie Library, Pittsburgh; and Oklahoma A. & M. College, Stillwater. The oldest patents likely to be present in these collections, however, date from about 1870.

A free three-page circular "Obtaining Information from Patents," issued by the U.S. Patent Office, Washington, D.C. 20231 is written with the would-be patentee in mind, but certain lists and microfilms that are perhaps pertinent to technological history are described and priced.

A copy of any U.S. printed patent can be obtained from the Patent Office; foreign patents and other matter can be photostated at a reasonable cost.

U.S., Patent Office (Benjamin Butterworth, Commissioner), *The Growth of Industrial Art* (Washington, 1892, 200 pp.), is an outsize-folio picture book that may appeal to an adman in search of quaint illustrations. The pictures are poor, and the text is worthless. A misguided effort.

An "Outline History of the United States Patent Office" is in a centennial number of the *Journal of the Patent Office Society, 18* (July 1936), 1–251. Included are biographical sketches of commissioners, summaries of patent acts, and descriptions of homes of the Patent Office.

(S. C. Gilfillan) *Invention and the Patent System* (Washington: U.S. Government Printing Office, 1964 [Joint Committee Print, Joint Economic Committee, 88 Cong., 2 sess.], 247 pp.). Although prepared as an appraisal of the patent system and its role in encouraging invention, the 671 "Citational Notes" contain references that should be helpful in any study of the history of the patent system.

2. *Manuscript Patent Records (U.S.)*

Except for some original documents in private hands and in manuscript collections, existing official records of patents before 1836 are those that were prepared in the Patent Office after the fire. All such records are in the **National Archives, Business Economics Branch.** After the fire of 1836, a conscientious attempt was made to reconstruct the records of the Patent Office by borrowing from an inventor or his heirs the original patent specification. A manuscript copy of the specification was taken and entered in a series of volumes arranged chronologically. Where possible, a "restored" drawing was made by the patent office; and these restored drawings constitute an attractive as well as important part of the records. A key to the restored drawings is the typed "List of Drawings for Name and Date Patents," prepared in 1949, in the Archives. It is hoped that an index of restored specifications will be available in the not-too-distant future.

For the years 1826 to 1836, the **Franklin Institute** *Journal* sometimes contains the only existing illustration or description of a patent not restored by the Patent Office.

For one year, 1828, **I. L. Skinner** edited the quarterly *American Journal of Improvements in the Useful Arts and Mirror of the Patent Office* (Vol. 1, Washington, 1828; 560 pp. and about 40 pls.).

An indispensable, critical survey of the extensive and complex patent records in the Archives and elsewhere is in **Nathan Reingold,** "U.S. Patent Office Records as Sources for the History of Invention and Technological Property (*Technology and Culture, 1* [Spring 1960], 156–167). It should be noted that the present remarks on manuscript patent records can be brief only because of the existence of Reingold's knowledgeable survey, which was written after he had been intimately concerned, as an archivist, with the accessioning and care of this group of records. The Restored Patent records were moved to the Archives after this article was published. It should be noted also that the 1790–1836 records form only a small portion of the manuscript patent records in the Archives.

The important point for the historian to recognize is that the printed record is only a part of the information available on many patents.

Fifty-seven original patent drawings have been reproduced in **Peter C. Welsh,** "United States Patents, 1790 to 1870: New Uses for Old Ideas" (Washington, 1965 [U.S.N.M., *Bulletin 241*]), paper 48, pp. 109–152. The author suggests that patents supply valuable information on changing contemporary interests of the public.

3. *Patent Models (U.S.)*

Until nearly 1880, a model of an invention was submitted with each patent application. All of the pre-1836 models were lost in the fire of 1836, and another fire in 1877 destroyed more of the models, but a large number have survived. Many significant models are in the collections of the **U.S. National Museum** (Smithsonian Institution), but many others are elsewhere. The present status of the models not in the National Museum is, as it has been for some years, uncertain. An interesting summary of the disposition of models is in **Donald W. Hogan,** "Unwanted Treasures of the Patent Office" (*American Heritage,* 9 [Feb. 1958], 16–19, 101–103).

While the patent models are unnecessary to an understanding of how patented devices operated—drawings and specifications will reveal all that one wants to know along this line—they are nevertheless of more than sentimental interest to the technological historian. Seldom is it possible to date a particular piece of machine work as accurately as one can date a patent model. What kind of screws, threads, and nuts were used in 1850? Look at several models and see. Casting and machining techniques, existence of materials that are difficult to document in other ways, and similar questions can be answered directly and unequivocally by careful study of the models. The antiquarians have thus far been largely responsible for the continued existence of any part of our industrial heritage. It is to be hoped that the day will come when the historian of technology can demonstrate the need for capturing materials like the patent models before they have been irretrievably lost.

4. *Searching of Patents (U.S.)*

Inventors or their representatives—usually patent attorneys' organizations—must employ a search process to establish the absence of "prior art" before submitting an application for a patent. Searching in patent records is a highly specialized art, involving intimate knowledge of the classification system in all its ramifications. However, the scholar should not be dismayed by this aspect of patent procedure because a great deal of useful information can be extracted from the records without involvement in the search apparatus, which of course is geared particularly to recent practice. There is the loose-leaf **De-**

partment of Commerce, United States Patent Office, *Manual of Classification* (no other information on title page), 4to, which enables one to select, in the Search Room, a stack of patents to be looked at; but one quickly becomes aware that this is merely the starting point for a search. Any serious attempt to locate "prior art" in the legal sense is beyond the scope of this bibliography.

Machine searching procedures, now being developed by the Patent Office, will necessarily involve a machine language. Thus the esoteric knowledge of search procedures will not be reduced.

5. *Foreign Patents*

Printed records of foreign patents are located in the Patent Office. A key to their use is **Belknap Severance,** *Manual of Foreign Patents* (Washington, 1935, 161 pp.). In this volume an introductory article by Arthur Worischek, a patent attorney, "Searching Foreign Patents," is worth careful study. Miss Severance's work was issued serially in seventeen installments in *Journal of the Patent Office Society* (Vol. 15, 1933, and Vol. 16, 1934).

Patent and Trade Mark Publications (New York: The New York Public Library, 1954; 43 pp.) This pamphlet lists the publications of and about patents of thirty-nine countries. Publications include patent lists, indexes, and compilations of patent abstracts, specifications, and drawings. Of interest in locating patents in countries other than the United States and Great Britain. The Library, according to this pamphlet, has the largest collection of patent and trade-mark literature in any public library in the United States.

British patents before 1852 (under the "old law") were made usable by the heroically diligent work of Bennet Woodcroft (1803–1879), who had printed the available patent specifications and drawings from 1617 to 1852 and who compiled several indexes. Complete sets of the 1617–1852 patents are in the Patent Office and in New York Public Library. The indexes are

Bennet Woodcroft, *Subject Matter Index . . . of Patents of Invention 1617–1852* (2 vols., London, 1857).

Bennet Woodcroft, *Alphabetical Index of Patentees of Inventions 1617–1852* (London, 1854).

Bennet Woodcroft, *Titles of Patents of Invention, Chronologically Arranged 1617–1852* (2 vols., London, 1854).

Bennet Woodcroft, *Reference Index of Patents of Invention 1617–1852, pointing out the Office in which each enrolled Specification of*

a Patent may be Consulted (and other references to published material) (2 vols., London, 1855).

Later indexes have been published; however, I can name only **Great Britain, Patent Office,** *Patents for Inventions: Fifty Years Subject Index, 1861–1910* (London, 1915), which I have not seen. Subject matter can be approached also through the abridgment series, whose description follows.

The British Patent Office has published a series of illustrated abstracts of individual patents under the title *Abridgment of Specifications*. The first series covered the period from 1617 to about 1866 and was divided into 103 subject-matter classes, each class comprising one to four volumes. Three later series of increasing bulk bring the record down to the present.

An index of the first series is in **London, Index Society,** *Report of the Second Annual Meeting* (1880), Appendix 2, pp. 71–76. More comprehensive is the listing in **Besterman** (p. 18), which also includes the several later series. Listing in Besterman is by subject, but in the index, cols. 7,906–7,915, the classes are all set out, and complete bibliographic references to individual volumes are given under the appropriate rubric. The several series of *Abridgments* are described in Besterman, Vol. 1, cols. 27–30. According to **Gregory,** *Foreign Government Serials* (p. 95), copies of the *Abridgments* are in Engineering Societies Library, New York; Grosvenor Library, Buffalo; Cleveland Public Library; Milwaukee Public Library; and Los Angeles Public Library.

H. Harding, *Patent Office Centenary: a Story of 100 Years in the Life and Work of the Patent Office* (London, H.M.S.O., 1953, 48 pp.). A brief but critical history, without further references.

Three helpful indexes, in English, of other foreign patents are the following:

U.S., Patent Office, *Subject-Matter Index of Patents for Inventions Granted in France from 1791 to 1876 inclusive* (Washington, 1883, 936 pp.).

U.S., Patent Office, *Subject-Matter Index of Patents for Inventions Granted in Italy from 1848 to January 1, 1886* (2nd ed., Washington, 1887, 175 pp.).

G. Doorman, *Patents for Inventions in the Netherlands during the 16th, 17th and 18th Centuries with Notes on the Historical Development of Technics* [abridged English version, Trans. Joh. Meijer] (The Hague: Nijhoff, 1942, 228 pp.). The history of patents in the Netherlands, pp. 11–36, is followed by sixteen short (1–8 pp.) chapters giving essential history of individual subjects, as follows: agricultural and medical inventions, grain milling, drainage, dredging, fire engine, gilt leather technique, pp. 47–54, gutting of herrings, horology, navigation, papermaking, perpetual motion, printing, saw mill, telescope, textiles, windmills. Short descriptions are given of approximately 1,000 patents, pp. 81–210. Finally, subject and name indexes complete this excellent volume. Full texts of patents are given in the Dutch volume, which I have not seen.

The printed patents of some thirty-three countries are listed in more or less detail in **Besterman** p. 18); see entry "Inventions, patent," col. 3,089.

M. Frumkin, "Les anciens brevets d'invention. Les pays du continent européen au XVIIe siècle," *Archives Internationales d'Histoire des Sciences*, 7 (1954), 315–323. The author refers to his own article, "The Early History of Patents for Inventions," in Chartered Institute for Patent Agents, *Transactions* (1947–1948), pp. 20–50, which I have not seen.

D. U.S. National Archives

The primary records of all kinds in the National Archives are much like the interior arrangement of the Archives building itself, which is utterly confusing unless one has a guide to lead the way until one becomes familiar with the main features of the general plan. Therefore, it should be emphasized at the outset that the staff expects to provide help to the searcher. It would be difficult to imagine a more competent or devoted group than that which maintains the collections for the future.

However, the scholar can learn a great deal about the records if he will take time to obtain and study some of the many finding aids. The manuscript patent records in the Archives are described in Part VII.C.2.

Nathan Reingold, "The National Archives and the History of Science in America" (*Isis, 46* [Mar. 1955], 22–28), should be read periodically. This well-knit survey of materials bearing upon technology as well as science mentions dozens of leads that might otherwise be overlooked.

The next reading assignment is **U.S. National Archives**, *Guide to the Records in the National Archives* (Washington, 1948, 684 pp.). This invaluable guide has been out of print for years; a new edition is being prepared for publication in 1969.

The next item required is a free leaflet **"Publications of the National Archives and Record Service"** (1966, 17 pp.). Here are listed, among other finding aids, 165 "Preliminary Inventories," each of which describes in considerable detail a particular record group. Ranging from a few pages to over 300, the Preliminary Inventories are for practical purposes preliminary in name only. They divide the large record groups into segments small enough to be analyzed intelligently and frequently mention the existence of unexpected materials that make visits to the Archives both enjoyable and profitable. Additional preliminary inventories are being constantly prepared. Of those in existence, the following are among those of probable interest to the technological historian:

- 10. Bureau of Yards and Docks
- 26. Bureau of Aeronautics
- 40. United States Mint at Philadelphia
- 53. Bureau of Agricultural Engineering
- 76. International Conferences, Commissions, and Expositions
- 91. Panama Canal (Cartographic Records)
- 105. Coast and Geodetic Survey
- 110. Public Building Service
- 133. Bureau of Ships
- 134. Bureau of Public Roads
- 153. Textual Records of the Panama Canal
- 161. Bureau of the Census

It should be noted that only a small fraction of the total record has been inventoried, even though there are unpublished lists, indexes, and studies for many record groups. A few examples of rich record groups for which guides are not yet available will be given shortly. First, however, another finding aid is to be mentioned. "Reference Information Papers" cut across record group lines and point to materials relating to—for example—Transportation and the Iron, Steel, and Tin Industries. These finding aids will perhaps become more numerous for technological subjects when a glimmer of interest is shown in their existence.

Not yet inventoried but eminently usable are such records as the industrial census reports for 1810 (part) and 1820, the latter now available on microfilm; the rich and voluminous papers of the offices of weights and measures from about 1830; papers of arsenals; Pacific railways surveys; and, of course, the patent records, mentioned earlier. Thousands of photographs and drawings on many subjects are scattered through the establishment. Staff help is required to locate appropriate indexes.

The National Archives are accessioning from state archives and other repositories copies of the 1850–1880 census schedules of manufactures, agriculture, social statistics, and mortality.

National Archives Accessions, quarterly 1940–1952, has been issued irregularly since June 1952 (e.g., No. 51, June 1954; No. 52, Feb. 1956; No. 53, Jan. 1957; No. 54, June 1958; No. 55, May 1960). Each issue contains a list of accessions and announcements of interest to searchers.

Meyer H. Fishbein, "Early Business Statistical Operations of the U.S. Government," *National Archives Accessions,* No. 54 (June 1958), 29 pp. On the industrial censuses.

Meyer H. Fishbein, "The Censuses of Manufactures: 1810–1890," *National Archives Accessions,* No. 57 (June 1963), 1–20. An analytical review of available schedules and reports of the censuses, published and unpublished, with much information on the qualifications and competence of the gatherers of information and on the adequacy of their procedures in reducing data and preparing reports. Of interest to anyone who uses, in any way, census figures or reports on manufactures. Very good.

Meyer H. Fishbein, "Business History Resources in the National Archives," *Business History Review,* 38 (Summer 1964), 232–257. A comprehensive review of records throughout the various collections in the National Archives. Many of the collections mentioned here are pertinent to the history of technology.

Harold B. Hancock, "Material for Company History in the National Archives," *American Archivist,* 29, 1 (Jan. 1966), 23–32. A description of the approach and detailed methods used in locating records pertaining to the DuPont Company by one of the most indefatigable and competent searchers in the field. Details are given of army, navy, diplomatic, legal, fiscal, social, economic, and legislative records. A useful case study of a peculiar talent in action.

Paul Lewison, "The Industrial Records Division of the National Archives," *National Archives Accessions,* No. 55 (May 1960), 7 pp.

Kenneth W. Munden and Henry P. Beers, *Guide to Federal Archives Relating to the Civil War* (Washington: National Archives, 1962, 721 pp.). Analysis of all agencies for 1860–1870 period; organized by departments and offices. Gives references to published books and articles summarizing subjects under discussion and to collections of manuscripts in other repositories. Subject index, pp. 605–721. Outstandingly good. Reviewed by David Donald in *American Historical Review* (Apr. 1963), p. 830; by Robert E. Schofield in *Technology and Culture,* 4 (Summer 1963), 356–357.

A definitive collection of the publications of the **National Archives,** 1936–1964, is available on seventeen rolls of microfilm. See the list that follows.

The Archives has issued at low cost several thousand rolls of microfilm, which are listed in the current (annual?) edition of *List of National Archives Microfilm Publications* (Washington: National Archives, 1966, 107 pp.). While most of the records thus far published are of little, or at best marginal, interest in the history of technology, the program is being steadily expanded, and technological records are becoming more numerous. One finds available on film such items as "Chief of Engineers, letters sent relative to internal improvements, 1824–1830"; "Letters sent by Director of U.S. Mint at Philadelphia, 1795–1817"; and "Records of the 1820 Census of Manufactures."

VIII. MANUSCRIPTS

(See also Early Manuscripts, Part IV.C.)

The collection of manuscript letters has a long and honorable tradition, of which too few engineers, scientists, and industrialists are aware. Attempts to inform those who have manuscript materials of their value for the future are being made by the Conference on Science Manuscripts (Secretary, Nathan Reingold, Smithsonian Institution), whose two-day organizing session was reported in *Isis*, 53 (Mar. 1962), 1–157; by the Case Institute Archives of Contemporary Science and Technology (Curator, Robert E. Schofield), and by individual curators in the Museum of History and Technology, Smithsonian Institution.

Although many repositories are aware of the value of drawings, it is important to emphasize and re-emphasize their importance as part of the manuscript record. A collection of U.S. Mint papers in the National Archives has a manuscript report of Franklin Peale's two-year visit to European mints, but the numerous working drawings that Peale made while in Europe are no longer present. These drawings may yet exist somewhere, but I did not find them. The Sellers drawings of machine tools were rescued by Robert Husband a few years ago, at the very last minute, and deposited in the Franklin Institute. Many unique drawings and photographs still exist, and a growing number of institutional and individual collectors are making heroic efforts to capture some of the primary evidence of our industrial heritage before it has been irretrievably lost.

A helpful discussion of manuscript sources is in the ***Harvard Guide*** (p. 21), pp. 79–88.

A. United States

Philip M. Hamer, Ed., *A Guide to Archives and Manuscripts in the United States* (New Haven: Yale University Press, 1961, 775 pp.).

The publication of this book has vastly simplified the problem of locating manuscript materials in some 1,300 repositories. Although it does not contain everything that it might, it does contain a very great deal. For example, references are given at the end of each part to printed lists and guides issued by individual repositories. These are important; until now it has often been difficult to learn just what was available. There is also a two-page "Note on Bibliographical Guides."

In this *Guide* are described extensive collections of industrial papers in the Baker Library, State Historical Society of Wisconsin (McCormick reaper papers and many others), Historical Society of Pennsylvania (Baldwin Locomotive Works and others), Library of Congress, Cornell University (Collection of Regional History), and University of California at Berkeley, to name only a very few. Browsers in the guide will become aware of such holdings as those of the University of Wyoming (500,000 items on barbed wire); of Harvard's Houghton Library (7 vols. in manuscript by Réaumur and others), apparently unpublished parts of the *Déscriptions des Arts et Métiers;* and the individual papers of G. M. Dodge (Iowa Archives), George Corliss (Brown University), A. S. Hallidie and Adolph Sutro (California Historical Society); and many, many others. Carefully planned, logically arranged, well indexed, and comprising a surprisingly large amount of detailed information about individual holdings, the *Guide* is highly recommended. Reference libraries may wish to assemble a collection of guides to individual repositories, listed throughout this work. This work is now reinforced and supplemented, but by no means superseded, by the *National Union Catalog*, immediately following.

Library of Congress, *National Union Catalog of Manuscript Collections.* Vol. 1, *1959–1961* (Ann Arbor: Edwards, 1962); Vol. 2, pt. 1, *1962,* Vol. 2, pt. 2, *Index 1959–1962* (Hamden, Conn.: Shoestring, 1964); Vol. 3, *1963–1964* (Washington: Library of Congress, 1965); Vol. 4, *1965* [and] *Index 1963–1965* (Washington: Library of Congress, 1966). In the words of its preface, this monumental undertaking is "a continuing series designed to bring under bibliographic control manuscript collections housed permanently in American repositories that are regularly open to scholars." More than 16,000 collections in 492 repositories have been described to date. Indexing is good, but as in any richly annotated listing, systematic browsing will be amply repaid. An annual volume can be expected in the future. A "Guide to Entries by Repository," new in the 1963–1964 volume, is a welcome innovation.

The following informal, incomplete, and unauthorized addenda to Hamer's *Guide* are noted here because of their particular bearing upon technological history. Numbers in parentheses refer to entries in the *National Union Catalog*.

The Franklin Institute Library (Philadelphia) has several hundred original drawings of machine tools, 1857–1890, from the design offices of William Sellers Co. (63-1,374); a collection of Wright brothers papers; the Lenthal collection on shipbuilding (63-155); the Frederick Graff Collection on the Philadelphia waterworks (100 ms. drawings from 1799–ca. 1860) (63-1,373); the James L. Woods Collection (cf. Greville Bathe, *Oliver Evans*, Philadelphia, 1935, p. 346); Cramp shipbuilding papers, 1830–1946 (63-153); and the manuscript records of the Franklin Institute's "Committee on Science and the Arts." While many of the reports of this committee were published in the *Journal of the Franklin Institute*, it is possible that unpublished correspondence, if it exists, would enhance our understanding of such developments as the Westinghouse air brake, Thomson's electric welder, Hollerith's tabulating device, and many others of like significance. See **Percy A. Bivins**, *Index to the Reports of the Committee on Science and the Arts of the Franklin Institute of the State of Pennsylvania, 1834–1890* (Philadelphia, 1890), which indexes some 1,500 reports. The records, not formally open to the public, are under control of the chairman of the still-existent committee.

The Smithsonian Institution, Museum of History and Technology, has important manuscript materials, including papers of Uriah Boyden and others of the Proprietors of Locks and Canals (Lowell, Mass.) (main body of the Proprietors' papers is in Baker Library); over 2,000 cloth tracings of mining equipment designed by Erasmus D. Leavitt (1836–1916); 75 volumes construction photographs of works designed by Cass Gilbert; a collection of drawings of Bessemer steelworks by Alexander Holley; some 200 mid-nineteenth-century locomotive drawings from Reading Railroad; and the De Forest Collection of early radio manuscript materials. The Charles S. Tainter (phonograph) collection is described in U.S. National Museum *Bulletin 218* (Washington, 1959) paper 5, pp. 70–79. The George H. Clark (radio) Collection, containing both printed and manuscript materials, is in the Division of Electricity. It is intended to issue a microfilm reel of portraits and illustrations in this collection in order to simplify the ordering of photographic prints by mail: thus students at a distance from the museum may first review in detail the available illustrative materials.

The Merrimack Valley Textile Museum, North Andover, Massachusetts, has over 1,000 lineal feet of business records of textile manufacturers. Nine firms are named in a list of holdings in American Historical Association, *Newsletter,* 6 (Oct. 1967), 14-15.

The Library of the DeGolyer Foundation (Southern Methodist University, Dallas, Texas) has important manuscripts and drawings of Baldwin Locomotive Works and of Samuel Vauclain (1856–1940), president of B.L.W. See note in *Technology and Culture*, 5, No. 3 (Summer 1964), 408–411.

The Eleutherian Mills Historical Library (Greenville, Wilmington, Del. 19807), in which the collections of the Hagley Museum and the Longwood Library were combined in 1961, will issue late in 1968 a complete guide to its manuscript holdings received through 1965. In addition to the rich collection of Du Pont Company manuscript records, admirably catalogued and calendared, the Library has collected industrial papers of many firms and families in the general vicinity. Particularly useful to scholars using the Library is a union list of manuscript holdings in nineteenth-century industrial and technological history located in other repositories within an 80-mile radius of Greenville.

The Library has the Lammot du Pont, Jr., Collection of Aeronautics, which contains between 3,000 and 4,000 items dealing with the growth of aeronautics, especially European.

The University of California at Davis acquired in 1959 the F. Hal Higgins Library of Agricultural Technology (175,000 items). While the collection is richest in trade catalogues, broadsides, and photographs, it includes manuscript materials of C. M. Eason, Frank N. G. Kranick, Philander M. Standish, and others. A three-page description of the Higgins collection has been prepared by A. Martinez of the University Library, Davis, Calif. See article by Wayne Rasmussen in *Technology and Culture*, 5, No. 4 (Fall 1964), 575–577.

The Library of Congress has 2,500 microfilm frames of the papers of Nikola Tesla (1856–1943), which are in Yugoslavia. Correspondence with George Westinghouse and others is included.

Papers of the American Institute of Aeronautics and Astronautics, including a group on Thaddeus S. C. Lowe (Civil War aeronaut) were presented to the **Library of Congress**. See note in *American Historical Review* (July 1965), p. 1,315.

An important collection of the papers (1823–1870) of John A. Dahlgren (1809–1870), of naval-ordnance fame, including letters, diaries, and so forth, has been recently acquired by the **University of Syracuse**. A description of this rich collection is in *National Union Catalog* (p. 114).

Frank B. Gilbreth (1868–1924), *Selected Papers, ca. 1910–1924, from the Gilbreth Library* (Lafayette, Ind.: Purdue University Libraries, 1964?). Four rolls microfilm of typed talks, annotated articles, notes, etc., by the motion study advocate of "the one best way." Included in roll one is an unpublished M.A. thesis on the life of Gilbreth by Ellen Hawley (U.C.L.A., 1928?, 215 pp.). Bibliography, pp. 208–215.

Ernst Posner, *American State Archives* (Chicago: The University of Chicago Press, 1964, 397 pp.). This book rests upon a thorough and perceptive survey of the archives of all the states. However, its purpose is to point out ways to improve the archives, not to display in any systematic way the contents of individual archives. That function is already served by **Hamer,** *Guide* (p. 113), insofar as the information was available to Hamer. Nevertheless, a great deal of additional specific information is incorporated in the book, and until the pertinent elements are recast in the form of a guide to the archives, the scholar will do well to be aware of these pages. This book serves also to show that archival work is not pursued because of monetary rewards. The situation is at the very least depressing.

Whitfield J. Bell, Jr. and Murphy D. Smith, *Guide to the Archives and Manuscript Collections of the American Philosophical Society* (Philadelphia: American Philosophical Society, 1966, 182 pp.). Holdings include voluminous records of the Society (1758–present); Benjamin Franklin Papers (more than half of those still in existence); rich collections of the Peale, Sellers, Bache, and other Philadelphia families; papers of the Carpenters Company (1683–1952); and of some shopkeepers and craftsmen. Microfilms of pertinent materials in other repositories are listed as, for example, Royal Society of Arts (London) selected materials relating to America (1754–1806), two reels. Holdings are strong in technology as well as science. The *Guide* is an attractive and informative book. Through it I have learned of two or three significant collections that I failed to find while doing research in the Library because I had not the wit to ask the proper questions of the card catalogue and of the curator. Guides of this kind will, in my opinion, continue in the age of computers to contribute to intelligent and informed scholarship.

Harry J. Carman and Arthur W. Thompson, *A Guide to the Principal Sources for American Civilization, 1800–1900, in the City of New York. Manuscripts* (New York: Columbia University Press, 1960, 453 pp.). This work "is designed to give the researcher a fair idea of what is available on *as many apsects* of nineteenth-century American civilization in as many depositories in the City of New York as possible." Arranged by subjects and, within subjects, by geographical areas. For example, a section on "Invention and Technology," pp. 228–231, includes general entries and entries for Illinois, Massachusetts, and New York. There are sections on description and travel, manufacturing and mining, transportation and communications, and so forth. Recommended.

Hubertis M. Cummings, *Pennsylvania Board of Canal Commissioners' Records . . . Descriptive Index* (Harrisburg, Pa.: State of Pennsylvania, Department of Internal Affairs, 1959, 235 pp.). Annotated list of a large and important body of manuscript materials on railroads and canals from *ca.* 1825.

Josephine L. Harper, *Guide to Manuscripts of the State Historical Society of Wisconsin. Supplement Number Two* (Madison, 1966). This volume describes 808 collections catalogued between 1956 and 1965. Earlier volumes in the series are listed in **Hamer,** *Guide* (p. 113).

Robert W. Lovett, *List of Business Manuscripts in Baker Library* (2nd ed., Boston: Baker Library, 1951, 213 pp.). Helpfully annotated. Addenda have appeared in *Business History Review* and its predecessor Business History Society, *Bulletin.* The business historians have been well ahead of historians of technology in systematic efforts to preserve industrial records; the collections in the Baker Library of Harvard Business School are very important. A new edition of this *List* is being prepared.

Nathan Reingold, "Manuscript Resources for the History of Science and Technology in the Library of Congress," Library of Congress, *Quarterly Journal of Current Acquisitions, 17* (May 1960), 161–169. Here is a survey by a persistent collector of manuscript materials that supplies the reader with facts, insights, and provocative suggestions. Recent accessions of manuscripts are mentioned from time to time in the *Quarterly Journal* and in American Historical Association, *Newsletter.*

John Stevens (1749–1838) family papers will be microfilmed by the **New Jersey Historical Society** under a grant from the National Historical Publications Commission. This project will make generally available a valuable but fragile collection, important in early transportation history (1770–1850). This note is based upon American Historical Association, *Newsletter, 5,* No. 4 (Apr. 1967), 3.

Reminiscences of "a number of executives and employees" of the Washburn and Moen Company (makers of wire since *ca.* 1830) are in the American Steel and Wire Co. Collection, **Baker Library.** See note in Business History Society, *Bulletin, 18,* No. 5 (Nov. 1944), 131–132.

Reminiscences (1826–1827) by Samuel W. Collins of the **Collins Company,** axmakers, are in the vault of the company. Excerpts are quoted in *ibid., 18,* No. 5 (Nov. 1944), 132–141.

Material relating to Ferdinand Hassler (1770–1843) and others is described in **Nathan Reingold,** "Research Possibilities in the U.S.

Coast and Geodetic Survey Records" (*Archives Internationales d'Histoire des Sciences*, 11 [1958], 337–346).

N.Y.P.L. Reference Department, *Dictionary Catalogue of the Manuscript Division* (2 vols., Boston: G. K. Hall, 1967). Approximately 25,000 cards are here reproduced. See note on p. 26.

Charles R. Schultz, *Inventory of the Lawrence & Co. Papers, 1822–1904* (Mystic, Conn.: Marine Historical Assn., 1966, 31 pp.). The papers of this firm of shipowners and agents are in the G. W. Blunt White Library of the Marine Historical Association.

Charles R. Schultz, "Manuscript Collections of the Marine Historical Association, Inc. (Mystic Seaport)," *American Neptune*, 25 (Apr. 1965), 99–111. Describes collections totaling some 70,000 pieces relating to maritime and naval history.

Microfilming of several manuscript collections of technological interest in **University of Washington Libraries** has been completed under the direction of Richard C. Berner through a grant from the National Historical Publications Commission. Collections include, for example, Washington Mill Company Papers, 1857–1888, on lumbering in the Puget Sound area; Isaac I. Stevens Papers, 1831–1862, on the northern route railroad survey; and several others. For each microfilmed collection a printed guide has been prepared. The Washington Mill Company Papers, on three rolls of microfilm, are described in a four-page guide, which includes historical sketch of the firm, provenance of the papers, and an inventory of the nature and quantity of the various components of the collection. Microfilms may either be purchased or borrowed through interlibrary loan. Inquiries should be addressed to Curator of Manuscripts, University of Washington Libraries, Seattle, Wash. 98105.

Elisabeth Coleman Jackson and Carolyn Curtis, *Guide to the Burlington Archives in the Newberry Library, 1851–1901* (Chicago: Newberry Library, 1949, 374 pp.). This and the following title are models of annotated finding aids for large collections of primary materials that are available to qualified scholars.

Carolyn Curtis Mohr, *Guide to the Illinois Central Archives in the Newberry Library, 1851–1906* (Chicago: Newberry Library, 1951, 210 pp.).

The Library of Rensselaer Polytechnic Institute, Troy, N.Y., has preserved the undergraduate theses of engineering students from 1856 to 1950. In 1857 three notable theses were mentioned by Alexander Holley in his *Railway Advocate* (and pointed out to me by Robert M.

Vogel): Washington A. Roebling, "Design for a Suspension Aqueduct"; Hezekiah Watkins, "Review of the Niagara Railway Suspension Bridge" (built by Washington Roebling's father John); and G. F. Kirby, "Review of the Hydraulic Motor of the Messrs. Burden at the Troy Iron Works." This last is the only reliable description that I have seen of the great Burden water wheel, one of the world's notable prime movers.

More recently, at **Iowa State University,** I became aware of the college collection of undergraduate theses in engineering, from about 1885. This suggests the probability that there are many other such collections, dating from the 1860's and 1870's. While the general run of these probably would yield little of lasting interest, some attempt to inventory such collections should be made before the pressure for library shelf space causes their destruction.

B. Other Countries

The **Center for the Coordination of Foreign Manuscript Copying** was established in 1965; it is located in the Manuscripts Division of the Library of Congress, Washington, D.C. The Center is concerned with publishing information on materials copied, past and present; with coordinating the photocopying in foreign archives and collections to avoid duplication and to ensure adequate reporting; and with bringing order into the present chaos of uncoordinated and ill-planned copying. The success of the Center will depend ultimately on the cooperation of individual scholars. Thus, a scholar who expects to use foreign materials should let the Center know of his tentative plans, in order that the Center may help and advise him. This note is abstracted from American Historical Association, *Newsletter* (Dec. 1966), pp. 34–36.

For manuscript materials relating to the United States in other countries, see the **Harvard Guide** (p. 21), pp. 87–88. To this listing should be added **B. R. Crick and M. Alman,** Eds., *A Guide to Manuscripts relating to America in Great Britain and Ireland* (London: Oxford University Press, 1961, 677 pp.). The "First List of Addenda" is in *American Studies* (Dec. 1962).

Alan Jeffreys, "Locating the Manuscript Sources of Science," *British Journal for the History of Science,* 2, No. 6 (1964), 157–161. Technology is noticed in this survey of British repositories, which includes the National Register of Archives and other clearinghouses and finding agencies.

"The National Register of Archives; Sources of Industrial History," *Industrial Archaeology,* 1 (May 1964), 59–65. A selected list of (281) collections of records of British industrial and commercial concerns, taken from a (much longer?) checklist in the Register's headquarters in London.

René Taton, "Projet de publication des registres des séances de l'Académie royale des Sciences de Paris (1666–1793)," International Congress on the History of Sciences, *Actes, 10,* pt. 1 (1964) 283–286.

Kenneth D. C. Vernon, "The Royal Institution's Collection of Manuscripts," *ibid.,* 311–314.

IX. ILLUSTRATIONS

A. General Guides

Helen Faye, Ed., *Picture Sources. An Introductory List* (New York: Special Libraries Assn., 1959, 115 pp.). Both the wealth of available original pictures and the difficulties of locating suitable pictures of a particular subject are suggested by this intelligently constructed guide. About 400 sources in the United States and Canada are given, including academic and commercial collections. Subject index, pp. 109–115. The list of "Picture-finding Tools," pp. 5–7, consists mainly of familiar titles, but these are the titles that we are not likely to think of in this context. Therefore, the list is a useful one. Collections are described only in general terms, and I hope the editor may discover a practical way to expand the descriptions in future editions.

G. W. A. Nunn, *British Sources of Photographs and Pictures* (London: Cassell, 1952, 220 pp.). This work, which I have not seen, is listed in **Ottley**, *Bibliography* (p. 237), item 7,871. It contained, according to Ottley, fifteen sources for railway illustrative materials.

John Weale (publisher), *Catalogue of Books on Architecture and Engineering, Civil, Mechanical, Military, and Naval: New and Old* (London, 1854, 111 pp.). Full titles are given, and in many cases lists of subjects of plates, of many fine illustrated books, including handbooks, guides, reports, and other compendia. For example, the following volumes are analyzed: *Public Works of the United States of America* (1841), 40 plates; Steel and Knowles's *Naval Architecture* (1832), 39 plates; chemin de fer d'Orléans à Bordeaux (1851–1853), 39 plates; Great Exhibition building (1851), 28 large folding plates; Corps of Royal Engineers, *Engineering Papers*, ten volumes; bridges, four volumes, 138 engravings, 92 woodcuts; civil and mechanical engineering, six volumes, over 200 plates. A photostat copy of this catalogue is in Smithsonian Museum of History and Technology Library.

Lucile E. Vance and Esther M. Tracy, *Illustration Index* (2nd ed., New York: Scarecrow, 1966, 527 pp.). This is a subject index of pictures in current popular magazines; it is not historically oriented.

B. Individual Collections

[Columbia University Library]. *The William Barclay Parsons Railroad Prints; an Appreciation and a Check List* (New York: Columbia University Library, 1935, 58 pp.). The prints are from the period 1820–1880.

Bella C. Landauer Collection of trade cards in New-York Historical Society. See (p. 68).

[Library of Congress]. Paul Vanderbilt, Comp., *Guide to the Special Collections of Prints and Photographs in the Library of Congress* (Washington: Library of Congress, 1955). A subject index has entries under engineering, railroads, aeronautics, mining, marine steam engines, Washington Aqueduct, water wheels, etc. This guide will repay careful study. Particularly notable are the Van Name collection of locomotive photographs (1890–1910) and water-power machinery (1890–1910) (see *Quarterly Journal of Current Acquisitions,* 9 [Nov. 1951], 45–47); an album of original brush drawings of merchant shops and craftsmen in China, from the late eighteenth century (*ibid.,* p. 48); lithographic prints of industrial establishments, exteriors and interiors (see *Quarterly Journal,* 17 [Nov. 1959], 59–60; 18 [Nov. 1960], 46–47; 19 [Dec. 1961], 54–55); and a group of twenty-six construction photographs of the U.S. Pension Office Building, 1883, designed by Montgomery C. Meigs (*ibid., 18* [Nov. 1960], 52). See also p. 88.

Illustrations of occupations in the **Kress Library,** Harvard Business School, are noted in the "Annual Report, 1963–64." In that year was added a "set of forty-three small paintings of East Indian trades by an unknown Lucknow artist, done about 1815 in gouache." The occupations portrayed include net weaver, laundress, goldsmith, stonecutter, and street trader. Already in the collections were illustrations of English, Chinese, German, Spanish, and Dutch trades.

Mariner's Museum, Newport News, Va. *Catalog of Marine Photographs* (5 vols., 72,700 cards) and *Catalog of Marine Prints and Paintings* (3 vols., 48,200 cards) have been published by G. K. Hall. See note on p. 26.

Boies Penrose, "Prints and Drawings in the Collections of the Historical Society of Pennsylvania," *Pennsylvania Mag. of History & Biog-*

raphy, 66 (1942), 140–160, 379–384. A few prints of railways, bridges, waterworks, and so forth; not limited to Philadelphia.

Harry T. Peters, *America on Stone* (New York: Doubleday, 1931, large 4to, 415 pp.). Profusely illustrated.

A portion (1,700 prints) of the Peters Collection, recently acquired by Smithsonian Institution, is reviewed in Anthony N. B. Garvan and Peter C. Welsh, *The Victorian American* (Washington: Smithsonian Institution, 1961, 30 pp.). The Currier & Ives prints in the Peters Collection are in Museum of the City of New York; the "California on Stone" collection in M. H. de Young Musuem, San Francisco.

Philip Lee Phillips, *A Descriptive List of Maps and Views of Philadelphia in the Library of Congress 1683–1865* (Philadelphia, 1926). About 500 items are enumerated.

Nicholas B. Wainwright, *Philadelphia in the Romantic Age of Lithography* (Philadelphia, 1958, 4to, 261 pp.). Covering the period 1828–1865, there are many subjects of technological interest—for example, ironworks, machine shops (exteriors), gasworks, fire engines, bridges, and the residence of Joseph Harrison, Jr., showing the influence of his sojourn in Russia, where he built locomotives for the Czar.

C. Other Sources

Science Museum, London, can furnish very excellent photographic prints of many thousands of pictures of exhibits, industrial objects, and plates copied from books. Public display books of available pictures are kept in the museum salesroom. Mimeographed subject lists (e.g., steam engines, glass, kinematography) are available by mail. A list of (60) available subject lists is entitled "Science Museum. Photographs and Lantern Slides."

Smithsonian Institution, Washington, holds rich collections of illustrative materials, but as yet there is no system of public display of available pictures. Individual requests, addressed to the Institution, are referred to appropriate curators for reply. See notes on Smithsonian holdings in Part VIII.A.

A number of commercial compendia of industries, published in the nineteenth century, are illustrated, showing chiefly exterior views of prominent works. For example, **Edwin T. Freedley**, Ed., *Leading Pursuits and Leading Men. A Treatise on the Principal Trades and Manufactures of the United States* (Philadelphia, 1856; about 600 pp.).

Industrial Chicago (4 vols., Chicago: The Goodspeed Publishing Co., 1891) is another heavily illustrated work. I have seen only one volume, "The Building Interests," which has 891 pages of text and 50 full-page plates. The plates include 16 steel-engraved portraits and about 25 buildings, some under construction. A few collotypes, many line drawings.

French industries of the nineteenth century are described and illustrated in **Julien F. Turgan**, *Les grandes usines de France* (9 vols., Paris: Michel Lévy Frères, 1860–1870). The illustrations, while not numerous, are fresh and unhackneyed.

Louis Figuier, *Les Merveilles de la Science* (6 vols., Paris, 1867–1891). Popular in approach, but illustrations are generally quite good. The emphasis is upon technology. See also the following title.

Louis Figuier, *Les nouvelles conquêtes de la Science* (4 vols., Paris, 1884).

Handwerk und Technik Vergangener Jahrhunderte. 124 graphische Blätter, ausgewählt von Hans Schmithals mit einer Einleitung und Erläuterungen von Friedrich Klemm (Tübingen: Verlag Ernst Wasmuth, 1958). An attractive selection of pictures of mechanical subjects, *ca.* 1475–1850.

In the preface of his *Medieval Technology and Social Change* (p. 11), **Lynn White, Jr.**, pays tribute to "the incomparable Princeton Index of Christian Art." He analyzes the Index on p. 157 of his article, "Technology and Invention in the Middle Ages" (p. 11).

Otto Bettmann, Ed., *Bettmann Portable Archive* (New York: Picture House, 1966; 299 pp., 3,669 illustrations). This is the first of a series of volumes intended to reveal the resources of the Bettmann Archive, a commercial supplier of pictures, and to "serve as an idea stimulator and image finder" for advertising writers. I have not seen the book.

Informative, often attractive pictures will be found in early source books and in most technical treatises up to about 1900. Until the advent of the halftone process, just before 1900, illustrations were generally sharp, clear, and readily reproducible. See also encyclopaedias, handbooks, compendia such as **Knight's** *Mechanical Dictionary* (p. 65), society proceedings, books of travel, trade catalogues, vanity works such as **Van Slyck** (p. 82), local histories such as **Scharf and Westcott** (p. 84), and especially periodical and serial publications.

D. Motion Pictures

Aside from those that I list here I am unaware of scholarly films in the history of technology. The usual "educational" film, which reviews the history of the Industrial Revolution or the history of transport, for example, is negligibly slight. On the other hand, it is immediately obvious that a large and probably very important body of unique information exists in news and documentary films. For example, **Felix Green's** recent film on China includes a short but deeply impressive sequence showing countless sampans and junks, each carrying a pitiful cargo of rocks, engaged in constructing in deep water a very long causeway from mainland China to one of the offshore islands.

A guide to films that contain source materials in technology would be a most valuable work. A collection of pertinent film segments would, I think, be more than occasionally useful in teaching the history of technology.

In support of its program in industrial archaeology, several films are planned by the **Museum of History and Technology, Smithsonian Institution,** to record vanishing handcraft and manufacturing techniques. (*Technology and Culture*, 8 [July 1967], 384, 386.)

A short film showing the Otto and Langen atmospheric engine in operation has been prepared by **Klöckner-Humboldt-Deutz,** of Deutz, Germany.

A film showing the collapse in 1940 of the Tacoma Narrows suspension bridge may be purchased from **The Camera Shop,** 1003 Pacific Avenue, Tacoma, Washington 98402.

A short (5 minutes) sequence in color, taken after 1955, of primitive ironmaking in Africa is at the beginning of "One Hoe for Kalabo," a promotional film of the **National Machine Tool Builders Association,** Washington, D.C.

An anthropological film, "The Hunters," produced by **Peabody Museum of Harvard University,** is a sensitive and moving record of hunting (and living) techniques of the Bushmen of the Kalahari, the people who are described in Elizabeth Marshall Thomas, *The Harmless People* (New York: Knopf, 1959). This is easily one of the most memorable films I have ever seen.

A Catalogue of Shell Films in International Circulation (London: Shell International Petroleum Company Limited, 1963, 147 pp.). I have seen "The Cornish Engine," a 33-minute sound film on the development of the steam engine. Historical information was provided by H. W. Dickinson; the film includes sequences of actual operation of the great pumping engines still extant in Cornwall. In addition to this, there is a series of (6) short (5–10 minutes) films of individual engines in operation. The sounds of the engine at Cook's Kitchen were recorded in one of these films.

I have seen also the perfectly delightful Shell film "The Flintknappers." This witty and imaginative review of Britain's oldest industry is only 5½ minutes in length, but it contains much information. This film is one of a series of (7) short films on handcrafts. The others in the series are the wheelwright, blacksmith, cooper, glassblower, chairbodger, and the tidemiller.

The Shell catalogue has also a section on motor races, which includes a series of four half-hour films covering the history of motor racing, 1902–1939. Presumably these include contemporary shots, but I have not seen the films.

"Airport," a 17-minute film of a day at Croydon Airport in 1935, is a reissue of a 1935 film. Thus, as the catalogue states, "this film shows the beginnings of the modern air transport system."

X. TRAVEL AND DESCRIPTION

Contemporary descriptions of engineering works are disappointingly meager before the advent of the technical press around the middle of the nineteenth century. However, remarks and occasional extended comments in books of travel will provide insights and a surprising amount of solid information to the student who has the patience to read through many long and often tedious narratives.

Travel accounts were used extensively by **Victor S. Clark** in his monumental *History of Manufactures in the United States* (p. 256). He listed thirty travel books, from 1744 to 1860, in his bibliography. **Louis C. Hunter** used at least seventy-five titles in his *Steamboats on the Western Rivers* (p. 230), duplicating only four of those used by Clark (Clark's were generally earlier). Although Hunter's list, which must be extracted from his footnotes, is too long to print here, many (but not all) of its titles are mentioned in the various guides and bibliographies enumerated.

The accounts listed under "Individual Travels" are chiefly concerned with technology. Emphasis is upon the United States. The catalogue of the New York Public Library is particularly good in geographical listings of works of travel and description.

Diaries, which have preserved contemporary descriptions and attitudes that are nowhere else recorded, can be located through the remarkable series of bibliographies by **William Matthews** (p. 79) and one by **Louis Kaplan** (p. 79). For several references to Canadian canal and railway construction and mining, for example, see Matthews's volume on Canadian diaries and autobiographies.

A. Finding Aids and Bibliographies

Max Berger, *The British Traveller in America 1836–1860* (New York: Columbia University Press, 1943, 239 pp.). A sequel to **Mesick** (p. 131). Well-constructed guide to a great mass of material. Berger has included a number of accounts published in periodicals. A critical bibliography, pp. 189–229, will help the reader to avoid irrelevant works. There is a list of "Tourist and Emigrant Guidebooks" on pp. 220–222.

Thomas D. Clark, Ed., *Travels in the Old South* (3 vols., Norman: Oklahoma University Press, 1956–1959). The three volumes cover in detail the period from 1527 to 1860. All citations are annotated fully, both critically and as to content, and all volumes are indexed. Quotations are given in many of the critical annotations. Technological interests include references to agriculture, canals, clocks, conditions of travel, cotton gins, inland waterways, inventions, ironworks, manufactures, meteorology, and rifle manufacturing. This work is an excellent tool to be consulted by anyone doing serious work on any aspect of southern history before 1860. [This note by Peter C. Welsh.]

Thomas D. Clark, Ed., *Travels in the New South* (2 vols., Norman: Oklahoma University Press, 1962). The South as used here includes Maryland, West Virginia, Kentucky, and Missouri, plus eleven states that seceded in 1860–1861. Vol. 1 covers "The South in Reconstruction, 1865–1880"; "The New South, 1880–1900." Vol. 2 covers "The Twentieth Century South, as Viewed by English Speaking Travelers, 1900–1955"; and "Foreign Language Accounts by Travelers in the Southern States, 1900–1955." All titles cited are annotated fully, both critically and as to content, and quotations are included in many of the annotations. Technological interest includes references (by state and city) to textile mills, manufacturing, iron industry, agriculture, railroads, architecture, expositions, and so forth. Both volumes are indexed. Excellent bibliography; useful guide for any aspect of post-Civil War history. [This note by Peter C. Welsh.]

Marvin Fisher, *Workshops in the Wilderness. The European Response to American Industrialization, 1830–1860* (New York: Oxford University Press, 1967, 238 pp.). A "Checklist of European Comment on America, 1830–60," pp. 223–231, includes nearly 200 travel accounts. The nature and significance of industrial growth in the United States before the Civil War is assessed through study of diaries, accounts, and reports of Europeans who visited America during this period.

Michael W. Flinn, "The Travel Diaries of Swedish Engineers in the Eighteenth Century as Sources of Technological History," Newcomen Society, *Transactions*, 31 (1957–1959), 95–109. The author gives the nature and quality of information available. He cites the following title:

Sven Rydberg, *Svenska Studieresor till England under Frihetstiden* (Uppsala, 1951). Flinn calls this "a most comprehensive guide . . . through which it is possible in a great many instances to locate to the exact page the reference in a journal to a visit to a particular place or works . . . . The book has a short summary of each chapter in English at the end."

Robert R. Hubach, *Early Midwestern Travel Narratives. An Annotated Bibliography 1634–1850* (Detroit: Wayne State University, 1961, 4to, 149 pp.). A large number of published and unpublished narratives, from a few pages to several volumes in length, are listed. For any specific purpose, however, the short annotations are of little help, except in giving the geographical location of places described. There is a chronological list that may be helpful in an intensive study of a particular region. The index gives place names and personal names. The Midwest includes Ohio.

Esther E. Larson, *Swedish Commentators on America 1638–1865: An Annotated List of Selected Manuscript and Printed Materials* (New York: New York Public Library, 1963, 139 pp.). Listed chronologically are 683 entries; index of names. The subjects include Axel Adelswärd, an engineer, 1855, in New York and Philadelphia; S. G. Hermelin's report of mines, 1783; anonymous report of mines and foundries, 1732, Pennsylvania and Maryland; and several untranslated travel accounts that appear to be significant.

Library of Congress, *Guide to the Study of the United States* (p. 32). A section on "Travel and Travelers," pp. 529–548, includes (7) guides—Mesick, Berger, and so forth—and reasonably full notes on the books of fifty travelers, 1743–1849. Very little on technology.

Jane L. Mesick, *The English Traveller in America 1785–1835* (New York: Columbia University Press, 1922). A taxonomic analysis of books of travel. Under manufactures and industry, the author mentions or gives pertinent passages from John Bradbury (1809), Isaac Candler (1824), Henry Fearon (1819), James Flint (1822), Adam Hodgson (1824), Charles J. Latrobe (1832), Harriet Martineau (1837), John Melish (1806–1811), Tyrone Power (1833–1835), Patrick Shirref (1835), and Henry Wansey (1794). List of books, pp. 347–352. See also **Berger** (p. 129).

Frank Monaghan, *French Travellers in the United States 1765–1931* (New York: New York Public Library, 1933). Revised version of the bibliography published serially in New York Public Library, *Bulletin*, 36 (1932), 163–189, 250–261, 427–438, 503–520, 587–596, 637–645, 690–702. The book has been reprinted, with a fifteen-page supplement by Samuel J. Marino (New York: Antiquarian Press, 1961).

Allan Nevins, Ed., *America through British Eyes* (New York, 1948). A book of extracts from early travel accounts. There is added an extensive list of British books of travel.

Antoine de Smet, *Voyageurs Belges aux Etats-Unis du XVIIe Siècle à 1900. Notices Bio-bibliographiques* (Brussels, 1959, 201 pp.). A bibliography of the writings of Belgian travelers in the United States as published in books, serials, and reports, many of which are relatively easy to find in American libraries. At least fifteen references are made to visits by engineers, 1849–1893, many of whom were commissioners to the Exhibitions of 1853, 1876, and 1893. The following names should be consulted: Beco (1876), Canon (1893), Deby, Dewilde, Gobert, and Jottrand (1876), Kaiser (1889), Lagasse (1876), Lambert (1853), Laveleye (1878), Le Hardy (1849–1855), Marlin (1876), Schorn (1871), Trasenster (1884), and Van Bruysell (1876). Very good.

Judith Williams, *Guide* (p. 25), has a short section on travels in England in the 1750–1850 period: Vol. 1, pp. 87–97.

See "Books of Travel and Description" in the **Harvard Guide** (p. 21), pp. 149–161. On p. 150 are listed a dozen guides and summaries (Mesick, Berger, etc.). A series of several alphabetical lists of books, pp. 151–161, is arranged chronologically. Annotations are usually one or two words in length.

B. Individual Travels

1. *United States*

Michel Chevalier, *Histoire et Déscription des Voies de Communication aux Etats-Unis* (3 vols., Paris, 1840–1841; 2 vols. text, 4to, atlas vol., large fo.). Descriptions and illustrations of Schuylkill Canal, Morris Canal, B. & O. locomotive and passenger cars, Georgetown aqueduct over the Potomac; description only of Pennsylvania Works; much other material.

Christopher Colles (Walter W. Ristow, Ed.), *A Survey of the Roads of the United States of America 1789* (Cambridge, Mass.: Harvard University Press, 1961). Ristow's sketch of Colles's life occupies pp. 3–95. Strip maps by Colles are reproduced.

Karl Culmann, "Wooden and Iron Bridges in the United States and England," *Allgemeinen Bauzeitung*, 16 (1851), 69–129, pls. 387–397 [in separate atlas of pls.]; 17 (1852), 163–222, pls. 478–487. Article is in German. Excellent illustrations of many American bridges by an

Austrian engineer who visited the United States in 1849–1850 to study civil works.

Charles Dickens, *American Notes for General Circulation* (2 vols., London, 1842). This account, reprinted many times, is notable for the descriptions of railroad, steamboat, stagecoach, and canal travel. Particularly good on the Pennsylvania works and the Boston-to-Lowell railroad.

William Ferguson, *America by River and Rail* (London: Nisbet, 1856). Described in Clark (p. 130), Vol. 3, pp. 357–358.

Franz Anton, Ritter von Gerstner, *Die Innern Communicationen der Vereinigten Staaten von Nordamerica* (2 vols., Vienna, 1842–1843, 34 pls.). Chiefly on railroads. Described in Clark, *Travels* (p. 130), Vol. 3, pp. 133–134.

A series of ten letters on the same subject by Gerstner were translated by L. Klein and printed in Franklin Institute, *Journal, 28* (1839), 145 ff.; *30* (1840), 217–227, 289–301, 361–369; *31* (1841), 3–11, 73–82, 165–173, 247–255, 300–308. See also *30* (Aug. 1840), 89–102. Meyer Weinberg has kindly informed me that at least one of these articles came from Gerstner's *Berichte aus den Vereinigten Staaten von Nordamerica* (Leipzig: C. P. Melzer, 1839).

Carlo de Ghega, *Die Baltimore-Ohio Eisenbahn* (2 vols., Vienna, 1844; 1 vol. text, 8vo, atlas of map and 19 pls., fo.). Description based upon visit.

Joshua Gilpin, "Journal of a Tour from Philadelphia through the Western Counties of Pennsylvania in the Months of September and October, 1809," *Pennsylvania Magazine of History and Biography, 50* (1926), 64–78, 163–178, 380–382; *51* (1927), 172–190, 351–375; *52* (1928), 29–58. Gilpin, a Philadelphia merchant and industrialist, was keenly aware of the technological implications of whatever he saw in his travels. See also his trip to Bethlehem, Pa., *ibid., 46* (1922), 15–38, 122–153. See also his trip to the British Isles (p. 136).

Great Britain, Parliament, House of Commons, *Sessional Papers,* 1854, Vol. 36: New York Industrial Exhibition. General Report of the British Commissioners (5 pp.) and special reports of George Wallis (85 pp.), Joseph Whitworth (44 pp.), Charles Lyell (50 pp.), Mr. Dilke (108 pp.), and Professor Wilson (133 pp.). Because the commissioners arrived in New York before the 1853 Exhibition had opened, they made independent tours which are reported here. Whitworth, for example, visited builders of machinery and machine tools in New England. Very good.

Basil Hall, *Forty Etchings, from Sketches Made with the Camera Lucida, in North America, in 1827 and 1828* (Edinburgh, 1892), includes good illustrations of Lake Cayuga bridge, Mississippi steamboat at wooding station, and an American stagecoach.

Samuel Gustaf Hermelin (Amandus Johnson, Trans.), *Report about the Mines of the United States of America, 1783* (Philadelphia, 1931). The introduction of this volume mentions other pertinent travel accounts. Appeared serially in Swedish-American Historical Society, *Bulletin*.

Franklin D. Scott, Trans., *Baron Klinkowström's America, 1818–1820* (Evanston, Ill., 1952). The Swedish edition, Axel Leonhard Klinckowström, *Bref om de Fortenta Staterna . . .* (Stockholm, 1824), was accompanied by an atlas of plates. Observations on Speedwell Iron Works, ships and boats, bridges, canals, the DuPont powder mills on the Brandywine, and other works.

Guillaume Lambert, *Voyage dans l'Amérique du Nord en 1853 et 1854, avec Notes sur les Expositions Universelles de Dublin et de New-York* (2 vols., Brussels, 1855; text vol., 320 pp., atlas vol., 32 double pls., 8vo). Extended discussion of the New York Crystal Palace Exhibition; descriptions of coal mines, ironworks, and machinery in the Pennsylvania anthracite region and in Pittsburgh. Describes a belt conveyer carrying 200 tons of coal per day over a distance of 100 yards. Pertinent material on stone dressing, tunneling, and mining machinery. Very good.

Emile Malézieux, *Travaux Publics des Etats-Unis d'Amérique en 1870 (rapport de mission)* (2 vols., Paris, 1873–1875; text vol., 572 pp., atlas vol., 61 double pls.). Malézieux, professor in Ecole des Ponts et Chaussées, and a senior student spent 100 days in America, visiting Niagara Falls, San Francisco, and numerous other places. Excellent plate of Roebling's Niagara Bridge; many plates and maps of canals, bridges, elevated railways, and so forth.

Jean Baptiste Marestier, *Mémoire sur les Bateaux à Vapeur des Etats-Unis d'Amérique* (2 vols., Paris, 1824; 1 vol. text, 4to, 1 vol. pls., large fo.). Primarily on ships and engines, with a few added notes on tools and equipment in American shipyards. Parts of this work were translated in Sidney Withington, *Memoir on Steamboats of the United States of America* (Mystic, 1957), and some of the plates were reproduced.

(Philip H. Nicklin) Peregrine Prolix, pseud., *A Pleasant Peregrination through the Prettiest Parts of Pennsylvania* (Philadelphia, 1836). Letters of a bookseller describe travels on canals and on the Portage

Railroad in the summer of 1835. I learned of this title from Miss Pearl I. Young.

Guillaume Tell Poussin, *Chemins de fer Américains* (Paris, 1836).

David Stevenson, *Sketch of the Civil Engineering of North America* (London, 1838; 320 pp., illustrated). This remarkably informative account was based upon a three-months' tour by the twenty-three-year-old uncle of Robert Louis Stevenson. Many clear engravings of bridges, steamboats, and so forth.

Henri Stucklé, *Voies de Communication aux Etats-Unis: Etude Technique et Administrative* (Paris, 1847, 470 pp.).

Edward W. Watkin, *A Trip to the United States and Canada: in a Series of Letters* (London, 1852, 149 pp.). This English railroad promoter, involved also in the Channel Tunnel, wrote *Canada and the States; Recollections, 1851 to 1886* (London, 1887, 524 pp.). Described in **Clark** (p. 130), 3, 329-330.

Elkanah Watson, *Men and Times of the Revolution; or, Memoirs of Elkanah Watson, 1777–1842* (2nd ed., New York, 1857). Watson (1758–1842), promoter of the Erie Canal, has useful descriptions of canal towns, Cleveland, and Detroit in 1818.

Avrahm Yarmolinsky, *Picturesque United States of America 1811, 1812, 1813, being a Memoir on Paul Svinin* (New York: Rudge, 1930; xviii + 46 pp. + 52 plates, fo.). Svinin was a Russian consular officer who with enthusiasm observed steamboats, public works, and many other American objects. The text summarizes parts of the Russian book that Svinin published in 1815. His water-color views include steam and sailing vessels, buildings, landscapes, horse-drawn vehicles, views of the "Colossus" bridge and Philadelphia waterworks, and a chain bridge over the Merrimack River.

Oscar Handlin, Ed., *This Was America . . . as Recorded by European Travelers to the Western Shore in the Eighteenth, Nineteenth, and Twentieth Centuries* (Cambridge, Mass.: Harvard University Press, 1949; 602 pp., illustrated). Reprinted (Harper Torchbook, 1964). This picture of United States is unhackneyed, being composed of accounts by obscure travelers, some of which are here for the first time published in translation.

2. Other Countries

Zachariah Allen, *The Practical Tourist, or Sketches of the State of the Useful Arts* [etc.] *in Great-Britain, France, and Holland* (2 vols.,

Providence, 1832). Disappointing, superficial work in both reports of foreign works and comparisons drawn with the United States.

G. D. Amery, "The Writings of Arthur Young," Royal Agricultural Society of England, *Journal*, 85 (1924), 175–205. A description of the travels and agricultural writings of Arthur Young (1741–1820) is followed by an exhaustive bibliography of his very numerous works. His books of travels, covering much of England, Wales, and parts of Ireland and France, describe roads, towns, estates, occasional industrial works, and workhouses for the poor. It was in Young's travels, for example, that I discovered the astonishing opulence of the English manor houses in the eighteenth century. His emphasis was upon agriculture, but he saw and commented upon much else.

A list of his articles in the forty-six volumes of *Annals of Agriculture*, which he edited from 1784 until 1809, demonstrates his interest in farm implements, agricultural education, effects of electricity upon plants (1790), manufactures in America (1797), and all aspects of the techniques and economy of agriculture.

A few American railroad builders of the 1820's can be followed to England in **Robert E. Carlson's** "British Railroads and Engineers and the Beginnings of American Railroad Development" (*Business History Review*, 34 [Summer 1960], 137–149).

Harold B. Hancock and **Norman B. Wilkinson**, "Joshua Gilpin, an American Manufacturer in England and Wales, 1795–1801," Newcomen Society, *Transactions*, 32 (1959–1960), 15–28; 33 (1960–1961), 57–66. Summaries of some sixty-two notebooks kept by Gilpin.

W. O. Henderson, *J. C. Fischer and his Diary of Industrial England 1814-51* (London: Cass, 1966; 184 pp., illustrated). The background of the diaries of Johann C. Fischer (1773–1854), the Swiss ironmaker, including information on his career and on the people and places visited, are carefully set out in the first part of this book, pp. 1–94. The actual diaries, pp. 125–167, and a series of letters and other documents, pp. 95–124, are followed by (7) pertinent maps and a short bibliography.

Patricia James, Ed., *The Travel Diaries of Thomas Robert Malthus* (Cambridge: University Press for Royal Economic Society, 1966; 316 pp., illustrated). Occasional, rather thin, notes on arts and industries; many details of transport facilities used by Malthus, particularly in Scandinavia in 1799. Short sections on a continental tour, 1825, and a Scottish holiday, 1826.

Gabriel Jars, *Voyages métallurgiques, ou recherches et observations* *sur les mines et forges de fer* [etc.] (3 vols., Paris, 1774–1781). Highly informative account of mining and manufacturing, principally in England. Notes on the book and on the travels of Jars will be found in Newcomen Society, *Transactions, 26* (1947–1949), 57–68.

Michel de Montaigne (E. J. Trechmann, Trans.), *The Diary of Montaigne's Journey to Italy in 1580 and 1581* (New York: Harcourt, Brace, 1929, 297 pp.). Conditions of travel in the sixteenth century as seen by an acute observer.

Sister St. John Nepomucene, "Franklin Peale's Visit to Europe in the U.S. Mint Service," *Journal of Chemical Education,* Vol. 32 (Mar. 1955). Peale's report of his two-year study of European mints and techniques of assaying and minting (1833–1835) is in the National Archives.

George Escol Sellers (p. 86) visited England in the fall of 1832. His descriptions of machine shops and metalworking techniques are precise, well written, and generally unique.

John Smeaton's Diary of his Journey to the Low Countries 1755 (Leamington Spa: Newcomen Society, Extra Pub. No. 4, 1938).

Two parts of **Joshua Field's** diary of an 1821 English tour are in Newcomen Society, *Transactions, 6* (1925–1926), 1–41; *13* (1932–1933), 15–50.

The travel diary of **John G. Bodmer** in England, 1816–1817, is in Newcomen Society, *Transactions, 10* (1929–1930), 102–114.

William Strickland, *Reports on Canals, Railways, Roads and Other Subjects Made to "The Pennsylvania Society for the Promotion of Internal Improvement"* (Philadelphia, 1826; 51 pp. + 57 pls., part colored, part folding; 11 in. high × 18 in. wide). The significant and attractive result of Strickland's visit to England in 1825.

W. Howard White, "European Railways — As They Appear to an American Engineer," American Society of Civil Engineers, *Transactions, 3* (1874), 61–66. A general but informative comparison of philosophies of construction in the United States and Europe.

The following (13) titles were generously supplied by Robert P. Multhauf, who used them in a study of European chemical and metals industries. Additional titles by some of these authors may be found in the British Museum Catalogue.

John Aiken, *A Description of the Country from Thirty to Forty Miles round Manchester* (London, 1785).

Anon., *Tournée faite en 1788 dans la Grande-Bretagne par un Français* (Paris, 1790).

John Brand, *History and Antiquities of the Town and Country of Newcastle upon Tyne* (2 vols., London, 1789).

J. J. Ferber, *Beiträge zur Mineralgeschichte Böhmens* (Berlin: C.F. Himburg, 1774; 162 pp., 2 pls.). Describes nineteen mining centers in 1768 and 1770.

J. J. Ferber, *Bergmännische Nachrichten der Herzog-Zwei . . . und Nassauischen Länder* (Meitau: J. F. Hinz, 1776, 94 pp.). More on mercury than on any other mineral product.

J. J. Ferber, *Mineralogische und metallurgische Bemerkungen in Neuchâtel, Franche Comté und Bourgogne* (Berlin: A. Mylius, 1789, 77 pp.). Trip was made in 1788. Description, pp. 49–77, of ironworks at Mt. Cenis in Bourgogne, established ten or twelve years since, by Wilkinson, an Englishman. Most of the eight plates are devoted to these works, which had eight steam engines.

J. J. Ferber, *Neue Beiträge zur Mineralgeschichte verschiedener Länder* (Meitau: J. F. Hinz, 1778, 462 pp.). Travels in Saxony, England, Scotland, and so forth.

J. J. Ferber (R. E. Raspe, Trans.), *Travels in Italy* (London: L. Davis, 1776, 377 pp.). Mostly mineralogical. Preface has an account of the career of Ferber, whom Raspe knew.

J. G. A. Forster, *Voyage philosophique et pittoresque en Angleterre et en France fait en 1790* (Paris, 1790).

C. F. Hollunder, *Tagebuch einer metallurgisch-technologischen Reise* (Nürnberg: J. G. Schrag, 1824; about 30 pls.). Travels in Poland, Germany, and Austria.

H. F. Link, *Bemerkungen über eine Reise durch das südwestliche Europa besonders Portugal* (Rostock and Leipzig, 1801).

H. F. Link, *Bemerkungen über eine Reise durch Frankreich, Spanien, & Portugal* (3 vols., Kiel, 1801–1804).

A. G. Monnet, "Voyages minéralogiques faits dans les années 1772, 1784 et 1785 en Auvergne," *Journal de physique, de chimie, d'histoire naturelle et des arts*, 32 (1788), 115–132, 179–199.

C. Travel Guides and Descriptive Works

H. S. Tanner, *A Description of the Canals and Railroads of the United States* (New York, 1840; many other eds.). Detailed descriptions and maps.

Wellington Williams was author of at least three travel guide books, which are described in **Clark** (p. 130), 3, 331–332. His *Appleton's Railroad and Steamboat Companion* (1847; many later eds.) is, according to Clark, one of the best available physical accounts of the United States.

There were many promotional but descriptive volumes issued in various cities in the United States before 1850. For example: **James Mease,** *The Picture of Philadelphia* (1811); **Charles Cist,** *Cincinnati in 1841: its Early Annals and Future Prospects* (1841); and **Samuel Jones,** *Pittsburgh in the Year Eighteen Hundred and Twenty-Six* (1826).

A careful descriptive and interpretive essay, based upon books of travel and other contemporary materials is **Ralph H. Brown,** *Mirror for Americans—Likeness of the Eastern Seaboard 1810* (New York, 1943). A chapter on "The Principal Occupations" brings together information from many sources.

Information on reports of British "productivity teams" that visited the United States after World War II will be found in **John E. Sawyer's** "The Social Basis of the American System of Manufacturing" (p. 299), which provides a striking demonstration of the comparative uses of travel accounts.

XI. PERIODICAL AND SERIAL PUBLICATIONS

Periodical publications of the last two centuries contain a vast quantity of primary and secondary information on technology. Unfortunately, most of this material is beyond the reach of existing indexes and bibliographies and can be recovered only through ingenuity, luck, or patient, page-by-page search. R. A. Peddie, a diligent English compiler of book lists, quite accurately called periodical literature "that extraordinary mass of fossilized knowledge." As an indication of the size of the mass, it may be noted that in 1897 — two generations ago — **Henry C. Bolton**, of the Smithsonian Institution, listed 8,600 titles of independent (i.e., not issued by a society) scientific and technical periodicals in Europe and America.

In addition to the countless thousands of articles and items (particularly after about 1850) that are useful as source materials, if they can be located, there are many hundreds of review articles, many in nontechnical magazines, that bring together scattered information and provide a point of departure for serious inquiry into all sorts of developments. Review articles, which provide a context for the latest developments, are still a welcome feature of many English and German technical magazines. In the United States, on the other hand, this class of article has nearly vanished.

A few methodical and dedicated individuals have made significant inroads into periodical literature in three limited fields. **G. F. Dow**, **A. C. Prime**, and **R. S. Gottesman** have copied trade advertisements from long runs of eighteenth-century newspapers (p. 170). **A. A. Jakkula** has put students of bridges in his debt and has suggested a format for similar studies in many other fields in his *A History of Suspension Bridges in Bibliographical Form* (p. 224). Finally, **Thomas James Higgins** (p. 78) has listed nearly 1,000 journal articles on the history of electrical engineering and electrophysics.

A. Finding Aids and Lists of Periodicals

Edna Brown Titus, Ed., *Union List of Serials in Libraries of the United States and Canada* (3rd ed., 5 vols., New York: H. W. Wilson, 1965, 4,694 pp.). This is the standard and indispensable guide, not only for (1) locating serials but also for (2) determining dates of origin and demise and (3) determining what individual indexes are available. There are 156,500 entries; 956 libraries are represented. This new edition will probably continue to be known as "Gregory," who was editor of the earlier editions (2nd ed., 1943; supplements 1941–1943 and 1944–1949). For new serials, see the following title.

New Serials Titles; a Union List of Serials Commencing Publication after December 31, 1949 is published monthly by the Library of Congress. One decennial cumulative volume (1950–1960) has been published; cumulative issues for shorter intervals after 1960 will be superseded by the second decennial volume (1961–1970).

British Union-Catalogue of Periodicals, James D. Stewart and others, Eds. (4 vols., Butterworth, 1955–1958). *Supplement to 1960* (1962). Lists some 140,000 titles of serials, from the seventeenth century to the present, in 440 British libraries. Many titles listed here are not to be found in **Titus**, *Union List* (above).

British Union-Catalogue of Periodicals Incorporating World List of Scientific Periodicals: New Periodical Titles, K. I. Porter, Ed. (London: Butterworth, 1964+). Published quarterly, cumulated annually.

World List of Scientific Periodicals Published in the Years 1900–1960, Peter Brown and George B. Stratton, Eds. (4th [and last] ed., 3 vols., Washington: Butterworth, 1963–1965). British locations are given for 59,404 titles. A list of periodic international congresses is given in 3, 1,789–1,824. This work has been effectively extended by the BUCOP *New Periodical Titles*, listed just previously.

Ruth S. Freitag, *Union Lists of Serials. A Bibliography* (Washington: Library of Congress, 1964, 151 pp.). Annotated list of some 1,200 union lists, 361 of which are in the United States, the rest throughout the world.

Eric H. Boehm and **Lalit Adolphus**, Eds., *Historical Periodicals. An Annotated World List of Historical and Related Serial Publications* (Santa Barbara: Clio Press, 1961, 618 pp.). This may furnish to the reader an entry to historical serials that would otherwise remain unknown to him.

Henry C. Bolton, *A Catalogue of Scientific and Technical Periodicals, 1665–1895, together with Chronological Tables* . . . (2nd ed., Washington: Smithsonian Misc. Colls., Vol. 40, 1897). Lists 8,600 titles of independent serials, not published by a society or association. Analytical subject index and a highly selective chronological table that relates volume numbers to dates. Very helpful to anyone who wants to learn something of the range of titles available in several languages.

David P. Forsyth, *The Business Press in America 1750–1865* (Philadelphia: Chilton, 1964, 394 pp.). Contains a large amount of fresh information on trade and technical publications of the period covered. Chapters on general technical magazines, such as Franklin Institute *Journal* and *Scientific American*, and on publications pertaining to railroads, mining and metallurgy, chemical industry, and so forth. Tables, bibliography. Very good. Reviewed in *Technology and Culture*, 6 (Summer 1965), 538–540.

Daniel C. Haskell, Comp., *A Check List of Cumulative Indexes to Individual Periodicals in The New York Public Library* (New York, 1942, 370 pp.). Much of the information given here is in **Titus**, *Union List* (p. 142).

Herman Haupt, *List of Periodical Engineering Literature (published in the English Language)* (Philadelphia: University of Pennsylvania, Rogers Engineering Library, 1879). Six hundred titles in ninety-one pages. Gives founding dates, title changes, and so forth, as in **Bolton** (above).

David A. Kronick, *History of Scientific and Technical Periodicals* . . . *1665–1790* (New York: Scarecrow, 1962, 274 pp.). Discusses early technical periodicals from standpoint of function: whether they contain ephemeral or substantive information; includes critical review of bibliographies. Has useful information about some individual series. Bibliography, pp. 241–260. For a critical review of the book by Nathan Reingold, see *Isis*, 54 (June 1963), 284–285.

Johannes Müller, *Die wissenschaftlichen Vereine und Gesellschaften Deutschlands im 19. Jahrhundert.* (3 vols., 1883–1917). A bibliography of the publications of societies. I found nothing on technology.

A. D. Roberts, *Guide to Technical Literature: Introductory Chapters and Engineering* (London: Grafton, 1939, 279 pp.). A very knowledgeable guide by the former librarian of the Technical Library, Birmingham. Although chiefly concerned with current literature, Roberts discusses finding aids that can be of help in running down serials of the last 300 years. There is a chapter on societies and institutions, which lists indexes, similar to the **Thompson** and **National Academy of Sciences** titles (pp. 174, 37), of societies in Great Britain, Germany, France, Holland, Denmark, and Switzerland. Historical books and monographs, when available, are listed under appropriate subject headings.

In **Sarton**, *Guide* (p. 17), see particularly national academies and national scientific societies, pp. 111–114; international congresses, pp. 290–302; and annotated list of journals and serials concerning the history of science, pp. 194–248.

Samuel H. Scudder, *Catalogue of Scientific Serials . . . 1633–1876* (Cambridge: Harvard University Library, 1879, 358 pp.). Reprint (New York: Kraus, 1965). There are 4,390 titles, arranged by geographical location. Index of titles, pp. 315–354; subject index, pp. 355–358.

Finding aids may be located through **Bates**, *Scientific Societies in the United States* (p. 173), pp. 245–293.

Current publications of American societies are listed in **NAS–NRC**, *Scientific and Technical Societies* (p. 37). British societies are listed in *Scientific and Learned Societies of Great Britain* (p. 37).

Thompson, *Learned Societies* (p. 174) treats exhaustively the publishing history, through 1906, of technical societies in North and South America.

Seventeenth-century serials are described in **Ornstein**, *Scientific Societies* (p. 174).

B. Indexes of Articles in Periodicals

British Technology Index (London: Library Assn., 1962+). Subject guide to some 400 British technical journals. The first annual volume has 28,000 entries (*Industrial Arts Index*, following, has about 80,000). There are no annotations, but inclusive page numbers are stated, and most one-page articles have been omitted. I found very few historical articles, but history is not excluded from the compilers' view.

Engineering Index (New York). Annual. Vol. 1, 1884–1891, was called *Descriptive Index of Current Engineering Literature*. First issued serially in Association of Engineering Societies, *Journal*, the index was prepared as "largely a labor of love" by John B. Johnson (1850–1902), professor of civil engineering at Washington University, St. Louis. The present name was assumed with Vol. 2, 1892–1895, and publication was taken over by *The Engineering Magazine* (New York); Vol. 3, 1896–1900; Vol. 4, 1901–1905; annual from

1906. Control was transferred to A.S.M.E. in 1919 and later to Engineering Index, Inc.

An "Engineering History" rubric, with cross indexing, was discontinued in the 1958 issue of *Engineering Index;* hence the current issues are of little use for history, although some historical articles are still indexed under subject headings. In the 1956 *Index,* for example, it was possible to locate within a half hour about three dozen historical articles. A "history" index like **Poole** (below) would make the *Engineering Index* through 1957 vastly more useful to a historian; the restoration of the "Engineering History" rubric should be considered a minimum hope for the future.

Francis E. Galloupe, *An Index to Engineering Periodicals* (2 vols., Boston, 1888–1893). Vol. 1 (1883–1887, 294 pp.); Vol. 2 (1888–1892, 396 pp.). No more published.

Industrial Arts Index (New York, yearly, 1913–1957). Since 1958, *Applied Science & Technoolgy Index,* a continuation of the series. See following entry.

Mary Elizabeth Poole, Comp., *"History" References from the Industrial Arts Index 1913–1957* (Raleigh, N.C.: D. H. Hill Library, N.C. State College, 1958, 119 pp.). This most useful index, which lists perhaps 3,000 titles, reduces drastically the time required to use the *Industrial Arts Index.* Available as a University Microfilms (Ann Arbor) "O-P Book."

Informationsdienst Geschichte der Technik (Technische Universität Dresden, Institut für Philosophie, Abt. III, Dresden A27, Liebigstrasse 30). Published six times a year, each issue approximately thirty-six pages. A recent typical issue (No. 3, 1966), contained ninety article entries, each with brief annotation. Eighteen articles were five pages or longer, and of these seven were in *Technikgeschichte;* eleven book reviews were listed, of which eight were in *Technikgeschichte;* all other entries (about sixty) were one- to four-page "Kurzgeschichte" and commemorative articles. Thus, the bibliography is quite marginal. On the other hand a recent issue — Vol. 5, No. 5 (1965) — was devoted to **E. Herlitzius,** "Technik und Philosophie" (p. 208), a listing of works on the philosophy of technology. There is also a list of eleven other East German institutes for the history of science and technology on the back cover of each issue.

William F. Poole, *Poole's Index to Periodical Literature 1802–1881* (rev. ed., 2 vols., Boston, 1891). Very thin on technology. Five supplements were issued, covering the years 1882–1906. Entire series reprinted (New York: Peter Smith, 1938).

Reader's Guide to Periodical Literature, from 1900 (New York, 1905+). Fairly good for technology in the general press.

For several additional titles of general periodical indexes, see *American Historical Association's Guide* (p. 18), p. 27.

Jeremias D. Reuss, *Repertorium commentationum a societatibus litterariis editarum* (16 vols., Göttingen, 1801–1821). Reprint (New York: Burt Franklin, 1962). Subject index to learned journals before 1800 in English, French, German, Swedish, and so forth. See particularly Vol. 3, chemistry and metallurgy; Vol. 4, physics; Vol. 6, agriculture and domestic economy; and Vol. 7, mathematics, mechanics, hydrostatics, hydraulics, aerostatics, engineering, architecture, naval and military technology, and so forth. Author index in each volume.

Royal Society of London, *Catalogue of Scientific Papers, 1800–1900* (19 vols., London, 1867–1902, and Cambridge, 1914–1925). A surprisingly comprehensive index of papers in the nineteenth-century literature of science and technology; listing is by author.

Royal Society of London, *Catalogue of Scientific Papers, 1800–1900, Subject Index* (3 vols. in 4, Cambridge, 1908–1914). An important but apparently not well-known bibliography, covering mathematics, mechanics, and physics. Contains hundreds of references to papers on measurement, kinematics of machinery, steam engines, electrical technology, and so forth. Sections on history and biography. This subject index, intended to encompass all science, was never completed.

[Royal Society]. *The International Catalogue of Scientific Literature* (254 vols., 1902–1916) grew out of the Royal Society *Catalogue*. Yearly volumes in seventeen subject fields were published until World War I. There is less emphasis upon technology than in the earlier *Catalogue*. For a convenient summary of the series, see **Sarton,** *Guide* (p. 17), pp. 98–99. The entire series has been reprinted in 238 volumes. E. Wyndham Hulme, in his *Statistical Bibliography in Relation to the Growth of Modern Civilization* (London: the Author, 1923, 49 pp.), derived certain statistical data to support his thesis by counting titles in the *International Catalogue*.

C. Current Serials Having Articles on Technical, Scientific, and Economic History

Historical articles and items of interest to the historian of technology have appeared with reasonable frequency in the periodicals listed in this section. To cover current periodicals thoroughly, however, one must also work one's way through the monthly journals of technical societies and a considerable number of trade magazines. A list of society publications follows in a separate section.

The journals of state and local historical societies contain occasional solid articles, but with few exceptions (*Delaware History, Pennsylvania Magazine of History and Biography,* for example) the rate of occurrence of articles of more than marginal interest is rather low. Such articles may be located through Griffin, *Bibliography of American Historical Societies* (p. 29) and the annual *Writings on American History* (p. 29). A few articles can be recovered from the annual bibliographies in *Economic History Review* (from 1927). Current articles are listed also in *American Historical Review* and *American Quarterly*. The serious student will want to work his way systematically through the annual bibliographies in *Technology and Culture* and in *Isis*.

Abhandlungen und Berichte (Munich, Deutsches Museum, 1929+). Six numbers per year to 1940 (?), irregular through 1949, three issues per year since 1950. Each issue consists of a single article on technical history. Titles of individual issues, 1948–1962, are listed in *Technology and Culture,* 6 (Winter 1965), 102.

Agricultural History (1927+). A quarterly journal having significant articles on various aspects of technology on the farm. Occasional bibliographical articles and listings.

American Historical Review (1895+). Very few articles on technology; but see the exceptions, such as Lynn White, Jr., "Tibet, India, and Malaya as Sources of Western Medieval Technology" (65 [1960], 515–526), and Frederic C. Lane, "The Economic Meaning of the Invention of the Compass" (68 [1963], 605–617). There are many book reviews, extensive bibliographies, and news items regarding acquisitions and publications of Library of Congress, National Archives, and other repositories.

Additional news of this nature appears in the *AHA Newsletter* (1962+), which is published five times during the academic year.

American Neptune (1941+). Historical articles on shipbuilding and maritime affairs. Current bibliography of maritime literature.

American Quarterly (1941+). Recent interest in technology is reflected by occasional articles on technology (e.g., "American Technology and the Nineteenth-Century World," Summer 1958) and notices of current articles in an annual bibliography (from 1959).

Annals of Science (London: Taylor & Francis, 1934+). Quarterly review of the history of science since the Renaissance. Vol. 1, articles on Chilean nitrate industry; Vol. 4, soap; Vol. 5, origins of thermometer; Vols. 6, 7, 11, rise of tinplate industry.

Archives Internationales d'Histoire des Sciences (1947+). Now sponsored by UNESCO. Vol. 5, articles on gas industry; Vol. 7, early patents; Vol. 11, continental influences on British civil engineering to 1800; Vol. 12, recording instruments. In **Sarton**, *Guide* (p. 17), pp. 200–201, see entries for this journal and for *Archivio di Storia della Scienza* (1919–1941), issued under title *Archeion* (1927–1941).

Die BASF (Ludwigshafen am Rhein, 1951+). A house organ of the Badische Anilin und Soda Fabrik A.G., of outstanding quality, nicely printed on coated paper; engraving and color work are excellent. Illustrations are unhackneyed and generally striking. A typical issue has two or three short articles on history in addition to articles on the firm's chemical products. A supplement (35–50 pp.), containing summaries of articles in five languages (English, French, Spanish, Italian, Portuguese), is included with each issue.

Beiträge zur Geschichte der Technik und Industrie. See *Technikgeschichte.*

Blätter für Technikgeschichte (Vienna, 1932+). The first eight volumes (1932–1942) contained about 1,000 pages and 500 illustrations. Yearly since Vol. 9 (1947). Each issue has many short reviews of books and articles; an abbreviated cumulative list of contents is also to be found in every issue. For an analytical review of a typical recent volume (23) by R. S. Hartenberg, see *Technology and Culture, 4* (Summer 1963), 347–350.

The British Journal for the History of Science (1962+). Two issues a year; occasional articles on technology — for example, electrical communications, dyeing, and scientific instruments. Supersedes British Society for the History of Science, *Bulletin* (1949–1961).

Business History Review (1926+). Until 1954, Business History Society, *Bulletin*. An important source for articles on American technology. Articles on kerosene industry (Mar. 1955); British influence on American railroads (Summer 1960); Niagara Falls power (Spring 1961). Index, Vols. 1–15 (1926–1941).

Ciba Review (Basle, 1937+). English edition of the attractive house organ of a chemical firm. The editors, literate and intelligent, exhibit a broad spectrum of interests. For example, evolution of the shirt (Sept. 1957); English wool trade (Jan. 1959); history of textile testing (Oct. 1959); artisans' dress (June 1960); and history of chlorine bleaching (Aug. 1960).

Daedalus. Tekniska Museets Arsbok (Stockholm, 1931+). One issue a year, about 160 pages. Profusely illustrated with pictures of high quality. Articles on Christopher Polhem, iron works, calculators, paper, chemical industries, photography, textiles, rock drilling, matches and match boxes, rockets, etc., chiefly of the eighteenth and nineteenth centuries. Cumulative author index, yearly from 1938. Contents page in English, from 1959. Very occasional articles in English; one each in 1960, 1961, and 1962. In 1965 there appeared a "Cumulative Index in English of *Daedalus* 1931–1965," pp. 264–282. Listing is by author; each article is described in a sentence or two.

Any serious work on the history of mechanization should include an appreciation of the contributions of Polhem (1661–1751) and his contemporaries in Sweden.

Early American Industries Association, *Chronicle* (Williamsburg, Va., 1933+). Frequent articles and notes of some importance to history of hand tools and techniques of the craftsman. See, for example, two articles on making wooden planes in America (8, 19 ff., 28 ff.). Index to Vols. 6–13 (?–1960) issued separately.

Economic History (supplement to *The Economic Journal*). Nos. 1–15 (Vols. 1–4, London, 1926–1939). See, for example, articles on the Albion steam flour mill (2, 380–395) and on American engineering competition in Britain (2, 292–311).

Economic History Review (Series 1, 1927+; series 2, 1948+). This English journal has many articles important to the history of technology. Bibliographic lists of current books and articles are in most volumes.

Economy and History. Published yearly by Institute of Economic History and the Economic History Association, University of Lund, Sweden, 1958–1961 (no more?). Nothing in the first four volumes seemed pertinent to history of technology.

The Edgar Allen News (Vols. 1–45, Sheffield, 1919–1966). The house organ of Edgar Allen & Co. Ltd., makers of alloy steels, published frequent articles on history. Discontinued after issue of December 1966.

Endeavour (1942+). Published quarterly by Imperial Chemical Industries, this distinguished magazine, whose tone and format were set by E. J. Holmyard, historian of science, has generally short (4–6 pp.) illustrated articles on the sciences and technology, reflecting always an awareness of a past, a development, and a heritage. Historical sketches of scientific societies, institutions, laboratories, colleges, and so forth. A publication of general interest, ever literate, with frequent suggestions of links between past and future. I can point to no periodical of equal quality sponsored by an industrial company in the United States.

Geschichtsblätter für Technik und Industrie (11 vols., Berlin, 1914–1927). No more. Edited by Klinkowstroem and Feldhaus. The entire volume (22 pp.) in 1923, a year of runaway inflation, consisted of the remarkable bibliography of early histories of technology (p. 15).

Glass Technology (Society of Glass Technology, Sheffield, 1960+). In the first six volumes there have appeared four articles on history: on development of automatic glass-bottle machine, on tools of the glassmaker, on a British seventeenth-century glassmaking manual, and on Frederick Carder (1863–1963), founder of Steuben Glass Works. Book reviews and abstracts appear in each issue; occasional historical books and articles are noticed.

History of Science (Cambridge: Heffer, 1962+). An annual review of literature, research, and teaching in the history of science, which is interpreted to include the history of technology and of medicine. Good essays on needs and opportunities; for example, Vol. 2 on the histories of scientific societies and on mathematical instruments and Vol. 5 on sources for early history of the Royal Society. Long and short reviews of books. Edited by A. C. Crombie and M. A. Hoskin.

Industrial Archaeology (1964+). A quarterly journal, until 1966 *The Journal of Industrial Archaeology*, edited by Kenneth Hudson, Bath University of Technology. This modest but attractive journal has an encouraging number of illustrated articles, describing machines and structures; book reviews and news of activities in the field of record and rescue, being cultivated by Hudson; and in each issue an up-to-date list of local societies for industrial archaeology.

International Congress on the History of Sciences, *Actes.* For the slightly complex publishing history, see **Sarton,** *Guide* (p. 17), p. 255, or *Isis, 48* (1957), 176. The *Actes* of the Eighth Congress, Florence–Milan, 1956, were published in three volumes (1,338 pp.); Ninth, Barcelona–Madrid, 1959, one volume (732 pp.); Tenth, Philadelphia–Ithaca, 1962 (2 vols., Paris: Hermann, 1964). *Actes* of Sixth Congress, for example, contain articles on patents in the seventeenth century, weights and measures, and British contributions to continental technology, 1600–1850. Interest in technology has increased sharply in recent years.

Isis (1913+). Infrequent articles on technology. "Critical Bibliography," now published annually but more frequently in earlier years, contains sections on technology. A table of contents and index of Vols. 1–20 is in 21 (1934), 502–698. A fifty-year index is being prepared. *Isis* is the official quarterly of the History of Science Society.

Journal of American History (1914+). Occasional reference to technological subjects. Technical and economic history book reviews. Until 1964, *Mississippi Valley Historical Review*. Published by Organization of American Historians, formerly Mississippi Valley Historical Association.

Journal of Economic and Business History (Vols. 1–4, Cambridge, Mass., 1928–1932). In this very valuable though short-lived journal there are major articles on the Wedgwoods, ironmaking in Pennsylvania, ropemaking and hemp trade in fifteenth-century Venice, the rise of industrial consciousness in the United States, 1760–1830, and many other technical subjects.

Journal of Economic History (1941+). An American journal in which technological implications are often recognized. See, for example, articles on the "American system" of manufacturing (1954, pp. 361–379); engineering of railways (Summer 1951); American automobile standardization (Winter 1954).

Journal of the Patent Office Society (1918+). Many historical articles on patents and the Patent Office have been published in the *Journal*. A centenary issue is mentioned on p. 105. Historical patent surveys on aeronautics (3, 478), alcohol (5, 107), anaesthesia (7, 128), beds (7, 157), caustic soda process (3, 368), coal-cutting machinery (4, 364), glass manufacturing (4, 155), railway signalling (5, 620), sound recording (3, 193; 10, 120), illumination (4, 7), sewing machines (4, 206), woodworking (3, 512), papermaking (3, 457), pavements (9, 246), road building (9, 60), refrigeration (7, 51), automatic telephone (4, 251), automatic loom (3, 412), tide and wave motors (8, 321).

Indexes for the *Journal* are Index, Vols. 1–20 (1918–1938), bound separately; 11–20 in Vol. 20 (1938); 21–30 in Vol. 30 (1948); 31–40 in Vol. 40 (1958). Analytical indexes have "history" headings in 11–20 and 21–30 but not in earlier or later indexes.

Journal of the Printing History Society (St. Bride Institute, Bride Lane, London, E.C. 4, 1965+). Yearly. The first number has an illustrated article on Mackie's steam type-composing machine, *ca.* 1867; the second has illustrated articles on a 150-year-old engraving firm, on Augustus Applegath (1788–1871), and the Albion Press of 1825. It is to be hoped that historians of printing will maintain this quickened interest in the machinery of the industry. The Society has also published at low prices one or two titles a year (type specimens and historical works) in addition to the *Journal*.

Journal of Transport History (1953+). Published twice yearly by Leicester University Press. Articles on many aspects of British transportation, including canals, roads, railroads, and vehicles; articles on sources, including information on records of British Transport Commission, Public Record Office, Parliament, and so forth; reviews and notices; and (in Vol. 4) a list of transport theses in British university libraries. From 1962, there is a short annual "Transport Bibliography."

Le Machine (Milan, Dec. 1967+). The "Bolletino dell' Istituto italiano per la storia della tecnica," whose first issue appeared late in 1967, is published at Via S. Vittore, 19, Milan 20123. Articles on the reasons for the history of technology, nineteenth-century Catalan iron industry, and medieval accounting; book reviews and news of the Istituto.

Mariner's Mirror (London, 1911+). Articles on the history of shipbuilding and maritime affairs. Literature reviews and book reviews.

Mississippi Valley Historical Review. See *Journal of American History.*

Newcomen Society for the History of Engineering and Technology, Transactions (1920+). Annual, except for a few irregular volumes after 1942. Much detailed history of the steam engine and a great deal more. At least forty articles on the origins of commonplace but generally ignored basic elements of technology: wood screws, spring balance, chain, rope, drafting instruments, wire and sheet gauges, glass, pencils, vacuum cleaner, and the like. Another forty or more biographical articles on technical men, such as Bramah, Drebbel, Huntsman, Weston, and Bodmer. Articles such as these supply many of the blocks needed to construct a sound foundation for the history of technology. Indexes to Vols. 1–10 (1920–1929) and Vols. 11–20 (1930–1940) also have topical indexes to the "Analytical Bibliography of the History of Engineering," which appeared in Vols. 2–22 and 25. *General Index to Transactions Volumes I–XXXII 1920–1960* (1962) does not include the bibliographical index. I reviewed this 111-page general index in *Technology and Culture,* 5 (Winter 1964), 136–137.

Five numbered "Extra Publications" have been issued: (1) M. Triewald, *Description of the Atmospheric Engine,* 1724; (2) R. d'Acres, *Art of Water Drawing,* 1659; (3) George Cayley, *Aeronautical and Miscellaneous Note-Book;* (4) *John Smeaton's Diary of his Journey to the Low Countries 1755;* (5) *Catalogue of the Civil and Mechanical Engineering Designs, 1741–1792, of John Smeaton.* Also published by the Society are three unnumbered volumes: **Rhys Jenkins,** *Collected Papers* (1936); **H. W. Dickinson,** *The Watersupply of Greater London* (1954) (p. 226); and **Joseph Needham,**

Development of Iron and Steel Technology in China (1958) (p. 260).

The Newcomen Bulletin, issued thrice yearly to members, contains news, brief reviews of books, and short lists of publications received.

The Newcomen Society in North America grew out of, but has grown very far away from, the parent English society. The present tie seems to be merely legal. The garish brochures issued by the American society are ubiquitous and extremely numerous. Comprising the after-dinner speeches of (principally) industrialists, some 630 titles, most of them of no value historically, were published before August 1959. A list of titles, 1933-1959, is given in Newcomen Society in North America, *Annual Report 1958-1959* (New York, 1959, 60 pp.).

Osiris (1936+). Each volume is dedicated to a historian of science, with a biography and bibliography. Some long articles of particular value. Most volumes contain cumulative "Alphabetical List of Memoirs."

Physis. Rivista di Storia della Scienza (Florence, 1959+). Articles relating to technology are published in this journal in English, Italian, French, and German.

Revue d'Histoire de la Sidérurgie (Nancy, 1960+). Published by Centre de Recherches de l'Histoire de la Sidérurgie, Nancy.

Revue d'Histoire des Sciences et de leurs Applications (Paris, 1947+). Articles on history of moment of inertia (3, 315-336); utilization of hydraulic energy (8, 53-72); Réaumur's contributions to thermometry (11, 302-329). Vol. 4, Nos. 3-4 (July-Dec. 1951) is devoted to *l'Encyclopédie* of Diderot and d'Alembert. Starting in 1963 this journal has devoted one issue each year to *Documents pour l'Histoire des Techniques* (p. 19).

Scientific American. Historical articles, some quite good, have appeared in this magazine since 1948, when the present publisher brought it back to life. A *Cumulative Index 1948-1963* was published separately in 1964.

The Smithsonian Journal of History (1966+). A quarterly journal of general history, designed for scholarly but heavily illustrated articles. The first several issues have emphasized technology.

Society of Architectural Historians, Journal (1941+). Notes on early elevators (Oct. 1951, May 1952); "Harbingers of Eiffel's Tower" (Dec. 1957). Sound articles. The first four volumes were published under the title: American Society of Architectural Historians, *Journal.*

Tasks of Economic History (10 vols., 1941–1950). Supplements to *Journal of Economic History*, comprising papers presented at annual meetings. Function was taken over by the Journal.

Technikgeschichte (Düsseldorf, Verein Deutscher Ingenieure, 1909–1941, 1965+). Until 1932 *Beiträge zur Geschichte der Technik und Industrie*. Thirty annual volumes of papers of permanent value were edited by Conrad Matschoss, from 1909 to 1941. Vols. 6, 9–20 have cumulative indexes from Vol. 1–date. The journal is now a quarterly. Vol. 31 is a cumulative index (see next entry); regular issues resumed with Vol. 32 (1965). In *34*, No. 3 (1967), 273–280, a "Literaturübersicht 1966" provides a checklist of articles in the history of technology from about forty-five German, English, French, and American journals. I hope this will be an annual feature.

Technikgeschichte, "Register zu Bd. 1 bis 30" (Index to Vols. 1–30). Published at Düsseldorf in 1965 as Vol. 31 of *Technikgeschichte* (90 pp.). Cumulative index of *Beiträge zur Geschichte der Technik und Industrie*, Vols. 1–21 (1909–1932), which was followed by *Technikgeschichte*, Vols. 22–30 (1933–1941). This valuable index is in seven parts: volume-by-volume lists of contents, pp. 1–21; index of personal names, pp. 23–41; index of firms, institutions, and organizations, pp. 43–47; subject index, pp. 49–79; industrial monuments, pp. 81–83; index of museums referred to in articles, pp. 85–86; author index, pp. 87–90.

Techniques et Civilisations (5 vols., Saint-Germain-en-Laye, 1950–1956). No more. This was a revival of *Métaux et Civilisations*, Vol. 1, Nos. 1–6 (1945–1946).

Technology and Culture (International quarterly of the Society for the History of Technology, 1959+). ". . . concerned not only with the history of technological devices and processes, but also with the relations of technology to science, politics, social change, economics, and the arts and humanities." — *1*, No. 1, 1. An annual current bibliography was inaugurated in 1964; the works of 1962 were listed, with annotations, in *5* (Winter 1964), 138–148. A cumulative subject index of the first four annual bibliographies, 1962–1965, appeared in *8* (Spring 1967), 291–309.

Textile History. David and Charles, Newton Abbot, Devonshire have announced a once-yearly journal, the first issue of which may be expected near the end of 1968.

Tradition; Zeitschrift für Firmengeschichte und Unternehmerbiographie (Munich: Verlag F. Bruckmann K. G., 1956+). A bimonthly journal of industrial history and biographies of entrepreneurs, reviewed annually (I have seen two reviews only) in *Business History Review*, beginning in Autumn 1965. A continuing bibliography of monographic literature in its field is expected to commence in 1967.

Transport History. David and Charles, Newton Abbot, Devonshire have announced a new (thrice yearly) journal devoted to British transport, published in association with the University of Strathclyde. The first issue appeared in March 1968.

D. Independent Technical and General Magazines (chiefly nineteenth century)

The independent technical press was established in Great Britain just before 1800 and in the United States about a generation later. Before 1850, there were only a few journals capable of reporting adequately on technical developments. There were, however, dozens of short-lived mechanics' magazines, journals for progress in agriculture and the arts, and polytechnic journals, most of which published only a volume or two. Generally, these are disappointingly superficial. With one or two exceptions, they have been omitted in the list that follows.

Throughout the nineteenth century some nontechnical magazines paid attention to technical works. These magazines are now frequently the best and sometimes the only contemporary sources available.

I have made no attempt to review the rapidly burgeoning technical press of the last quarter of the nineteenth century. In **Bolton**, *Catalogue* (p. 143), the final 3,600 entries cover a period of only ten years, from 1885 to 1895. See, for example, the entries "Electric" and following, "Industrial" and following.

Allgemeine Bauzeitung (Vienna, 1836–1918). Concerned principally with design and construction of buildings, but with some attention to bridges. Large, fine engraved plates.

American Engineer. See *American Railroad Journal.*

American Journal of Science (Silliman's, 1818+). In the 1820's there were occasional notes on steam engines, internal-combustion vacuum engines, surveying of the Erie Canal, and notice of textbooks at Rensselaer. Index, Vols. 1–49 (1847), and decennial indexes thereafter.

American Mechanics' Magazine (2 vols., New York, 1825–1826). Not very substantial. Succeeded by the much superior Franklin Institute *Journal* (p. 165).

American Railroad Journal (now *Railway Locomotives and Cars*) (New York 1832+). Founded by D. K. Minor, publisher of New York *American*. Very important for history of public works in the United States. Provided good coverage when technical publications were very few and continued solid and informative throughout the nineteenth century. Two series of articles should be noted: one on locomotive working stresses (1898–1900) and Octave Chanute on progress in flying machines (1891–1893). There are probably many others. Title was changed frequently after 1886: *Railroad and Engineering Journal*, 1887–1892; *American Engineer and Railroad Journal*, 1893–1911; *American Engineer*, 1912–1913; *Railway Age Gazette, Mechanical Edition*, 1913–1915; *Railway Mechanical Engineer*, 1916–1949; and so forth.

Annals of Electricity. See William Sturgeon (p. 161).

The Annual of Scientific Discovery: or Year-Book of Facts in Science and Art, David A. Wells, Ed. (21 vols., Boston, 1850–1871). In the preface of the first volume, it is noted that this series was suggested by "similar works, which have been published in Europe for several years past." The section on "mechanics and the useful arts," about 100 pages in length, gives short summaries of events with references to journal sources.

Annual Record of Science and Industry for 1871, Spencer F. Baird, Ed. (New York: Harper, 1872, 634 pp.). The first of eight annual volumes; similar in scope to the preceding title. Agricultural and rural economy, pp. 313–354; household economy, pp. 355–388; mechanics and engineering, pp. 389–424; technology, pp. 425–534. Each article, generally a single paragraph in length, ends with a reference to a periodical in the United States, the United Kingdom, France, or Germany.

Appletons' Mechanics' Magazine and Engineers' Journal (3 vols., New York, 1851–1853). No more. The coverage ran to steamboat engines, but an alert if not very expert eye was peeled for all technical developments. Should certainly be checked by anyone concerned with engineering of the period. Contains a running controversy on hot-air engines between John Ericsson and army Maj. John G. Barnard.

Atlantic Monthly (Boston, 1857+). Occasional articles on railroads, iron and steel, mining, and the like; but neither as frequent or as good as similar articles in *Harper's* (p. 158).

Cassier's Magazine (New York, 1891–1913). One of the very best sources for ten- to twenty-page review articles of contemporary practice. Written for the intelligent engineer and informed general reader. Profusely illustrated (unfortunately with halftones). Detailed, informative articles on Columbian Exposition (1893); Niagara Falls hydropower (special issue, July 1895); special issues also on electricity (Jan. 1896), marine engineering (Aug. 1897), electric railways (Aug. 1899), mining and metallurgy (July 1902), machine shops (Nov. 1902); a series on technical colleges, another on technical societies, and a third on technical journals. Many biographical sketches, with portraits, of prominent living engineers. It should be noted that this magazine is listed in *Union List of Serials* (p. 142) as Vols. 1–44 of *Mechanical Handling*. A cumulative index has been published separately. See next entry.

Cassier's Magazine, Engineering Monthly 1891–1913: Subject Index, Eugene S. Ferguson, Comp. (Ames, Iowa, 1964 [Iowa State University Bulletin, Vol. 63, No. 10], 46 pp.). In addition to the subject index, a "biography and portrait" list locates short sketches and portraits of over 300 engineers, most of whom were alive at the time of publication and are now quite obscure.

Civil Engineer and Architects' Journal (London, 1837–1868). Not very substantial, but this journal was published before there were many others of high quality.

De Bow's Review (New Orleans, 1846–1870). A compendium of southern commercial and industrial progress, recently made more readily available by Willis D. Weatherford, *Analytical Index of De Bow's Review* (n.p., 1952). See also Otis C. Skipper, *J. B. D. De Bow, Magazinist of the Old South* (Athens, Ga., 1958), and James A. McMillen, *Works of James B. D. De Bow. A Bibliography of De Bow's Review* [etc.] (Hattiesburg, 1940).

Dingler's Polytechnisches Journal (Stuttgart, 1820–1931). Described and illustrated a selected group of patents in the early years. Later published original papers on the developing engineering sciences.

Emporium of the Arts and Sciences (Philadelphia, 1812–1814). A few original articles on steam engines and chemical arts.

The Engineer (London, 1856+). Excellent coverage of civil and mechanical engineering, broadly interpreted. Well illustrated. *The Engineer* published in 1956 a special "Centenary Number," which contained 125 folio pages of serious and mature editorial matter and 325 pages of advertising. There is in this issue of brief history of the magazine.

'The Engineer' Index 1856–1959, C. E. Prockter, Comp. (London: Morgan Bros., 1964, 4to, 212 pp.) makes a real contribution to the

historical record by providing a compressed but very expertly made key to the riches of this journal. Name index, pp. 1–159; subjects, pp. 161–212.

Engineering (London, 1866+). *The Engineer* (previous listing) and *Engineering* should be consulted first for balanced, illustrated accounts of nearly all engineering projects and developments of general interest to technical readers. Unfortunately, there is no cumulative index of *Engineering*.

Engineering Magazine (New York, 1891–1916). Usually catalogued under *Factory and Industrial Management*. Full articles on the several branches of engineering. A section called "Industrial Sociology" provides a running commentary on management techniques such as Taylorism. Thick volumes. No index, but contents pages, divided by fields, can be readily scanned.

Engineering News-Record (New York, 1874+). Originally issued as *Engineering News* (1874–1917, with minor variations). A full, informative record of engineering in the United States to around the end of the nineteenth century. Selective indexes, 1874–1890 (118 pp.), and 1890–1899 (324 pp.).

Franklin Institute, Journal. See p. 165.

[Gill's] *The Technical Repository* (Vols. 1–11, London, 1822–1827); superseded by *Gill's Technological Repository* (Vols. 1–11, 1827–1830). No more. A journal of invention and improvements in the "useful arts," edited by Thomas Gill.

Harper's Magazine (originally *Harper's New Monthly Magazine*) (1850+). A considerable number of illustrated descriptive articles on technology, ten to fifteen pages long, appeared in the early years. For example: Novelty Iron Works, Vol. 2; Springfield Armory, Vol. 5; nail-making, Vol. 21; shipbuilding, Vol. 24; description of new Harper's building, Vol. 32; watches made by machinery, Vol. 39; industries of Pittsburgh, Vol. 42; oil refining, Vol. 45; coke-making, Vol. 47; cotton manufacture, Vol. 50. Analytical index, Vols. 1–85 (1850–1892).

[Hunt's] *Merchants' Magazine and Commercial Review* (New York, 1839–1870). Good editorial coverage of selected technical developments. No general index.

Illustrated News Magazines. *The Illustrated London News* (1842+), *L'Illustration* (Paris, 1843+), and other general magazines that followed shortly in considerable profusion must be recorded as last resorts, although I have found in them some technical pictures that are unique. Generally, the results of page-by-page search are quite disappointing. *Ballou's Pictorial Drawing Room Companion* (1850–1859; 1851–1854 as *Gleason's*); *Leslie's Illustrated Weekly Magazine* (1855–1922); and *Harper's Weekly* (1857–1916) were American contributions to pictorial reporting. *Harper's Weekly* published some good pictures of the Centennial Exhibition and Brooklyn Bridge.

Iron Age (New York, 1859+). A continuation of *Hardware-Man's Newspaper and American Manufacturers' Circular* (1855–1859). In its early years, *Iron Age* was concerned particularly with machines, tools, and hardware, although it carried also iron and steel statistics. Some good articles on details of manufacture of various hardware items.

The Journal of Science, and Annals of Astronomy, Biology, Geology, Industrial Arts, Manufactures, and Technology (22 vols., London, 1864–1885). Quarterly 1864–1879, monthly 1879–1885. Many good articles on metals and chemical industry, public works, electricity, beer, and air pollution; however, there were other good journals by the time this one appeared. Edited by James Samuelson and William Crookes.

Mechanics' Magazine (London, 1823–1858). Usually catalogued under *Iron*, its successor. Very good for mechanical technology. Many good illustrations.

Mechanics' Magazine, and Journal of Public Internal Improvement (Boston, 1830–1831). No more.

Mechanics' Magazine and Register of Inventions and Improvements (New York, 1833–1837). Published by D. K. Minor. Much secondhand material, but some significant original articles. Absorbed in 1838 by Minor's *American Railroad Journal*. Not to be confused with *American Mechanics' Magazine*, which was succeeded by Franklin Institute *Journal*.

Merchant's Magazine. See [Hunt's].

Nature (London, 1861+). News notes of technological interest in the early years.

La Nature (Paris, 1873+) [since 1961 subtitled *Science-progrès*]). Occasionally, in the nineteenth century, wood engravings of industrial subjects; few, but good when present. Band-saw mill in Brooklyn (2, 45); cutting ice on Hudson River (4, 136–137). Four decennial indexes, 1873–1912.

Newton's London Journal of Arts and Sciences (London, 1820–1866). A patent repertory. See also *Repertory of Arts, Manufactures, and Agriculture.*

[Nicholson's] *Journal of Natural Philosophy, Chemistry, and the Arts* (London, 1797–1813). See S. Lilley, "Nicholson's Journal 1797–1813" (*Annals of Science*, 6, [1948], 78–101).

Niles' Weekly Register (Baltimore, 1811–1849). Many short news paragraphs and valuable editorial comment on civil works and manufacturing. There is no general index, but see Norval Neil Luxon, *Niles' Weekly Register: News Magazine of the Nineteenth Century* (Baton Rouge: Louisiana State University Press, 1947, 337 pp.), particularly chapter 10, "Roads, Rivers, Canals, and Railroads."

North American Review (1815–1940). Occasional comment in the early years on the problems of public works. A series of articles on the alternating- vs. direct-current controversy (1889, pp. 321–325, 625–634, 653–664). Index for Vols. 1–125 (1815–1877).

Popular Science Monthly (New York, 1872+). A series on "Development of American Industries since Columbus" in Vols. 38–42 (1890 ff.) includes W. F. Durfee on iron and steel, nine parts; and shorter articles on pottery, musical instruments, leather, shoes, and glass. A modest contribution.

Port Folio (Philadelphia, 1801–1827). A few technical articles and occasional editorial comment on current civil and private works. For example, Finley's chain bridge (1809–1810, pp. 440–453); Schuylkill Permanent Bridge (1808, pp. 168–171, 182–187, 200–204, 222–224); National Road (1822, pp. 121–125; 1823, pp. 61–65); S. H. Long on railways and canals (1825, pp. 265–280, 353–360).

Railroad Gazette. See *Railway Age.*

Railway Age (New York). Published 1870–1908 as *Railroad Gazette;* in 1908 absorbed *Railway Age* (Chicago); became *Railroad Age Gazette* 1908–1909; *Railway Age Gazette,* 1910–1917; *Railway Age* since 1917. Both *Railroad Gazette* and *Railway Age* were good during the first generation. There were occasional reminiscences and articles on earlier events.

Railway Locomotives and Cars. Present (1967) name of *American Railroad Journal* (q.v.).

Repertory of Arts, Manufacture, and Agriculture (London, 1794–1862). A patent journal, in which specifications and illustrations of patents were frequently commented upon editorially. See also *Newton's London Journal*.

Scientific American (New York, 1845+). Although founded by Rufus Porter, this weekly magazine was edited, 1846–1907, by Orson Munn (1824–1907). For many years the magazine was an organ of the patent-soliciting firm of Munn & Co., but it was sufficiently spiced with news and editorial comment to make it important even in the early years. After about 1860, its coverage of technical events (with many outstanding folio-size wood engravings) made it the leading technical journal in the United States for half a generation or more. An interesting illustrated series on "American Industries," comprising more than eighty-five articles, each with a full page of fair-to-good views, appeared in 1879–1882.

Scientific American Supplement, a weekly, 1876–1919, was published concurrently with *Scientific American* by the same publishers but was a distinct and separate publication. There is an index to the *Supplement*, 1876–1910.

Scribner's Monthly (New York, 1870–1881). Illustrated articles on Hoosac Tunnel (Dec. 1870), Alvan Clark's telescope (Nov. 1873), and a trip to Niagara, viewing canals and bridges (May, June 1872). Index, Vols. 1–10 (1870–1875).

[William Sturgeon]. *Annals of Electricity, Magnetism, and Chemistry* (London, 1836–1843). Very good while it existed.

United States Magazine of Science, Art, Manufactures, Agriculture, Commerce and Trade (1854–1858). There are several well-illustrated major articles on industrial works: R. Hoe printing-press and saw shops (*1*, 337–347); Norris locomotive works (*2*, 129, 151–167); Salamander Safe Co. (*2*, 328–334); writing-pen manufacture (*4*, 348–357); Colt's armory (*4*, 221–248).

Van Nostrand's Engineering Magazine (New York, 1869–1886). As implied in its original title — *Van Nostrand's Eclectic Engineering Magazine* — this journal was a compendium of technical literature of the period. Some French and German articles were translated for this magazine. A few original articles appeared in the later years. Absorbed by *American Railroad Journal*.

E. Technical and Scientific Society Publications

This list consists chiefly of technical societies founded before 1910 in the United States and Great Britain. It exists because current directories, while listing surviving societies and their present publications, do not indicate the dates or titles of early transactions and proceedings. Moreover, the directories require systematic search in order to separate technical societies from the more numerous groups concerned with other fields.

Two dozen or more "Railroad Clubs" and electric light associations, founded between 1880 and 1900, have been omitted, as have most trade associations. Those that I am aware of can be located, however, through the finding aids listed at the beginning of this part. A very few scientific societies — Royal Society, Académie des Sciences, American Philosophical Soceity, and one or two others — are listed because of their deep commitment to the useful arts in their early years. For scientific societies in general, see **Sarton,** *Guide* (p. 17). and **Roberts,** *Guide* (p. 143).

Several engineering schools in the United States and Canada published serious engineering magazines and yearbooks during the period *ca.* 1890–1910; e.g., Stevens, Purdue, Colorado, Iowa, Minnesota, Toronto, Princeton, Rensselaer, and so forth.

There is a vast quantity of source materials in the nineteenth-century publications of many of the societies. Additionally, historical articles and reminiscences of value will be found in the earlier U.S. transactions and journals, and in those of Great Britain down to the present. The publications of state and local societies are generally slight, but the early transactions of the Engineers' Clubs in Philadelphia and Pittsburgh and of the Western Society of Engineers, in Chicago, stand up well when compared to transactions of the national societies.

Similar lists of journals of technical societies of other countries would in my opinion be useful. I should hope that eventually we may have a world checklist of technical publications that will approach the bibliographic completeness of **Thompson's** *Handbook* of 1908 (p. 174), to which the present list makes no distant pretense.

Indexes to individual serials can be located through the *Union List of Serials* (p. 142) and **Haskell,** *Check List* (p. 143).

1. *United States*

American Concrete Institute. Org. 1905. *Proceedings* [*Journal*], 1905+.

American Institute of Chemical Engineers. Org. 1908. *Transactions,* 1908+.

American Institute of Electrical Engineers. Org. 1884. *Transactions,* 1884+; *Electrical Engineering,* 1905+. Merged, 1962, with Institute of Radio Engineers. See Institute of Electrical and Electronics Engineers.

American Institute of Mining Engineers (now American Institute of Mining, Metallurgical, and Petroleum Engineers). Org. 1871. *Transactions*, 1871–1873+.

American Institute of the City of New York. Org. ?. *Journal*, 1836–1837, 2 vols. (no more?); *Annual Report* (many vols. entitled *Transactions*), 1841+.

American Iron and Steel Association. Org. 1855; reorg. 1864. *Bulletin*, 1866–1912.

American Philosophical Society Held at Philadelphia for Promoting Useful Knowledge. Org. 1769. *Transactions*, 1769–1809 (6 vols., O.S.); N.S., 1818+. *Proceedings*, 1838+; *Memoirs*, 1935+. See *Classified Index to the Publications of the American Philosophical Society . . . 1769–1940* (Philadelphia, 1940, 173 pp.). A subject index.

American Society for Engineering Education (until 1946, Society for the Promotion of Engineering Education). Org. 1893. *Proceedings*, 1893–1952; *Journal*, 1910+.

American Society for Testing Materials. Org. 1898. *Proceedings*, 1898+.

American Society of Agricultural Engineers. Org. 1907. *Transactions*, 1907–1935; *Agricultural Engineering*, 1920+.

American Society of Civil Engineers. Org. 1852; reorg. 1867. *Transactions*, 1868–1871+. *Civil Engineering*, 1930+. Index to *Transactions*, Vols. 1–83 (1867–1920).

American Society of Heating and Ventilating Engineers. Org. 1894. *Transactions*, 1895+.

American Society of Mechanical Engineers. Org. 1880. *Transactions*, 1880+; *Proceedings*, 1906–1908; *Journal*, 1908–1918; *Mechanical Engineering*, 1919+. The best index is that for Vols. 1–45 (1880–1923), which refers, however, to more complete indexing of early informal discussion periods in Index, Vols. 1–25, bound with Vol. 27 (1906), pp. 849–999.

American Society of Refrigerating Engineers. Org. 1904. *Transactions*, 1906+.

American Society of Sanitary Engineering. Org. 1906. *Proceedings*, 1906+.

American Water Works Association. Org. 1881. *Proceedings*, 1881+.

American Welding Society. Org. 1919. *Welding Journal*, 1922+.

Association of Engineering Societies, *Journal*, Vols. 1–55 (1881–1915). Formed in 1880 to secure joint publication of papers and transactions of the constituent societies. Papers of the following societies are in the volumes indicated. Starred (*) societies also published other works. See separate entry.

Boston Society of Civil Engineers (Vols. 1–51, 1881–1913)*
Engineers' Club of St. Louis (Vols. 1–55, 1881–1915)*

Western Society of Engineers (Vols. 1–15, 1881–1895)*
Cleveland Engineering Society (Vols. 1–39, 1881–1907)*
Engineers' Club of Minneapolis (Vols. 4–38, 1884–1907)*
Civil Engineers' Society of St. Paul (Vols. 4–55, 1884–1915)
Engineers' Club of Kansas City (Vols. 6–16, 1886–1896). Disbanded 1896; reorg. 1911.
Montana Society of Engineers (Vols. 7–54, 1888–1915)
Wisconsin Polytechnic Society (Vols. 11–12, 1892–1893)
Denver Society of Civil Engineers (Vols. 14–21, 1895–1898). Disbanded 1898.
Association of Engineers of Virginia (Vols. 14–21, 1895–1898). Disbanded 1898.
Technical Society of the Pacific Coast (Vols. 14–53, 1895–1914)*
Detroit Engineering Society (Vols. 18–51, 1897–1913)
Engineers' Society of Western New York (Vols. 20–37, 1898–1906)*
Louisiana Engineering Society (Vols. 21–53, 1898–1914)*
Engineers' Club of Cincinnati (Vols. 23–31, 1899–1903)*
Toledo Society of Engineers (Vols. 32–40, 1904–1908)
Engineers' Society of Milwaukee (Vols. 40–45, 1908–1910)*
Utah Society of Engineers (Vols. 40–53, 1908–1914)*
Oregon Society of Engineers (Vols. 47–55, 1911–1915)

Association of Railway Superintendents of Bridges and Buildings. Org. 1891, *Proceedings*, 1891–1930.

Boston Society of Civil Engineers. Org. 1848. *Proceedings*, 1879–1881; *Journal*, 1914+. For 1881–1913, see Association of Engineering Societies.

Cleveland Engineering Society. Org. 1880. *Journal*, 1908–1919; *Cleveland Engineering*, 1917+. For 1881–1907, see Association of Engineering Societies.

Connecticut Society of Civil Engineers. Org. 1884. *Proceedings*, 1885–1903; *Transactions*, 1903+.

Deutsch-Amerikanischer Techniker-Verband. Org. 1874. *Mittheilungen*, 1886–1895; *The Technologist*, 1896–[?].

Engineering Association of the South. Org. 1889. *Proceedings*, 1889/90–1916.

Engineering Society of Buffalo. See Engineers' Society of Western New York.

Engineering Society of Cincinnati. See Engineers' Club of Cincinnati.

Engineers' Club of Baltimore. Org. 1905. *Baltimore Engineer*, 1926+.

Engineers' Club of Cincinnati (now Engineering Society of Cincinnati). Org. 1888. *Cincinnati Engineer and Scientist*, 1942+. For 1899–1903, see Association of Engineering Societies.

Engineers' Club of Minneapolis. Org. 1883. *Minnesota Engineer*, current. For 1884–1907, see Association of Engineering Societies.

Engineers' Club of Philadelphia. Org. 1877. *Proceedings* [*Engineers and Engineering*], 1879–1932; *Announcer*, 1928+.

Engineers' Club of St. Louis. Org. 1868. *Transactions*, 1871, 1874; *Journal*, 1916+. For 1881–1915, see Association of Engineering Societies.

Engineers' Society of Milwaukee. Org. 1904. *Milwaukee Engineering*, 1921+. For 1908–1910, see Association of Engineering Societies.

Engineers' Society of Western New York (now Engineering Society of Buffalo). Org. 1894. *Transactions*, 1895–1897. For 1898–1906, see Association of Engineering Societies.

Engineers' Society of Western Pennsylvania. Org. 1880. *Proceedings*, 1880+.

Franklin Institute of the State of Pennsylvania for the Promotion of the Mechanic Arts. Org. 1824. *Journal*, 1826+. Easily the outstanding technical periodical in the United States before 1850. Patent records in the *Journal* before 1836 are important because official records were destroyed in the 1836 Patent Office fire. Records of the Steam Boiler Explosion Committee, in the early 1830's, are significant. Reports of the Committee on Science and the Arts provide descriptions and considered criticism of many inventions submitted to it for approval. There is a good index to Vols. 1–120 (1826–1885). The "New Series" and "Third Series" confuse the volume numbers. To convert from "New Series" to whole volume numbers, add four to the N.S. volume number; to convert from "Third Series" to whole volume numbers, add thirty to the 3rd Ser. volume number. Preceded by *American Mechanics' Magazine* (p. 155).

Illinois Society of Engineers and Surveyors. Org. 1886. *Annual Report*, 1886–1915.

Illuminating Engineering Society. Org. 1906. *Illuminating Engineer*, 1906+.

Indiana Engineering Society (now Indiana Society of Professional Engineers). Org. 1881. *Proceedings*, 1881–1930; *Indiana Professional Engineer*, 1937 (?)+.

Institute of Electrical and Electronics Engineers. Org. 1962 by merging of American Institute of Electrical Engineers and Institute of Radio Engineers. *IEEE Spectrum*, 1964+; *IEEE Transactions* are issued in several series, classified by subject.

Institute of Radio Engineers. Org. 1912. *Proceedings*, 1913+. Merged, 1962, with American Institute of Electrical Engineers. See Institute of Electrical and Electronics Engineers.

Institute of the Aerospace Sciences. Org. 1932. *Aero Space Engineering Index* (a bibliography), 1947+. Merged with the American Rocket Society (1963) to form American Institute of Aeronautics and Astronautics (AIAA).

Iowa Engineering Society. Org. 1889. *Proceedings*, 1889–1940; *Exponent*, 1940+.

Lake Superior Mining Institute. Org. 1893. *Proceedings*, 1893–1939.

Louisiana Engineering Society. Org. 1898. *Louisiana Engineer*, 1915+. For 1898–1914, see Association of Engineering Societies.

Maryland Institute for the Promotion of the Mechanic Arts. Org. 1825; reorg. 1847. *Annual Report*, 1851–1907; *Book of the . . . Annual Exhibition*, ca. 1848–1859 (no more?).

Massachusetts Charitable Mechanic Association. Org. (?). *Proceedings*, 1881–(?).

Master Car Builders' Association. Org. 1867. *Report of the Proceedings*, 1867–1918.

Michigan Engineering Society. Org. 1880. *Michigan Engineer*, 1880+.

Mining and Metallurgical Society of America. Org. 1908. *Proceedings* [*Bulletin*], 1908+.

National Association of Stationary Engineers (now National Association of Power Engineers). Org. ?. *National Engineer*, 1897+.

National Electric Light Association. Org. 1885. *Proceedings*, 1885–1932. Succeeded by Edison Electric Institute. These are trade associations.

New England Water Works Association. Org. 1882. *Transactions*, 1882–1885; *Journal*, 1886+.

[New York] Electrical Society. Org. 1881. *Transactions*, 1888–1916. Special papers, read before the society by Crocker, Elihu Thomson, Steinmetz, and others, separately published, 1887–1905.

Ohio Engineering Society. Org. 1880. *Proceedings*, 1880–1930.

Pacific Northwest Society of Engineers. Org. 1902. *Proceedings*, 1902–1917.

Society for the Promotion of Engineering Education. See American Society for Engineering Education.

Society of Automotive Engineers. Org. 1904. *Transactions*, 1907+; *Journal*, 1917+.

Society of Naval Architects and Marine Engineers. Org. 1893. *Transactions*, 1893+.

Technical Society of the Pacific Coast (San Francisco). Org. 1884. *Transactions*, 1884–1896. For 1896–1914, see Association of Engineering Societies.

Utah Society of Professional Engineers. Org. 1907. *Journal*, 1915–1922; *Utah Engineer*, current. For 1908–1914, see Association of Engineering Societies.

Western Society of Engineers (Chicago). Org. 1869. *Proceedings*, 1870/76–1881; *Journal*, 1896–1948; *Midwest Engineer*, 1948+. For 1881–1895, see Association of Engineering Societies.

2. *Great Britain*

British Institution of Radio Engineers. Org. 1925. *Journal*, 1934–1940+.

Builders' Society. See Institute of Builders.
Cold Storage and Ice Association. See Institute of Refrigeration.
Illuminating Engineering Society. Org. 1909. *Transactions*, 1936+; *Light and Lighting*, 1908+.
Institute of Builders (formerly Builders' Society). Org. 1834. Founded "to promote excellence in the construction of buildings . . ." *Proceedings*, 1887+; *Journal*, 1937+.
Institute of Marine Engineers. Org. 1889. *Transactions*, 1889+.
Institute of Metals (nonferrous). Org. 1908. *Journal and Metallurgical Abstracts*, 1909+. The abstracts section lists history articles and reviews books of history. Good.
Institute of Refrigeration (was Cold Storage and Ice Association). Org. 1900. *Proceedings*, 1900+.
Institute of Welding. Org. 1923. *Proceedings*, 1923–1934. *Transactions*, 1938–1953; *British Welding Journal*, 1954+.
Institution of Automobile Engineers. Org. 1906 (?). *Proceedings*, 1906/07–1946/47. Superseded by Institution of Mechanical Engineers, Automobile Division, *Proceedings*.
Institution of Chemical Engineers. Org. 1922. *Transactions*, 1923+.
Institution of Civil Engineers. Org. 1818. *Transactions*, 1836–1842, 3 vols.; *Minutes of the Proceedings*, 1837–1935; *Journal*, 1935–1951; *Proceedings*, 1952+. In addition to those on civil engineering subjects, major papers on steam engines and internal-combustion engines appeared in these publications throughout the nineteenth and early twentieth centuries.
Institution of Electrical Engineers. Org. 1871. *Proceedings*, 1872+.
Institution of Engineers and Shipbuilders in Scotland. Org. 1857. *Transactions*, 1857+.
Institution of Gas Engineers. Org. 1863. *Proceedings*, 1863–1881; *Transactions*, 1882+; *Journal*, current.
Institution of Heating and Ventilating Engineers. Org. 1897. *Journal*, 1898+.
Institution of Mechanical Engineers. Org. 1847. *Proceedings*, 1847+; *Chartered Mechanical Engineer*, 1954+. Indexes, 1847–1950 and 1937–1957. *Chartered Mechanical Engineer*, a monthly, publishes frequent articles of history and biography and has an annotated list of new books in each issue. Two compilations of the historical articles have appeared thus far under the title *Engineering Heritage* (2 vols., London: Heinemann, 1963–1966). Quality of articles varies, but some are quite good.
Institution of Mining and Metallurgy. Org. 1892. *Transactions*, 1892+; *Bulletin*, 1904+.
Institution of Mining Engineers. Org. 1889. *Transactions*, 1889+. Principally concerned with coal and iron ore.
Institution of Municipal Engineers. Org. 1873. *Chartered Municipal Engineer* (formerly *Journal*), 1873+.
Institution of Naval Architects. Org. 1860. *Transactions*, 1860+.

See K. G. Barnaby, *The Institution of Naval Architects 1860–1960* (London, 1960, 4to, 645 pp.), for year-by-year review of *Transactions* and table of contents of each volume.

Institution of Water Engineers. Org. 1896. *Transactions*, 1896–1945; *Journal*, 1947+.

Iron and Steel Institute. Org. 1869. *Journal*, 1871+. Very good. Monthly abstracts generally include several historical articles. Annual author-subject index as separate volume, 1962+.

Junior Institution of Engineers. Org. 1884. *Journal*, 1884+.

Liverpool Engineering Society. Org. 1875. *Transactions*, 1877+; *Bulletin*, 1928+.

Mining Institute of Scotland. Org. 1878. *Mining Engineer* (includes *Transactions*), 1879+.

North-East Coast Institution of Engineers and Shipbuilders. Org. 1884. *Transactions*, 1884+.

North of England Institute of Mining and Metallurgical Engineers. Org. 1852. *Transactions*, 1852+.

Radio Society of Great Britain. Org. 1913. *T. & R. Bulletin*, 1925+.

Royal Aeronautical Society. Org. 1866. *Annual Report*, 1866–1893; *Journal*, 1897+.

Royal Institute of British Architects. Org. 1834. *Transactions*, 1835–1892; *Journal*, 1893+.

Royal Institution of Great Britain. Org. 1799. *Journal*, 1802–1803, 2 vols.; *Quarterly Journal of Science, Literature, and Art*, 1816–1830; *Journal*, 1830–1831; merged into *London, Edinburgh and Dublin Philosophical Magazine and Journal of Science*.

Royal Society of Arts, Manufactures, and Commerce. Org. 1754. *Transactions*, 1783–1851; *Journal*, 1852+. Good illustrations of current inventions in *Transactions*; good summary articles of particular industries and branches of technology in *Journal*.

Royal Society of London. Org. 1660. *Philosophical Transactions*, 1665+. Abridgments of early *Transactions*, 1665–1750 (10 vols., 1705–1756). In the fifth edition, 1749, Latin papers were translated into English.

Society of Chemical Industry. Org. 1881. *Journal*, 1882+.

Society of Engineers. Org. 1854. *Journal & Transactions*, 1857+.

South Wales Institute of Engineers. Org. 1857. *Transactions*, 1857–1893; *Proceedings*, 1894+.

West of Scotland Iron and Steel Institute. Org. 1892. *Journal*, 1892+.

3. *Canada*

Canadian Mining Institute (now Canadian Institute of Mining and Metallurgy). Org. 1895 (?). *Transactions*, 1895+; *Canadian Mining and Metallurgical Bulletin*, 1908+.

Canadian Society of Civil Engineers (now Engineering Institute of Canada). Org. 1887. *Transactions,* 1887–1930; *Engineering Journal,* 1918+.

Mining Society of Nova Scotia. Org. 1892. *Journal,* 1892/93–1915.

4. *France*

Académie des Sciences. Org. 1666. Until about 1700, many papers of the Académie were published in *Journal des savants*. After 1700 there was a confusing mass of publications, which has been partly untangled in a brief summary in **Russo,** *Bibliographie* (p. 17), pp. 54–55.

Paris. Société d'Encouragement pour l'Industrie Nationale. *Bulletin,* 1801+. See also journals of similar societies in Mulhouse (1826+), Lille (1827+), Rouen (1873+), and so forth.

Société des ingénieurs civils de France. *Mémoires,* 1848–1965. No more. The society is an unspecialized one for all engineers; therefore, articles are often general and some are historical.

F. Newspapers

The first three titles are standard finding lists. The next is the result of a systematic search (by Brayer) for indexes. The three final titles are a remarkable series of gleanings, in which advertisements and pertinent news articles concerning the mechanical trades have been painstakingly copied out of long runs of early newspapers.

Clarence S. Brigham, *History and Bibliography of American Newspapers, 1690–1820* (2 vols., Worcester, 1947). A full and informative union list, showing holdings in great detail. Essential points are given of the history of individual newspapers.

Winifred Gregory, Ed., *American Newspapers 1821–1936. A Union List of Files Available in the United States and Canada* (New York: H. W. Wilson, 1937, fo., 789 pp.). A union list.

Library of Congress, Union Catalog Div., *Newspapers on Microfilm* (5th ed., Washington: Library of Congress, 1963, 4to, 305 pp.). Lists holders of negatives and of prints in the United States, pp. 1–230, and in other countries, pp. 231–305. With corrections and additions through June 1, 1963.

Herbert O. Brayer, "Preliminary Guide to Indexed Newspapers in the United States, 1850–1900," *Mississippi Valley Historical Review,* 33 (Sept. 1946), 237–258. A checklist of indexes — long, short, good, indifferent — of newspapers in about thirty-five states.

George F. Dow, *The Arts & Crafts in New England 1704–1775* (Topsfield, Mass., 1927).

Rita S. Gettesman, *The Arts & Crafts in New York 1726–1776* (New-York Historical Society, *Collections*, LXIX, 1936). Also issued separately (New York, 1938).

Alfred C. Prime, *The Arts & Crafts in Philadelphia, Maryland, and South Carolina 1721–1800* (2 vols., Topsfield, Mass., 1929–1932).

G. Miscellaneous Series

There are several series of papers and monographs, pertinent to the history of technology, that cannot be conveniently classified under a single subject.

The Monograph Series of the Society for the History of Technology, for example, is one to be aware of. The first three titles are Robert S. Woodbury, *History of the Lathe to 1850* (Cambridge: The M.I.T. Press, 1961); A. W. Richeson, *English Land Measuring to 1800* (Cambridge: The M.I.T. Press, 1966); and Frederick B. Artz, *The Development of Technical Education in France 1500–1850* (Cambridge: The M.I.T. Press, 1966).

Greville Bathe (1883–1964) published ten volumes of biography, monographs, and collected papers as follows:

Oliver Evans (1935)
An Engineer's Miscellany [collected papers] (1938)
Jacob Perkins (1943)
Citizen Genêt, Diplomat and Inventor (1946)
Horizontal Windmill, Draftmill and Other Air-Flow Engines (1948)
Ship of Destiny [Merrimac] (1951)
An Engineer's Notebook [collected papers] (1955)
The St. Johns Railroad (1959)
Three Essays [on] *Mechanical Transport in America before 1800* (1960)
Rise and Decline of the Paddle-wheel (1962)

All volumes are heavily illustrated.

Bern Dibner, under the imprint of Burndy Library, Norwalk, Conn., has published since 1945 an annual monograph or equally valuable alternative in the history of science and technology. Because a complete and current list is available from the Burndy Library, only a few titles, particularly concerned with technology, are mentioned here. An outstanding feature of these interesting works is the superb quality of reproduction of illustrations. Halftone screens are used only when unavoidable and then are very fine (160-line). Few modern publications exhibit the excellence of craftsmanship that these do. *Moving the Obelisks* (p. 45); 24 plates from **Stradanus,** *Nova Reperta* (p. 47); *Heralds of Science* [200 Books] (p. 40); fine reproduction of Schussele's 1862 engraving, *American Scientists and Inventors* (1956); *Agricola on Metals* (1958); *The Atlantic Cable* (1959); *The Victoria and the Triton* (1963).

Ostwald's Klassiker der exakten Wissenschaften (248 vols., Leipzig, 1889–1960. Thin 12mo.). German texts of all or parts of classics of science; some volumes combine several short texts on a single subject. Most (230) were published before 1930. I have found no list of the titles except in library card catalogues and in the publisher's catalogue, the latter kindly furnished to me by H. Drubba, of Hannover. Works of authors such as Euclid, Archimedes, Huygens, Guericke, Monge, Davy, Seebeck, Maxwell, Doppler, and so forth.

I. Bernard Cohen, Charles C. Gillispie, Sir Harold Hartley, Duane H. D. Roller, and Edward Rosen, Eds., *Landmarks in Science* (New York: Readex Microprint, 1967——). An ambitious and promising effort to make available on microcards (opaque cards, 6 x 9 inches) a very wide range of printed source materials in the history of science. Some attention will be given to the history of technology. The publishers expect, over a ten-year period, to reproduce approximately 20,000 separate publications totaling over three million pages of text. Authors of all periods from classical times through the twentieth century are represented, and a series of scientific journals published before 1800 may also be included. The entire series is available by subscription. I have seen no indication that individual titles may be purchased separately.

Deliveries over the first six months (from May 1967) totaled 1,682 cards comprising 127 titles. Galileo's works, for example, cover 124 cards, Descartes' works, 89 cards, John Harris's *Lexicon Technicum* (2 vols.), 20 cards, and Biringuccio's *Pirotechnia*, 4 cards. In addition to the books of Harris and Biringuccio, the first group of titles include technological works by Branca, Guericke, and Accademia del Cimento of Florence. The rest of the works are in science.

A comprehensive guide and index may be expected ten years hence. Meanwhile, a short-title checklist accompanies each shipment of microcards. Perhaps the most urgent need is for the production of library catalogue cards in order that readers in subscribing libraries may learn of the existence of microcard reproductions of individual titles for which they are searching.

XII. TECHNICAL SOCIETIES, EDUCATION, AND EXHIBITIONS

A. Technical and Scientific Societies

Directories of societies contain more or less historical information. In fact, the distinction that has been made here between histories and directories of societies was in several cases an arbitrary one. Therefore, the reader should also review Part III, Directories.

Ralph S. Bates, *Scientific Societies in the United States* (3rd ed., Cambridge: The M.I.T. Press, 1965, 326 pp.). A general narrative history of societies, technical and scientific; finding lists of publications can be located through the bibliography, pp. 245–293. A section of the bibliography on "Histories of Scientific Societies" lists one or more references to printed histories or historical articles about the following technical societies: A.I.E.E., A.S.C.E., A.S.M.E., Boston Soc. of C.E., Engineers' Clubs of Philadelphia and St. Louis, Engineers' Soc. of Western Pennsylvania, Federated American Engineering Societies, Iowa Engineering Soc., National Academy of Engineering, Purdue University Engineering Societies.

Monte A. Calvert, *The Mechanical Engineer in America 1830–1910* (Baltimore: The Johns Hopkins Press, 1967; 296 pp., illustrated). The sometimes conflicting aims of various interest groups within this single branch of engineering are explored. Bibliography, pp. 283–287.

Clark, *Age of Newton* (p. 13) has a chapter on the rise of scientific societies in the seventeenth century.

G. B. Goode, "The Origin of the National Scientific and Educational Institutions of the United States" (U.S. National Museum, *Annual Report*, 1897, pp. 263–354). Includes information and documents on the proposed national university, Barlow's proposed National Institution (1806), Columbian Institute (1819), Columbian College (1821),

Smithsonian Institution, Coast Survey, Naval Observatory, state academies of science, and land-grant colleges.

Edwin T. Layton, "The American Engineering Profession and the Idea of Social Responsibility," unpublished dissertation (Los Angeles: U.C.L.A., Dec. 1956, 299 pp.). Bibliography, 260–299. Despite the concern of American engineering societies over their public "image," this and **Calvert** (p. 173) are the only historical and critical works I know of on the philosophy of the profession. Layton's provocative work, recast in a much more mature form, will be published in 1968 or 1969.

Martha Ornstein, *The Role of Scientific Societies in the Seventeenth Century* (2nd ed., Chicago, 1928). Reprint (Hamden, Conn.: Shoe String Press, 1963). This careful monograph traces the history and influence of the earliest scientific societies, nearly all of which encouraged pursuit of science that would lead to practical results. Chapters on Italian societies; Royal Society of London; Académie des Sciences, Paris; German societies; and on individual serials, including *Acta eruditorum* (1682+) and *Journal des savants* (1665+). Very good.

Robert E. Schofield, "Histories of Scientific Societies: Needs and Opportunities for Research" (*History of Science,* 2 [1963], 70–83), has detailed information on published histories and on sources for studies not yet commenced.

The following list of histories is merely indicative of the extensive printed historical record of societies. I made no systematic search for such books and articles. Useful if generally uncritical historical information can be obtained from the serial publications of a society: see Part XI, Periodicals. Many of the older societies have published one or more volumes on their own history.

J. David Thompson, Ed., *Handbook of Learned Societies and Institutions — America* (Washington, 1908 [Carnegie Institution Pub. No. 39], 592 pp.). Fortunately for our purpose, the interpretation of learned societies is very broad in this volume. Engineers' clubs and trade associations are exhaustively treated. A thorough bibliography for each society gives publishing history through about 1906. Large sections are included on Canada, Mexico, and South America. Publications of engineering societies in Mexico, Cuba, Brazil, Argentina, Colombia, Peru, and Chile are listed. The National Academy of Sciences directory will answer some questions about publishing events after 1906, but this handbook is not likely to be superseded in the near future. Very, very good.

A historical sketch of the **American Philosophical Society** will be found in the Society's *Yearbook* (1964), pp. 36–106.

Derek Hudson and Kenneth W. Luckhurst, *The Royal Society of Arts 1754–1954* (London: Murray, 1954; 411 pp., illustrated). An intelligent, critical work.

Thomas Martin, "Early Years at the Royal Institution," *British Journal for the History of Science,* 2 (Dec. 1964), 99–115, 4 illustrations. Covers the period 1800–1802.

Thomas Martin, "Origins of the Royal Institution," *British Journal for the History of Science,* 1 (June 1962), 49–63.

Robert E. Schofield, *The Lunar Society of Birmingham: A Social History of Provincial Science and Industry in Eighteenth-Century England* (Oxford: Clarendon, 1963; 481 pp., illustrated). A first-rate study of the scientific, technical, and social society whose roster included such illustrious men as Priestley, Erasmus Darwin, Watt, Boulton, and many others.

Joseph Bruce Sinclair, " 'Science with Practice; Practice with Science': A History of the Franklin Institute, 1824–1837," unpublished dissertation (Cleveland, Ohio: Case Institute of Technology, 1966, 241 pp.). The development of the Franklin Institute and its relationship to American society, from 1824 to 1837. Formed in the flush of enthusiasm for mechanics' institutes, the Franklin Institute became, in succession, an agency for promoting internal improvements and domestic manufacture and later, under the leadership of Alexander D. Bache, a center for the professionalization of science and technology. Sinclair is currently (1967) extending his research to 1865. [This note by Edwin T. Layton.]

K. D. C. Vernon, "The Foundation and Early Years of the Royal Institution," Royal Institution of Great Britain, *Proceedings,* 39, No. 4 (1963), 364–402. A critical history covering the years 1799–1803. See also the two articles by **Martin,** preceding.

Ethel Mary Wood, *A History of the Polytechnic* (London: Macdonald, 1965; 176 pp., illustrated). The London Polytechnic Institution was founded in 1838. Bibliographic references.

Sydney L. Wright, *The Story of the Franklin Institute* (Philadelphia: The Franklin Institute, 1938; 105 pp., illustrated). Generally slight; a few early pictures are of interest.

B. Technical Colleges and Institutes

The development of technical education in technical institutes and in the universities of the United Kingdom is reviewed by Sir Eric Ashby in his chapter, "Education for an Age of Technology," in **Singer** (p. 2), Vol. 5, pp. 776–798. The reference notes of this chapter identify the principal reports of the several commissions that made formal studies of various aspects of the problem.

Michael Argles, *South Kensington to Robbins: An Account of English Technical and Scientific Education since 1851* (London: Longmans, 1964, 178 pp.). "Robbins" refers to the report on higher education made for the government in 1963. Bibliography, pp. 151–161, is a good point of entry into British technical education.

W. H. G. Armytage, *Civic Universities. Aspects of a British Tradition* (London: Benn, 1955). While technology is not the author's central concern, he treats among other matters the polytechnics of the latter part of the nineteenth century.

Eric Ashby, *Technology and the Academics* (London: Macmillan, 1959, 118 pp.). Ashby, an educational statesman and one of the most perceptive students of education, considers the current problems in the light of historical analysis.

D. S. L. Cardwell, *The Organisation of Science in England. A Retrospect* (London: Heinemann, 1957, 204 pp.). Two thirds of the book is concerned with technical education from about 1800 to 1918. Cardwell observes, p. 189, that industrial demand for technical graduates did not exist until after such graduates were already on the market. Extensive reference notes.

Charles P. Daly, *Origin and History of Institutions for the Promotion of Useful Arts: Discourse Delivered at the Thirty-fifth Anniversary of the American Institute of the City of New York* (Albany, 1864, 35 pp.). A few facts and a point of view are supplied in this work. A serious study of mechanics' institutes in the United States is much needed.

Great Britain, Royal Committee on Technical Education, *Second Report* (3 vols., London: Eyre, 1884). Comparative information is given on engineering education in several countries, including the United States.

Thomas Kelly, *George Birkbeck: Pioneer in Adult Education* (Liverpool: The University Press, 1957; 380 pp., illustrated). Seventy pages on the general development of the mechanics' institute movement are employed to place Birkbeck's contributions in a proper perspective. Bibliography, pp. 340–363.

Thomas Kelly, *A History of Adult Education in Great Britain* (Liverpool: The University Press, 1962, 352 pp.). The mechanics' institutes are traced in this book. Footnote references to further sources.

John Scott Russell, *Systematic Technical Education for the English People* (London, 1869, 437 pp.) is the response of a leading technologist, long interested in technical education, to the lessons of the 1867 Paris international exhibition.

Mabel Tylecote, *The Mechanics' Institutes of Lancashire and Yorkshire before 1851* (Manchester: Manchester University Press, 1957, 346 pp.). A thorough work which inquires into the aims and achievements of the early mechanics' institutes in the textile manufacturing district. Bibliography, pp. 312–331.

Frederick B. Artz, *The Development of Technical Education in France 1500–1850* (Cambridge: The M.I.T. Press, 1966 [Monograph Series, Society for the History of Technology, No. 3], 274 pp.). This book rests upon a series of articles by Artz that appeared in *Revue d'histoire moderne,* 12 (1937), 467–519; 13 (1938), 361–407; and *Revue historique,* 196 (1946), 257–286, 385–407. A chapter has been added, the articles have been translated, and the result is a major contribution by a man who is himself a great teacher.

Antoine Léon, *Histoire de l'éducation technique* (Paris: Presses universitaires de France, 1961 [Que sais-je? No. 938], 126 pp.). A brief summary history covering 200 years. A few of the references may be useful, but I have included this title chiefly in order to ensure that the reader will know of the "Que sais-je?" series of paperback works, which summarize briefly but intelligently a very wide variety of subjects.

René Taton, Ed., *Enseignement et diffusion des sciences en France au XVIIIe siècle* (Paris: Hermann, 1964; 780 pp., 20 pls.). A major collaborative work, in which adequate attention is accorded to technology. Long, documented articles; lists and tables; extensive bibliographies; name index, pp. 717–756; index of institutions, pp. 757–769. Pt. 1: science in colleges and universities; pt. 2: medicine and pharmacy; pt. 3: Le Collège Royal et le Jardin du Roi; pt. 4: technical schools, including ponts et chaussées, mining and metallurgy, hydrography, ship construction, "design," baking, and so forth; pt. 5: military schools, including artillery, naval, and engineers' schools; pt. 6: scientific cabinets and observatories. An article of great significance by Taton is on the engineers' school at Mézières, precursor of l'Ecole Polytechnique and thus of modern engineering colleges.

Fritz Schnabel, "Die Anfänge des technischen Hochschulwesens," on the origins of German engineering colleges, is in Karlsruhe, Technische Hochschule, *Festschrift anlässlich des 100-jährigen Bestehens* (Karlsruhe, 1925), pp. 1–44.

Henri Schoen, "L'enseignement supérieur technique en Allemagne: ses origines, son organisation . . . ," *Revue internationale de l'enseignement supérieur,* 57 (1909), 214–233. An article on technical education in Germany. References are bibliographic in scope.

John L. Bruce, Ed., *A Quarter Century of Technical Education in New South Wales* (Sydney: William Applegate Gullick, Government Printer, 1909; 4to, 319 pp., illustrated). On trade instruction in the secondary schools.

Archibald Liversidge, *Report upon Certain Museums for Technology, Science and Art, also upon Scientific, Professional, and Technical Instruction* (Sydney, N.S.W.: Legislative Assembly, 1880, 237 pp.). The author points to British technical institutes, the Massachusetts Institute of Technology, Science Museum, and other existing institutions.

B. E. Lloyd and W. J. Wilkin, *The Education of Professional Engineers in Australia* (2nd ed., n.p.: Association of Professional Engineers, Australia, 1962, 358 pp.). Emphasis is upon the present, but earlier experiences are mentioned. One chapter is specifically historical, and reference is made to a paper, A. H. Corbett, "The First Hundred Years of Engineering Education, 1861–1961" (Institution of Engineers, Australia, *Journal* [Apr./May 1961]).

John Nicol, *The Technical Schools of New Zealand: An Historical Survey* (n.p.; New Zealand Council for Educational Research, 1940, 250 pp.). On schools at the secondary level; technical institutes of the 1850's are mentioned.

Stephen P. Timoshenko, *Engineering Education in Russia* (New York: McGraw-Hill, 1959, 47 pp.). Written by a noted American engineering teacher who was born in Russia.

Tuge, *Science and Technology in Japan* (p. 7) describes the importation in the late nineteenth century of Western technical education.

Minoru Wantabe, "Japanese Students Abroad and the Acquisition of Scientific and Technical Knowledge," *Cahiers d'histoire mondiale*, 9, No. 2 (1965), 254–293. Emphasis is upon the period 1860–1900.

W. E. Dalby, "The Training of Engineers in the United States," Institution of Naval Architects, *Transactions*, 45 (1903), 37–72. An enthusiastic report of a visit to the United States by an English teacher of engineering.

James G. McGivern, *First Hundred Years of Engineering Education in the United States* (Spokane, Wash.: Gonzaga University, 1960, 171 pp.). While all data and conclusions in this book must be approached very cautiously, the footnote references can be of help in locating source materials.

Charles R. Mann, *A Study of Engineering Education* (New York, 1918 [Carnegie Fund for the Advancement of Teaching, Bull. No. 11], 139 pp.). The assumptions were (1) that engineering students were being prepared for "industrial production" and (2) that successful engineers would be managers. This report is much shorter but also more perceptive than the S.P.E.E. report, following. The Mann report was reviewed in *Engineering News-Record*, 81, No. 17 (Oct. 24, 1918), 741–751.

Earle D. Ross, *Democracy's College* (Ames: Iowa State College Press, 1942, 267 pp.), is still the best history of the land-grant movement. Bibliography, pp. 231–254.

Society for the Promotion of Engineering Education (now American Society for Engineering Education), *Report of the Investigation of Engineering Education 1923–1929* (2 vols., Pittsburgh, 1930–1934). Extensive but, like successive reports in each decade to the present, not very profound.

Robert H. Thurston, "Technical Education in the United States," American Society of Mechanical Engineers, *Transactions*, 14 (1893), 855–1,013. This is an important policy statement because Thurston's point of view still largely prevails in engineering education.

J. A. L. Waddell, *The Principal Professional Papers of J. A. L. Waddell* (New York: Hewes, 1905). Waddell (1854–1938) was a consulting civil engineer who had taught in Japan and had thought about the problem of technical education. This collection of papers includes one on "Civil Engineering Education," pp. 87–184, and an address delivered in 1886 to the Japanese Engineering Society.

Marjorie Carpenter, *The Larger Learning. Teaching Values to College Students* (Dubuque, Iowa: Wm. C. Brown Company, 1960, 78 pp.). The problem of teaching (not preaching) values in technical as well as nontechnical courses is considered in this well-reasoned book.

[Cooper Union]. *The Humanistic-Social Stem of Engineering Education* (rev. ed., New York: The Cooper Union, 1955 [Engineering and Science Ser. Bull. No. 33], 56 pp.). A classified bibliography having 1,340 entries. While the emphasis is upon the subject in the title, a large number of entries — certainly over 100 — are directly pertinent to any study of the history of engineering education in the United States. Earliest titles are from around 1876; all but a few dozen are later than 1900. Over 150 serials are represented.

Edwin J. Holstein and E. J. McGrath, *Liberal Education and Engineering* (New York: Columbia University Teachers College, 1960, 132 pp.). An inquiry into the problem of acquiring a broad education while pursuing specialist studies. Selected bibliography, pp. 129–132. An exhaustive bibliography, mimeographed, was available in 1963.

Histories of individual technical schools and colleges are very numerous and readily located. The following nine titles suggest the nature of available works.

Russell H. Chittenden, *History of the Sheffield Scientific School of Yale University 1846–1922* (2 vols., New Haven: Yale University Press, 1928; 610 pp., illustrated). Chittenden was director of the school.

Ecole Polytechnique. Livre du centenaire 1794–1894 (3 vols., Paris: Gauthier-Villars et fils, 1895–1897). A handsome set of books, heavily biographical; many fine portraits.

James K. Finch, *A History of the School of Engineering, Columbia University* (New York: Columbia University Press, 1954, 138 pp.). Finch, after retiring as dean of engineering, wrote extensively on the history of engineering.

Sidney Forman, *West Point* (New York: Columbia University Press, 1950, 255 pp.). A reasonably thorough study of the U.S. Military Academy, which was founded in 1802. Bibliography, pp. 229–242.

Franklin Furman, Ed., *Morton Memorial. A History of the Stevens Institute of Technology* . . . (Hoboken, N.J.: Stevens Institute of Technology, 1905; 4to, 641 pp., illustrated). Includes history of the institute, pp. 1–77; biographies of faculty members, pp. 165–284; biographies of alumni, pp. 287–641.

T. J. N. Hilken, *Engineering at Cambridge University 1783–1965* (Cambridge: University Press, 1967, 277 pp.). Bibliography and notes, pp. 262–266. The first engineering degree was awarded in 1894, one hundred years after the first mechanical lectures of William Farish had been delivered.

Richard H. Kastner, "Die Entwicklung von Technik und Industrie in Österreich und die Technische Hochschule in Wien," *Blätter für Technikgeschichte*, 27 (1965), 1–181.

Samuel C. Prescott, *When M.I.T. Was "Boston Tech" 1861–1916* (Cambridge: The Technology Press, 1954; 350 pp., illustrated). Dean Prescott was associated with the institute for more than sixty years.

Thomas T. Read, *The Development of Mineral Industry Education in the United States* (New York: American Institute of Mining and Metallurgical Engineers, 1941, 298 pp.). The several schools of mining are considered in turn.

Palmer C. Ricketts, *History of Rensselaer Polytechnic Institute 1824–1934* (3rd ed., New York: Wiley, 1934, 293 pp.). Lists of courses are given for 1854, 1866, and 1933. Bibliography, pp. 213–219.

Russo, *Bibliographie* (p. 17), p. 118, lists a number of histories of the principal technical colleges of Paris.

C. Technical Museums

There are two principal ways in which technical museums are valuable to the student in his pursuit of the history of technology. First, the museum preserves objects in order that they may be studied; second, nearly every good museum supports and encourages scholarly work and publishes the results.

1. *Museums, Objects, and Lists of Museums*

If a historian has never seen, at first hand, at least a fair sample of the objects about which he writes, it is likely when he applies the yardstick of preconceived notions that his judgments will be wildly in error; for he will be ignorant of the character of the objects that he judges. Form, size, finish, and texture are frequently as important to an understanding and appreciation of a technical accomplishment as the mechanical configuration and details of operation.

For example, the Watt engine is frequently described as "crude" and "inefficient." Surely an author who uses such adjectives has not taken the trouble to learn about eighteenth-century technological capabilities or to look closely and critically at the Watt engines in the Science Museum, London, or the Henry Ford Museum, Dearborn, Michigan. He would find the engines far from clumsy and uncouth; indeed, he would find that the Watt engine is a highly sophisticated machine, as advanced in its day as the latest space vehicle of current interest.

To mention another example, the Cugnot steam carriage of 1769 is often thought of as a comical teakettle on wheels, obviously cobbled together by a simple-minded tinker. However, a few minutes spent in the presence of the actual Cugnot carriage, which is exhibited in the Conservatoire des Arts et Métiers, Paris, will give any but the most obtuse observer the distinct impression of a careful, workmanlike approach to a problem, even if the solution was not a successful one. The quality of craftsmanship is of an astonishingly high order. Just as a successful modern designer must spend a considerable part of his time in the shop, observing that which does not appear in reports, drawings, and photographs, so the student of technical developments must spend a part of his time examining carefully the artifacts of the period with which he is concerned.

There is no general guide to objects in technical museums. A student must be diligent and resourceful even to learn which objects still exist and where they are preserved. A start has been made on an inventory of scientific instruments (*History of Science*, 2 [1963], 21, 32, note 8), but in the absence of an organizing genius that can perceive a useful approach to indexing and describing the many thousands of diverse and miscellaneous objects, other than instruments, scattered in hundreds of museums all over the world, the great bulk of technical artifacts must remain in the realm of individual expert knowledge.

The list of museums of science, technology, and industry in **Sarton**, *Guide* (p. 17), pp. 260–289, is very helpfully annotated. Histories of individual museums, their chief fields of interest, and some of their publications are mentioned. I know of only one important technical museum, in existence in 1952, not listed by Sarton: namely, Vienna's Technische Museum für Industrie und Gewerbe, founded in 1908. The Museo Nazionale della Scienza e della Tecnica, in Milan, is a major museum that was first opened in 1953. The Smithsonian Institution's Museum of History and Technology opened its new building in 1964; the U.S. National Air Museum will occupy a new building on the Mall opposite the National Gallery.

American Association of Museums, *Museums Directory of the United States and Canada* (Washington: American Association of Museums and Smithsonian Institution, 1965, 1,039 pp.). General, often cryptic, notes, reflecting the quality of information supplied to the editors by museum directors, describe the collections of 4,956 museums of all kinds. Nevertheless, within its geographical territory, this is the best guide available.

Dorothy and M. V. Brewington, "The Marine Museums in Italy, Southern France, and Spain," *American Neptune* (Oct. 1960).

John J. Brown, "Museum Census: A Survey of Technology in Canadian Museums," *Technology and Culture*, 6, No. 1 (Winter 1965), 83–98. The author drove 30,000 miles in making a one-man survey of 120 museums in which he found significant materials in the history of technology. I find encouraging Dr. Brown's painstakingly personal approach to gathering information that is to be incorporated into a machine-retrieval system, even though I disagree with his use of the U.S. Patent Office classification system.

Laurence V. Coleman, *The Museum in America. A Critical Study* (3 vols., Washington: American Assn. of Museums, 1939). One of several books on various aspects of museum operation by the longtime director of the American Association of Museums.

Lucius Ellsworth, "A Directory of Artifact Collections," in **Hindle,** *Technology in Early America* (p. 22), pp. 95–126. This is the first serious attempt I know of to locate objects in U.S. technical museums. The field work leading to this directory was limited in the time that could be devoted to it. Arrangement is by subject matter. Hindle, in his essay, discusses the problem of recording artifacts in some comprehensible way, pp. 10–13. Perhaps, in the absence of an individual who will devote two or three years of his life to a systematic catalogue of museum objects, the format of the *National Union Catalog of Manuscript Collections* (p. 114) can suggest a feasible approach to a useful compilation.

Kenneth Hudson, "Company Museums," *Industrial Archaeology*, 1 (May 1964), 21–26. The author has compiled a list of about sixty-five collections in the United Kingdom, a number of which are not now open to the public.

Articles on museums have been a feature of *Industrial Archaeology*: for example, Open-air museum in Durham, 1 (May 1964), 3–8; Bristol City Museum, 1 (Aug. 1964), 103–105; Festiniog Railway Archives and Museum, 1 (Jan. 1965), 222–224; Glasgow Museum of Transport, 2 (Mar. 1965), 7–12; Wakefield Museum, 3 (Nov. 1966), 241–244; Museum of Industrial Locomotives in Caernarvonshire, 3 (Nov. 1966), 278–291.

The following East German technical museums are listed in *Informationsdienst* (p. 145): Polytechnisches Museum Dresden, 806

Dresden, Friedrich-Engels-Strasse 15; Kulturhistorisches Museum Magdeburg, Abteilung Polytechnisches Museum, 301 Magdeburg, Otto-von-Guericke-Strasse 68–73; Verkehrsmuseum Dresden, 801 Dresden, Augustusstrasse 1 (Johanneum); Zweitakt-Motorrad-Museum Augustusburg, 9382 Augustusburg über Zschopau (Sachsen).

Herbert and Marjorie Katz, *Museums, U.S.A. A History and Guide* (Garden City: Doubleday, 1965; 395 pp., illustrated). An attractive, popular, and, with respect to technical museums, uncritical work. The list of museums, pp. 267–375, is based upon the A.A.M. *Museums Directory* (p. 182) and information obtained directly by authors from subject museums. Of marginal utility.

"List of Museums with Horological Interests," National Assn. of Watch and Clock Collectors, *Bulletin,* 11, No. 116 (1965), 112–140. I have not seen this list.

Robert P. Multhauf, "European Science Museums," *Science,* 128 (Sept. 5, 1958), 512–519, gives a useful introduction to the nature of the collections of a number of European museums.

"Museums of Technology," *Technology and Culture,* 6, No. 1 (Winter 1965), 1–98. In the absence of a systematic history of technical museums, the articles in this issue of *Technology and Culture* will provide information and suggest sources for further inquiry. The pertinent articles are: Silvio A. Bedini, "The Evolution of Science Museums," pp. 1–29; Eugene S. Ferguson, "Technical Museums and International Exhibitions," pp. 30–46; Robert P. Multhauf, "A Museum Case History," pp. 47–58; Robert M. Vogel, "Assembling a New Hall of Civil Engineering," pp. 59–73; Bernard S. Finn, "The Science Museum Today," pp. 74–82; and John J. Brown, "Museum Census: A Survey of Technology in Canadian Museums," pp. 83–98.

"Museums of Science and Technology: A List Compiled by the Committee on Museums of Science and Technology (Chairman, Torsten Althin) of The International Council of Museums (UNESCO)," *Technology and Culture,* 4, No. 1 (Winter 1963), 130–147. A list of 900 museums in 44 countries. Brown (p. 183) notes, however, that of the 39 Canadian museums listed, 5 never existed, 9 are not museums of technology, 4 have been closed and put in storage, and that there are other errors in the list. VDI (p. 185) includes 35 German museums that are not on this list; conversely, 29 museums on this list are not in the VDI compilation. Hudson (p. 183) notes that only 3 of the company museums listed by him are on the list of U.K. museums. On the other hand, I find that even 50 per cent accuracy is much more helpful than no information at all. As the authors of this list freely allow, further refinement is called for.

Alex J. Philip, *An Index to the Special Collections in Libraries, Museums and Art Galleries in Great Britain and Ireland* (London: F. G. Brown, 1949, 190 pp.). A mixture of book, manuscript, and artifact collections is contained in this modest guide, which was intended to emphasize little-known collections. For example, one finds museum collections of decorative ironwork in Lewes and Newcastle, of weights and measures in Birmingham, Lancaster, and Winchester, of Japanese armor in Nottingham, trade cards in Cheltenham, and so forth. A scholar who plans to visit England would be repaid for the effort required to obtain and read this index.

Paul Sharp and E. M. Hatt, *Museums* (London: Chatto & Windus, 1964 [National Benzole Books]; 128 pp., illustrated). An attractive and inexpensive guidebook, sponsored by National Benzole Co., in which more than 200 museums in the United Kingdom, "idiosyncratic, humble, quirky, or touching," are located and briefly described. Included are museums of carriages, railways, canals, horology, motor transport, iron founding (at Coalbrookdale), textiles, rural life, and arts and crafts.

The locations of a number of specific objects are mentioned incidentally in **Singer**, Ed., *A History of Technology* (p. 2), Vols. 3–5. See index: "Museums."

Descriptions, by **F. Sherwood Taylor**, of the Museum of the History of Science, Oxford, and Science Museum, London, are in *Endeavour*, *1* (1942), 67 ff. and *10* (1951), 82 ff., respectively.

In *Technikgeschichte*, Index (p. ooo). pp. 85–86, approximately 160 articles and notes on museums in many countries are listed. Nearly all the references are to Vol. 17 and later.

Verein Deutscher Ingenieure, Hauptgruppe Technikgeschichte, Committee for Technical Museums, Industrial Museums and Archives, Chairman: Dipl.-Ing. Dr. techn. Gustav Goldbeck, "Technikgeschichtliche Sammlungen in Deutschland." An eight-page mimeographed list, dated Sept. 1, 1965, of sixty-eight museums, briefly identified. Thirty-four of the museums have been described in articles published since 1959 in *VDI-Nachrichten*, the weekly newspaper of VDI. References are given to issue and page number. This series is expected eventually to include technical museums in Switzerland and Italy.

Philip P. Wiener, "Leibniz's Project of a Public Exhibition of Scientific Inventions," *Journal of the History of Ideas*, *1* (1940), 232–240. The proposal was made in 1675.

2. *Museum Publications*

In addition to handbooks and catalogues of collections, the publications of technical museums cover fields as wide and varied as the interests of their curators and research associates. A reasonably comprehensive list of even major works having a direct bearing on technical history, which would consist of several hundred titles, is beyond my reach. Because the museum as publisher is seldom in the main stream seined by library order departments, a special effort is required to assemble a respectable collection of museum publications. The Library of Congress and British Museum printed catalogues, for example, have extremely spotty and generally unsatisfactory selections of titles in this category.

The Library of the Museum of History and Technology, Smithsonian Institution, is the only library in the United States that to my knowledge makes a systematic attempt to develop a strong collection of technical museum imprints.

I hope that a way will be found to assemble and publish a definitive list of publications, including, of course, titles that are out of print. A mere listing of "publications available," voluminous as this might be, would, in my opinion, be a sadly misguided effort. Jane Clapp, *Museum Publications* (2 vols., New York: Scarecrow, 1962), is such a compilation, listing currently available publications—U.S. only — in anthropology, archaeology, art, biological and earth sciences. Understandably, I found in this work a negligibly small number of titles pertaining to the history of technology.

I can only suggest, through the following highly selective listing, that the scholar pay due attention to titles that carry a museum imprint.

CARDIFF. **National Museum of Wales.** F. J. North, *Mining for Metals in Wales* (1962, 112 pp.). Prehistoric, Roman, and modern (through nineteenth century) periods.

DOYLESTOWN, PA. **Bucks County Historical Society.** Henry C. Mercer, *Ancient Carpenters' Tools* (2nd ed., 1951, 339 pp.). A standard work on hand tools through the nineteenth century. For other titles of this museum see **Sarton,** *Guide* (p. 17), p. 282.

LONDON. **Bryant and May Museum of Fire-Making Appliances.** Miller Christy, *Catalogue of the Exhibits* (2 vols., 1926–1928; illustrated). The second volume is a supplement; Bryant and May are makers of watches.

LONDON. **London Museum,** *London in Roman Times* (1946, 211 pp.).

LONDON. **Museum of the Worshipful Company of Clockmakers of**

London. G. H. Baillie, *Catalogue of the Museum* . . . (3rd ed., London, 1949, 137 pp.).

LONDON. *Science Museum.* Some of the publications of Science Museum will be found listed in many places. A list has appeared in most of the handbooks and descriptive catalogues (but a few of the later ones have omitted this feature in a self-conscious effort to make booklets more attractive). A mimeographed sheet obtainable from the Museum, "Publications, etc., on Sale, List 458," includes only titles in print. Some earlier and variant titles can be found in the Library of Congress printed catalogue, first series, 89, 588–591. The difficulty of finding, when wanted, a reasonably comprehensive list prompts the following, in which authors' names have for convenience been omitted. All of these titles were published by Her Majesty's Stationery Office, London. This list is not definitive.

Handbooks and Descriptive Catalogues

These are illustrated; most have bibliographies; length is generally between 50 and 100 pages. All are inexpensive.
Aeronautics — I. Heavier-than-air Aircraft
    Pt. I Historical Survey (1949)
    Pt. II Descriptive Catalogue (1949)
Aeronautics — II. Lighter-than-air Aircraft (1950)
Aeronautics — III. Propulsion of Aircraft (1936)
Agricultural Implements and Machinery (1930)
Applied Geophysics (1936)
British Fishing Boats and Coastal Craft
    Pt. I Historical Survey (1950)
    Pt. II Descriptive Catalogue (1952)
Chemistry
    Pt. I Historical Review (1962)
    Pt. II Descriptive Catalogue (1962)
Cycles
    Pt. I Historical Survey (1955)
    Pt. II Descriptive Catalogue (1958)
Electric Power
    Pt. I History and Development (1933)
    Pt. II Descriptive Catalogue (1933)
Heat and Cold
    Pt. II Descriptive Catalogue (1954)
Light Cars and Cyclecars. Historical Survey (1957)
King George III Collection of Scientific Instruments
    Outline Guide (1949)
    Descriptive Catalogue (1951)
Machine Tools (1966)

Marine Engineering
    Pt. I History and Development (1935)
    Pt. II Descriptive Catalogue (1953)
Mechanical Road Vehicles
    Pt. II Descriptive Catalogue (1948)
Merchant Steamers and Motor Ships
    Pt. II Descriptive Catalogue (1949)
Motor Cars
    Pt. II Descriptive Catalogue (1959)
Motor Cycles
    Pt. I Historical Survey (1956)
    Pt. II Descriptive Catalogue (1958)
Radio Communications
    Pt. I History and Development (1949)
Railway Locomotive and Rolling Stock
    Pt. I Historical Review (1947)
    Pt. II Descriptive Catalogue (1948)
Sailing Ships
    Pt. I Historical Notes (1951)
    Pt. II Descriptive Catalogue (1952)
Temperature Measurement and Control
    Pt. II Descriptive Catalogue (1955)
Time Measurement
    Pt. I Historical Review (1958)
    Pt. II Descriptive Catalogue (1955)
Typewriters (1964)

Monograph Series

D. B. Thomas, *The First* [photographic] *Negatives* (1964, 41 pp.)
K. R. Gilbert, *The Portsmouth Blockmaking Machinery* (1965, 42 pp.)

Other Publications

M. J. B. Davy, *Interpretive History of Flight* (1948)
H. W. Dickinson, *The Garret Workshop of James Watt* (1958)
C. St. C. B. Davison, *History of Steam Road Vehicles* (1953, 60 pp.)
G. R. M. Garratt, *One Hundred Years of Submarine Cables* (1950, 60 pp.)
C. H. Gibbs-Smith, *The Aeroplane* (1960, 375 pp.). See p. 242.
C. H. Gibbs-Smith, *Sir George Cayley's Aeronautics, 1796–1855* (1962, 268 pp.)
W. T. O'Dea, *A Short History of Lighting* (1958, 40 pp.)
H. T. Pledge, *Science Since 1500* (1939, 357 pp.)
The Science Museum. The First Hundred Years (1957, 85 pp.)

TECHNICAL SOCIETIES, EDUCATION, EXHIBITIONS

South Kensington Museum, *Handbook to the Special Loan Collection of Scientific Apparatus 1876* (London: Chapman & Hall, 1876, 339 pp.)

H. P. Spratt, *Outline History of Transatlantic Steam Navigation* (1950, 60 pp.)

E. G. Stewart, *Town Gas: Its Manufacture and Distribution* (1958, 52 pp.)

G. F. Westcott, *The British Railway Locomotive 1803–1853* (1958; 14 pp. + 33 pls.)

G. F. Westcott (revised by H. P. Spratt), *Synopsis of Historical Events; Mechanical and Electrical Engineering* (1960, 44 pp.). See p. 25.

A "Science Museum Booklet" series of illustrated articles includes one by Arthur Ridding, *S. Z. Ferranti* (1964, 32 pp.).

An attractive "Illustrated Booklet" series (6" × 6", 20 color pls. in each) includes three on ship models and one each on making fire, railways, and timekeepers.

MILAN. **Museo Nazionale della Scienza e della Tecnica.** Numerous guide books and informal catalogues have been issued during the ten years of the museum's existence: for example, *Gallerie di Leonardo da Vinci* (1956, 64 pp.); *Guida Breva* (ca. 1957, 90 pp.). The museum's periodical, *Museoscienza, notizario mensile,* is a popular rather than scholarly publication.

MUNICH. **Deutsches Museum,** *Abhandlungen und Berichte,* thrice yearly. See p. 147.

Conrad Matschoss, *Das Deutsche Museum, Geschichte, Aufgaben, Ziele* (Munich, 1933; 394 pp., illustrated).

NORTH ANDOVER, MASS. **Merrimack Valley Textile Museum,** *Wool Technology and the Industrial Revolution: An Exhibition* (1965; 100 pp., illustrated). Brief text but adequate to the purpose; no bibliography.

OXFORD. **Pitt Rivers Museum** publishes monographs in the series, "Occasional Papers on Technology."

Beatrice Blackwood, *Technology of a Modern Stone Age People in New Guinea* (No. 3)

H. H. Coghlan, *Notes on the Prehistoric Metallurgy of Copper and Bronze in the Old World* (No. 4, 1951; 131 pp. + 16 pls.)

Francis H. S. Knowles, *Stoneworker's Progress: A Study of Stone Implements in the Pitt Rivers Museum* (No. 6)

H. H. Coghlan, *Notes on Prehistoric and Early Iron in the Old World* (No. 8, 1956, 220 pp.)

PARIS. **Conservatoire National des Arts et Métiers.** A detailed catalogue of the collections, to be issued in an indefinite number of volumes, is in progress. The following list is from **Russo** (p. 17), pp. 15–16, with amendments based upon later information. The dated titles have been published.

A. Instruments et Machines à calculer (1942, 135 pp.)
B. Mécanique Essais des matériaux (1956, 278 pp.)
C. Machines motrices et réceptrices (1952, 326 pp.)
DA. Transports sur routes (1953)
DB. Transports sur rails (1952)
DC. Navigation marine, fluviale, aérienne (1954)
E. Electricité, magnétisme
F. Télécommunications
GA. Physique mécanique (1955)
H. Géodésie, levé des plans, photogrammétrie (1953)
J. Astronomie. Mesure du temps
JB. Horlogerie (1949, 329 pp.)
K. Poids et mesures, métrologie (1941, 226 pp.)
L. Photographie, cinématographie (1949, 216 pp.)
M. Arts graphiques
N. Verrerie
P. Chimie
R. Mines, métallurgie
S. Céramique (1943, 214 pp.)
T. Industries textiles, teintures et apprêts (1942, 224 pp.)
U. Machines et outillage agricole
V. Construction et matériaux de construction
X. Economie domestique et hygiène
Y. Mathématiques
Z. Automates et mécanismes à musique (1960).

READING, BERKS. **Museum of English Rural Life.** J. Geraint Jenkins, *The English Farm Wagon: Origins and Structure* (1961; 248 pp., illustrated). Based upon a survey of nearly 600 wagons, revealing 28 distinct regional types.

SHELBURNE, VERMONT. **The Shelburne Museum.**
 Frank H. Wildung, *Woodworking Tools at Shelburne Museum* (1957, 79 pp.)
 Gordon P. Manning, *Life in the Colchester Reef Lighthouse* (1958, 43 pp.)
 H. R. Bradley Smith, *Blacksmiths' and Farriers' Tools at Shelburne Museum* (1966, 272 pp.)
 Similar publications, which I have not seen, are *The Story of the Ticonderoga* and *The Carriages at Shelburne Museum*.

STOCKHOLM. **Tekniska Museet,** *Daedalus,* a yearbook. See p. 149.

WASHINGTON. U.S. National Museum, Museum of History and Technology (Smithsonian Institution). For general information on publications, see Part VII.B.2, Smithsonian Institution Publications.
The Smithsonian Journal of History. See p. 153.
The following catalogues of collections have been published:
Frank A. Taylor, *Catalog of the Mechanical Engineering Collections* (U.S.N.M., Bulletin 173, 1939, 203 pp.)
S. H. Oliver, *Automobiles and Motorcycles* (Bulletin 213, 1957, 157 pp.)
Howard I. Chapelle, *The National Watercraft Collection* (Bulletin 219, 1960, 4to, 327 pp.)

WORCESTER, MASS. John Woodman Higgins Armory, *Catalog of Armor* (1961, 4to, 128 pp.). Profusely illustrated. Reviewed in *Technology and Culture*, 3, No. 2 (Spring 1962), 182–183.

3. *Industrial Archaeology*

A vigorous movement is under way in England to preserve industrial monuments and artifacts or to record particulars of them if preservation is impossible or impracticable. The quarterly journal *Industrial Archaeology* (p. 150) is the best source for news, articles of lasting interest, and reviews of significant books. A list of about thirty "Local Societies for Industrial Archaeology" is printed in every issue.

A Committee on Industrial Archaeology, formed at the 1967 annual meeting of the Society for the History of Technology, is described in *Technology and Culture*, 9 (Apr. 1968), 349.

Kenneth Hudson, *Industrial Archaeology. An Introduction* (London: Baker, 1963; 179 pp., illustrated). A general handbook of the subject. Bibliography, pp. 164–172.

J. P. M. Pannell, *The Techniques of Industrial Archaeology* (Newton Abbot, Devon.: David & Charles, 1966; 192 pp., illustrated). A handbook for those who already understand the objectives of industrial archaeology.

E. R. R. Green, *The Industrial Archaeology of County Down* (Belfast: H.M. Stationery Office, 1964; 99 pp., 32 pls.). This is the first intense area study to appear. Sections on linen industry, brewing and distilling, and system of communication. Reviewed by Kenneth Hudson in *Technology and Culture*, 6 (Winter 1965), 123–125.

David & Charles, publishers, have undertaken a series of books of regional surveys; for example:

David Smith, *Industrial Archaeology of the East Midlands* (Newton Abbot: David & Charles, 1966).

D. B. Barton, *A Historical Survey of the Mines and Mineral Railways of East Cornwall and West Devon* (Truro, Cornwall: Barton, 1964; 102 pp., illustrated). A detailed guide to the region. This is one of a number of regional topical works of this publisher. Another work follows.

D. B. Barton, *A Guide to the Mines of West Cornwall* (Truro, Cornwall: Barton, 1963; 52 pp., illustrated).

Society for the Protection of Ancient Buildings, 55 Great Ormond St., London, W.C. 1, has published a series on water mills, of which the following is the eighth number:

David Luckhurst, *Monastic Watermills. A Study of the Mills Within English Monastic Precincts* (London: S.P.A.B., n.d.; 20 pp., illustrated). This pamphlet features attractive drawings of sections, perspectives, and plans.

Neil Cossons, *Industrial Monuments in the Mendip, South Cotswold and Bristol Region* (Bristol: Bristol Archaeological Research Group, 1966 [Field Guide No. 4], 32 pp.). A field guide to the Port of Bristol; logwood, sugar, snuff, glass, soap, textile, and paper industries; mines and quarries; wind and water mills; and roads, bridges, canals, railroads, and so forth. The B.A.R.G. may be reached c/o The City Museum, Bristol, 8.

National Park Service, Department of the Interior, administers the program of The Historic American Buildings Survey (HABS), which aims to produce and collect measured drawings, photographs, and technical details of architecturally significant buildings of all kinds in the United States. The collection is housed in Library of Congress, Division of Prints and Photographs. Two guides have been published: *Catalog of the Measured Drawings and Photographs of the Survey in the Library of Congress* (1941) and *HABS Catalog Supplement* (1959). It is expected that these will be replaced by a series of catalogues, arranged geographically.

D. Exhibitions

The thoroughly trivial world's fair is a twentieth-century phenomenon. During the latter half of the nineteenth century, international exhibitions not only recorded technological change but also, through interchange of ideas and artifacts, encouraged further change. Moreover, there is a considerable body of serious printed material concerning the earlier exhibitions that has no counterpart in the public relations "releases" about the absurd carnivals of today.

The "Great Exhibition of the Industry of All Nations," held in London in 1851, in the iron-and-glass building called the "Crystal Palace," had of all fairs the greatest influence upon subsequent events; but the Paris Exhibitions of 1867, 1878, 1889, and 1900 and the Vienna Exhibition of 1873 were not without lasting effects.

Local and regional industrial exhibitions have a much longer history than international exhibitions. In the United States, for example, mechanics' institutes in New York and Philadelphia were holding annual trade fairs before 1830. In Europe, the local fair can be traced at least to Roman times.

The following list is intended to suggest the nature of available source materials. Because I made no exhaustive search on exhibitions, the selection of local and regional exhibitions is haphazardly selective.

After completing this part, I learned that Dr. Josef Mayerhöfer of the National Library of Austria has prepared an extensive bibliography on international exhibitions (1851-1904), which I hope will be published in the Smithsonian Institution, U.S. National Museum *Bulletin* series. See also **Mandell**, *Paris 1900* (p. 198).

1. *Exhibitions in General*

Kenneth W. Luckhurst, *The Story of Exhibitions* (London: Studio, 1951; 224 pp., illustrated). A useful work on exhibitions in general. Some statistics of exhibitions, 1851-1939, are tabulated. There is a short bibliography.

Encyclopedia of the Social Sciences (15 vols., New York: Macmillan, 1930-1935). A good discussion by Guy Stanton Ford of the reasons for and results of international exhibitions is in 6, 23-27. A short bibliography is appended. See also, same volume, the article "Fairs," pp. 58-64.

[Chronology]. A list of approximately 350 exhibitions was published in H.T.W., "International Exhibitions, 1851-1907" (Royal Society of Arts, *Journal*, 55 [Nov. 8, 1907], 1,140-1,146). A fairly convenient chronology is imbedded in the article "Exhibition" in *Encyclopaedia Britannica*, eleventh edition. Unpublished supplementary lists and indexes in the library of the Royal Institute of British Architects are mentioned in **Luckhurst** (above), p. 217.

U.S., Superintendent of Documents, *Checklist* (p. 93), Section S6, "International Exhibitions and Expositions," pp. 949-953, gives a comprehensive listing of government publications for the following: Paris, 1867, 1878, 1889, 1900; Philadelphia, 1876; Vienna, 1873;

Melbourne, 1880, 1888; Cincinnati, 1888; Madrid, 1892–1893; Chicago, 1893; Nashville, 1897; Omaha, 1898; and St. Louis, 1904.

Merle Curti, "America at the World Fairs, 1851–1893," *American Historical Review,* 55 (July 1950), 833–856. A summary that should suggest many additional sources.

[U.S. National Archives]. H. Stephen Helton, *Records of United States Participation in International Conferences, Commissions, and Expositions, Preliminary Inventory No. 76* (Washington, 1955). A very uneven collection, transferred to the Archives from the State Department, of material pertaining to the exhibitions of 1867 and later are described in considerable detail on pp. 93–134. About 120 linear feet of records in all.

In *Records of the Public Building Service, Preliminary Inventory No. 110* (1958), entry 47 records approximately 650 photographs of buildings used for government exhibits, 1891–1909.

New York Public Library has "books, pamphlets, and official publications, catalogues, and guides relating to World's Fairs," according to **Brown,** *Guide to the Reference Collections* (p. 23). Materials for the Centennial (1876) and later exhibitions.

Manual of the Goss Library (p. 27). A few titles of published works, not listed here, on individual exhibitions are given on pp. 171–175.

2. *International Exhibitions*

LONDON, 1851.

Marcus Cunliffe, "America at the Great Exhibition of 1851," *American Quarterly,* 3 (Summer 1951), 115–126. A balanced view of England's reaction to American contributions.

Charles Wentworth Dilke, *Exhibition of the Works of Industry of All Nations, 1851. Catalogue of a Collection of Works on or Having Reference to the Exhibition of 1851, in Possession of C. Wentworth Dilke* (London, 1855, 116 pp.). About 500 titles.

France. La Commission Française sur l'Industrie des Nations, *Exposition Universelle de 1851. Travaux de la Commission . . .* (8 vols. in 12 parts, Paris, 1856–1858). No illustrations.

Charles H. Gibbs-Smith, *The Great Exhibition of 1851* (London: H.M.S.O., 1950, 1964; 142 pp., 216 illustrations). Brief but comprehensive, this book is the best secondary work on the Great Exhibition.

TECHNICAL SOCIETIES, EDUCATION, EXHIBITIONS   195

Christopher Hobhouse, *1851 and the Crystal Palace* (New York, 1937). A sound, popular account of the Exhibition.

Benjamin P. Johnson, *Great Exhibition of the Industry of All Nations, 1851. Report* (Albany, 1852, 193 pp.). Johnson was sent to London by the State of New York.

Lectures on the Results of the Exhibition, Delivered before the Society of Arts, Manufactures, and Commerce, at the Suggestion of H.R.H. Prince Albert, President of the Society (London, 1853, 463 pp.). Papers by Whewell, Playfair, Willis, Hensman, and Washington.

London. Great Exhibition of the Works of Industry of All Nations, 1851, *Official Descriptive and Illustrated Catalogue* (3 vols., London, 1851). A supplemental volume was issued also (1851?). Exhibits from the United States are described in 3, 1,431–1,469.

John Tallis, *Tallis's History and Description of the Crystal Palace, Exhibition of the World's Industry in 1851* (3 vols., London, 1852). Many steel engravings, but chiefly concerned with buildings and fine arts.

Tomlinson, *Cyclopaedia* (p. 63), has a description of the Exhibition in Vol. 1, pp. i–clx.

NEW YORK, 1853.

Earle E. Coleman, "The Exhibition in the Palace: A Bibliographical Essay," N.Y.P.L., *Bulletin*, 64 (Sept. 1960), 459–477. Any serious study of the New York "Crystal Palace" Exhibition should start with this article. A checklist of thirty-two books and pamphlets issued as a result of the fair is appended, as are sources of pictures, manuscripts, trade cards, and so forth.

Charles Hirschfeld, "America on Exhibition: The New York Crystal Palace," *American Quarterly*, 9 (Summer 1957), 101–116. Straightforward description.

National Union Catalog of Manuscript Collections (p. 114). Item MS 60–2,682 records 735 pieces (1851–1874) on the 1853 Exhibition in collections of the New-York Historical Society.

For the burning of the Crystal Palace, see *Harper's Weekly*, 2 (Oct. 16, 1858), 660–661.

LONDON, 1862.

Daniel K. Clark, *The Exhibited Machinery of 1862: a Cyclopaedia* of the Machinery Represented at the International Exhibition (London, 1864, 447 pp.). Many good pictures of individual objects.

DUBLIN, 1865.

Illustrated Record and Descriptive Catalogue of the Dublin International Exhibition of 1865 (London, 1866, 4to, 570 pp.).

PARIS, 1867.

J. E. and C. E. Armengaud, *Les Progrès de l'Industrie à l'Exposition Universelle* (2 vols., Paris, 1868–1869, 240 pls., many double-folio). One of several well-illustrated works on this exhibition.

Etudes sur l'Exposition de 1867 (8 vols., Paris, May 1867–June 1869, 258 pls.).

C. A. Opperman, *Visites d'un Ingénieur à l'Exposition Universelle de 1867* (Paris, 1867; 1 vol. text; 1 vol. pls.).

Reports of Artisans Selected by a Committee Appointed by the Council of the Society of Arts to Visit the Paris Universal Exhibition, 1867 (London: Royal Society of Arts, 1867 [?], 689 pp.). Title is from Hudson and Luckhurst (p. 175), p. 381. Excerpts were given by John Scott Russell in *American Journal of Education, 21* (1870), 394–400.

[U.S.] *Reports of the United States Commissioners to the Paris Universal Exposition, 1867* (6 vols., Washington, 1870). A series of monographs by individual commissioners, issued as H. Misc. Doc. (unnumbered), 40 Cong., 2 sess., Serial Nos. 1,351–1,356. In Vol. 1, see William P. Blake, "Bibliography of the Paris Universal Exposition of 1867." Over 300 titles are listed.

E. Saveney, "Les délégations ouvrières à l'Exposition Universelle de 1867; l'opinion des ouvriers sur l'industrie et sur eux-mêmes" (*Revue des Deux-Mondes* [Oct. 1, 1868] 586–622). Title is from **Landes** (p. 22), p. 970.

VIENNA, 1873.

Robert H. Thurston, Ed., *Reports of the Commissioners of the United States to the International Exhibition held at Vienna 1873* (4 vols., Washington, 1875–1876 [H. Ex. Doc. 196, 44 Cong., 1 sess., Serial Nos. 1,694–1,697]).

PHILADELPHIA, 1876.

National Union Catalog of Manuscript Collections (p. 114). Item

MS 60–1,654: 150 pieces souvenirs, such as printed form letters, invitations; and item MS 61–242: 3,000 pieces and 12 volumes, records of the General Director, plans of buildings, photographs, cards of visitors, stereoscopic views, and so forth. Both groups are in Historical Society of Pennsylvania. See also **Hamer**, *Guide* (p. 113), p. 543.

Oesterreichischen Commission für die Weltausstellung in Philadelphia 1876, *Bericht über die Weltausstellung in Philadelphia 1876* (4 vols., Vienna, 1877–1878). Many illustrations.

Franz Reuleaux, *Briefe aus Philadelphia* (Braunschweig, 1877, 98 pp.). Letters registering the author's disappointment upon comparing German exhibits with others. Most German products, he said, were "cheap and ugly." The book caused much discussion in Germany and probably helped the movement toward excellence in manufacturing. Excerpts were translated for *The Times*, London. I do not have dates of publication.

Reuleaux, who was a German commissioner to several international exhibitions, wrote *Die Anfänge des Ausstellungswesens* (Berlin, 1901, 71 pp.). (In H. Kraemer's *XIX-Jahrhundert*, Vol. 2.)

U.S. Centennial Commission, International Exhibition, 1876, *Reports* (11 vols., Washington, 1880–1884). An index to Vols. 1–8 is in Vol. 8. Only Vols. 10–11 have document serial numbers, which are 2,119 and 2,120.

Joseph M. Wilson, *The Masterpieces of the Centennial International Exhibition* (3 vols., Philadelphia: Gebbie & Barrie, 1876, large 4to. Vol. 3 includes history of this exhibition and description of exhibits in mechanics and science.

PARIS, 1878.

Reports of the United States Commissioners to the Paris Universal Exposition, 1878 (5 vols., Washington, 1880 [H. Ex. Doc. 42, pts. 1–5, 46 Cong., 3 sess., Serial Nos. 1,970–1,974]).

PARIS, 1889.

Reports of the United States Commissioners to the Paris Universal Exposition, 1889 (5 vols., Washington, 1890–1891 [H. Ex. Doc. 410, pts. 1–5, 51 Cong., 1 sess., Serial Nos. 2,753–2,757]).

CHICAGO, 1893.

James Dredge, *A Record of the Transportation Exhibits at the World's Columbian Exposition of 1893* (London: Engineering, 1894, fo., 779 pp.). Partially reprinted from *Engineering* but reset. Profusely illustrated.

International Engineering Congress (1893). In Library of Congress is a twenty-page pamphlet of "Announcements Relative to Publications" of the Congress. Div. A, Civil engineering, published in A.S.C.E., *Transactions,* July 1894 ff.; Div. B, Mechanical engineering, published in A.S.M.E., *Transactions,* Vol. 14; Divs. C and D, Mining and metallurgical engineering, published in two volumes (1894) by A.I.M.E.; Div. E, Engineering education, published in first volume of S.P.E.E., *Proceedings;* Div. F, Military engineering, to be published by G.P.O.; Div. G, Marine engineering and naval architecture, published in one volume (1893) by Wiley; separate conference on aerial navigation, to be published in *American Engineer* and in book form.

World's Columbian Commission, *Report of the Committee on Awards* (2 vols., Washington, 1901 [H. Doc. 510, 57 Cong., 1 sess., Serial Nos. 4,373-4,374]).

World's Columbian Exposition, *Report of the President to the Board of Directors* (Chicago, 1898, 497 pp.). A narrative and statistical report.

World's Columbian Exhibition Illustrated (2 vols., Chicago, 1892-1893). Exteriors of buildings, people, pomp, and celebration. Nothing on exhibits.

See also **Hamer,** *Guide* (p. 113), pp. 149, 151.

Paris, 1900.

Report of the Commissioner-General for the United States to the International Universal Exposition, Paris, 1900 (6 vols., Washington, 1901 [Sen. Doc. 232, pts. 1–6, 56 Cong., 2 sess., Serial Nos. 4,055–4,060]). Index to exhibits in Vol. 4, pp. 425 ff.

Richard D. Mandell, *Paris 1900. The Great World's Fair* (Toronto: University of Toronto Press, 1967; 173 pp., illustrated.). In addition to a thorough bibliographic coverage of the 1900 exposition (pp. 125–132) this book contains a bibliography of other nineteenth-century exhibitions (pp. 133–139). Details are given of an impressive series of mimeographed bibliographies of French international exhibitions published by the Conservatoire national des arts et métiers, Paris.

San Francisco, 1915.

Frank M. Todd, *The Story of the Panama-Pacific International Expo-*

sition 1915 (5 vols., New York, 1921, illustrated). See also **Hamer,** *Guide* (p. 113), p. 11.

3. *Local and Regional Exhibitions*

BALTIMORE. **Maryland Institute for the Promotion of the Mechanic Arts,** *Book of the . . . Fair* (Vols. 3–12, 1850–1859).

BRUSSELS. *La Belgique Industrielle, Compte Rendu de l'Exposition des Produits de l'Industrie en 1835* (Brussels, 1836, 419 pp.). The first Brussels fair was in 1798. See "Introduction," pp. xi–xliv.

CHICAGO. *Inter-State Exposition Souvenir; containing a Historical Sketch of Chicago; also a Record of the Great Interstate Exposition of 1873* (Chicago, 1873, 317 pp.). Full information on individual exhibits.

CINCINNATI. **Ohio Mechanics' Institute,** *Report of the . . . Annual Fair* (Cincinnati, 1838 ff.). The first was in 1838, the eighteenth in 1860.

Cincinnati Industrial Exposition 1874. Catalogue and List of Awards, 341 pp. This was the fifth exposition. The volume includes a report of tests made on Harris-Corliss and Cooper, Babcock & Wilcox steam engines, pp. 42–88.

Cincinnati Industrial Exposition 1883. "Special Report of Jurors on the Systems of Electric Lighting," 35 pp. Tests of Thomson-Houston and Edison systems.

LEIPZIG. *Die deutsche Industrie-ausstellung 1850 in der Central-Halle zu Leipzig* (Leipzig, 1850, 4to, 112 pp.). One hundred text figures, a few of machines.

NEW YORK. *American Railroad Journal,* 3 (1834), 636, 710, 726–727. Brief notices of the seventh Annual Fair of the American Institute, at Niblo's Gardens, Oct. 1834. See also American Institute, *Annual Reports,* for other exhibits. The interests of the American Institute were heavily agricultural at this time.

NEWARK. *National Union Catalog of Manuscript Collections* (p. 114), Item MS 60–333, seven volumes of records of Newark Industrial Exhibition, 1872, in collections of New Jersey Historical Society.

PARIS. F. B. W. Hermann, *Die Industrie Ausstellung zu Paris im Jahre 1839* (Nürnberg: J. G. Schrag, 1840). On pp. 18–26, "Geschichtliches über Industrieausstellungen, insbesondere die französischen."

PHILADELPIHA. *Franklin Institute.* Between 1824 and 1858, the Franklin Institute held twenty-six exhibitions. The twenty-seventh was held in 1874; the International Electrical Exhibition, in 1884; the Novelties Exhibition, in 1885; and the Franklin Institute was cosponsor of National Export Exposition, 1899. All were held in Philadelphia. A number of catalogues and other publications are in the Library of the Franklin Institute.

SAN FRANCISCO. Mechanics' Institute of the City of San Francisco, *Report of the Industrial Exhibition* (2nd, 1858; 8th, 1871).

E. Diffusion of Technology (geographic transfer of techniques)

The diffusion of culture has been going on for well over five thousand years, all over the world. Thus, almost any interregional study involves the diffusion of technology. Nevertheless, a number of studies, consciously aimed at measuring and assessing the transfer of techniques, can suggest useful formats for further studies of the phenomenon and its results.

In addition to the titles listed here, see **Lynn White's** studies (p. 11), all of which are largely concerned with transfer of technology. **Tuge**, *Science and Technology in Japan* (p. 7), treats the introduction of Western ideas, especially in the nineteenth century. See also Parts X, Travel and Description, and XII.D, Exhibitions.

Hindle, *Technology in Early America* (p. 22), discusses the transfer of technology from Europe to the United States, pp. 18–21. In Hindle's bibliography are listed a number of titles not given here; see his index, "transfer of technology."

Merle Curti and Kendall Birr, *Prelude to Point Four. American Technical Missions Overseas 1838–1938* (Madison: The University of Wisconsin Press, 1954, 284 pp.). "A Note on Materials," pp. 221–225; reference notes, pp. 226–267. The book was written apparently to test the assumptions upon which foreign aid was based: that technical assistance will help defeat poverty in all parts of the world, will win allies for the United States, and will produce universal amity.

H. W. Dickinson and A.A. Gomme, "Some British Contributors to Continental Technology (1600–1850)," International Congress on the History of Sciences, *Actes, 6,* pt. 1 (1950), 307–323.

S. B. Hamilton, "Continental Influences on British Civil Engineering to 1800," *Archives Internationales d'Histoire des Sciences,* 11 (1958), 347–355.

Herbert Heaton, "The Industrial Immigrant in the United States, 1783–1812," *American Philosophical Society, Proceedings,* 45 (1951), 519–527. A detailed study based upon federal records concerning aliens in the War of 1812.

Alan Moorehead, *The Fatal Impact; An Account of the Invasion of the South Pacific, 1767–1840* (New York: Harper, 1966, 230 pp.). On the results of Western influence.

Carroll W. Pursell, Jr., "Thomas Digges and William Pearce: An Example of the Transit of Technology," *William and Mary Quarterly,* 3rd Ser., 21 (Oct. 1964), 551–560. After the American Revolution, Digges encouraged Pearce and other English artisans to emigrate to the United States. This article also illuminates some unexplored aspects of the Washington, Jefferson, and Hamilton papers.

A. C. Todd, "Some Cornish Industrial Artifacts in the United States of America," *Industrial Archaeology,* 1 (Nov. 1964), 154–162. Almost 100,000 Cornish miners emigrated to the United States in the nineteenth century, taking with them techniques as well as objects.

Norman B. Wilkinson, "Brandywine Borrowings from European Technology," *Technology and Culture,* 4 (Winter 1963), 1–13. Several examples of European practices in mills along the Brandywine Creek, in the vicinity of Wilmington, Delaware.

XIII. TECHNOLOGY AND CULTURE

"It makes me groan to think o' the thousands of hours I've spent i' reading the wrong books."
—Graham Wallas
The Great Society (New York, 1915), p. 284.

    The literature of this subject is vast, which is another way of saying that the subject is diffuse and ill-defined, perhaps undefinable.
    There is unquestionably a conflict in men's minds between the technological marvels that their skill and enthusiasm have made possible and the undesirable side effects that seem always to be associated with the marvels. Leo Marx tells us that man has always had "the Machine in the Garden."
    We find a paradox when we think of Thoreau's rejection of the machine. Thoreau tells us how he turned his back on Concord and built his own house in the woods, buying only nails and some other items of hardware. To my mind he ignores the mining industry, the ironworks, the distributor, the hardware-store owner, and the whole freight of banking and credit that this simple act of buying nails carries with it. It is difficult to see how Thoreau might have expected that a well-stocked hardware store would be awaiting his fortuitous visit if the technological society had not already been grinding away for many generations.
    But if we cannot ignore the technological society, neither can we accept it uncritically. It has occurred to many thoughtful observers that technology has a way of becoming an end in itself and that human needs and human values are too often pushed aside. It is toward a reasonable ecological balance that we must continually bend our efforts because an ecological system that includes man must always, it seems, be unstable.
    The titles listed in this part are intended to be merely indicative of the nature of serious thought on the subject. I have made no attempt to assess the literature of sociology, nor have I plumbed the important body of fiction that can instruct us in ideas that cannot be learned through facts.

A. Interrelationships between Technology and Its Cultural Matrix

The title of *Technology and Culture* reflects the concern of the Society for the History of Technology with the problems arising from an imbalance between our technical capabilities and our human propensities. To name only a few of the expressions of this concern in the journal, the "Encyclopaedia Britannica Conference on the Technological Order" occupied the thickest issue yet published, in Fall 1962; the conference on "Technology and the Idea of Mankind" was reported in 5 (Winter 1964), 45–56; the conference on "Technology and Human Values" was reported in 5 (Summer 1964), 359–376; and each annual "Current Bibliography in the History of Technology" contains from a few to a score of titles on the general relationships between technology and culture.

W. H. G. Armytage, *The Rise of the Technocrats: A Social History* (London: Routledge & Paul, 1965, 448 pp.). A weird and eccentric work, according to Edwin Layton's review in *Technology and Culture*, 7 (Fall 1966), 532–534. Nathan Reingold demonstrates that it is a "significantly poor book" in *Isis*, 58 (Spring 1967), 111–113. Nevertheless, a great amount of material bearing on attitudes toward technology over the last 300 years may be pursued through the reference notes, pp. 359–421, which are bibliographic in scope.

Jacques Barzun, *Science: the Glorious Entertainment* (New York: Harper, 1964, 322 pp.). Technology is a part of the author's "Science." This intelligent appraisal of some of the misapprehensions of the scientist has been frequently rejected by scientists because Barzun is not a scientist and thus, they say, does not understand science. The basis for such rejection is itself another misapprehension.

J. Bronowski, *Science and Human Values* (New York: Harper, 1959, 94 pp.). Paperback reprint of lectures first published in 1956. Bronowski argues that we should do well to adopt the values of science — honesty, tolerance, and so forth — in all of our relationships. William Arrowsmith writes (in *Harper's Magazine*, March 1966, p. 51) that these values "are without exception the values of the master craftsman. They are impressive values, but they cannot help a man to live or die well. Only the humanities can do that."

J. Bronowski, *The Identity of Man* (Garden City, N.Y.: The Natural History Press, 1965, 108 pp.). In my opinion, the author here takes a long step beyond *Science and Human Values*, recognizing as he does the importance of subjective as well as objective modes of knowledge. He now finds science single valued: science reduces ambiguities; poetry, which is many valued, exploits ambiguities.

Bureau international de recherche sur les implications sociales du progrés technique, *Changements techniques, économiques, et sociaux: Etude théoretique* (Paris: UNESCO, 1958, 355 pp.). "Social, Economic and Technological Change: A Theoretical Approach." A collection of papers, some in French and some in English, on the relationships between technology and social change. Bibliography, pp. 337–355.

John G. Burke, Ed., *The New Technology and Human Values* (Belmont, Calif.: Wadsworth, 1966, 408 pp.). A balanced, intelligently selected book of readings for adult-education classes. Tolstoy, Bronowski, Krutch, Rickover, Hutchins, Diebold, Galbraith, Skinner, and many others are represented.

Roger Burlingame, *Backgrounds of Power: The Human Story of Mass Production* (New York: Scribner, 1949, 372 pp.). This book, says the author, is intended for those more interested in people than in machines. Essentially an optimist, Burlingame is here concerned because "mechanical processes and sequences . . . are becoming ends in themselves . . . in the minds of too many men and women."

Carlo M. Cippola, *Guns and Sails in the Early Phase of European Expansion 1400–1700* (London: Collins; New York: Pantheon Books, 1965; 192 pp., illustrated). The author recognized, "against his tastes and inclinations," the importance of the technological aspects of exploration and conquest. Thomas Esper suggests in a review in *Technology and Culture,* 8 (Apr. 1967), 220–222, that the guns were of overriding importance.

Jacques Ellul, *The Technological Society,* John Wilkinson, Trans. (New York: Knopf, 1964, 449 pp.). The book has been taken seriously, but reviews have been generally unfavorable; for example, Howard Falk's in *Technology and Culture,* 6 (Summer 1965), 532–535. Bibliography, pp. 437–449, "somewhat less extensive than the bibliography in the original French edition since it includes only those works which are readily available in American libraries."

Maurice Goldsmith and Alan Mackay, Eds., *The Science of Science; Society in the Technological Age* (London: Souvenir, 1964, 234 pp.). In this collection of essays written in honor of J. D. Bernal, Derek J. deSolla Price contributed the title piece, "The Science of Science." He notes the absence of scientific thought in considering problems about science and its cultural matrix. The science of science would treat such problems by methods scientific but as yet unknown. I continue to favor for such problem-solving the development of informed judgment, hopefully compassionate.

Carlton J. H. Hayes, *A Generation of Materialism 1871–1900* (New York: Harper, 1941; 390 pp., illustrated). The effects of technology upon European civilization are explored. A comprehensive "Bibliographical Essay," pp. 341–380, includes a rich section on the mechanization of work and thought.

Friedrich Georg Juenger, *The Failure of Technology. Perfection Without Purpose,* F. D. Wieck, Trans. (Hinsdale, Ill.: Regnery, 1949, 186 pp.). Translated from *Die Perfektion der Technik* (1946), "Begun in the spring, finished in the summer, 1939." A series of bleak assertions, reflecting time and place.

John A. Kouwenhoven, *Made in America: The Arts in Modern Civilization* (Garden City, N.Y.: Doubleday, 1962; 259 pp., illustrated). Paperback reissue, with a new preface, of a book first published in 1948. Exploration of the relationship of art — largely art in structures and industrial products — to American life. The author's thesis is that we cannot understand either the limitations or the achievements of the United States if we continue to think that its culture is solely the product of Western Europe. An original and widely known book.

Kranzberg and Pursell, *Technology in Western Civilization* (p. 5) is generally sensitive to the relationships between technology and culture. Notice, for example, Vol. 1, Chapter 31, on economic effects of technological change; Chapter 33, on the effects of technology on domestic life; and the bibliography on the social and cultural impact of industrialization, pp. 769–774.

Arthur O. Lewis, Ed., *Of Men and Machines* (New York: Dutton, 1963, 349 pp.). A paperback anthology of pieces by Bacon, Forster, Whitman, Emerson, Snow, Orwell, and many others. Many sides of the issues are touched upon.

C. S. Lewis, *The Abolition of Man* (New York: Collier, 1962, 121 pp.). A paperback reprint of a book first published in 1947. The author argues that "What we call Man's power over Nature turns out to be a power exercised by some men over other men with Nature as its instrument."

Leo Marx, *Machine in the Garden: Technology and the Pastoral Ideal in America* (New York: Oxford University Press, 1964, 392 pp.). The conflict between the machine and those who would have the garden undisturbed is traced in American literature. Reviewed by Harold D. Woodman in *Technology and Culture,* 6 (Fall 1965), 661–664.

Elting E. Morison, *Men, Machines, and Modern Times* (Cambridge,

Mass.: The M.I.T. Press, 1966, 235 pp.). A series of historical studies of nineteenth-century innovations is followed by the author's speculations on how to keep the power and complexity of modern technology within the bounds and on the scale of "all the human dimensions." The author does not deplore change, but neither does he deify it. A perceptive review by Dale Riepe is in *Technology and Culture*, 8 (Oct. 1967), 524–532.

Lewis Mumford, *The Myth of the Machine. Technics and Human Development* (New York: Harcourt, Brace, 1967; 342 pp., illustrated). In this very important book Mumford argues that man's toolmaking capacities are only a part of his cultural freight, and until very recently a subordinate part. The relationships among the organizational schemes that made possible the building of pyramids, military organizations of all ages, and the industrial corporation are explored in detail. The book ends about 1500 A.D. As in Mumford's other books, an extensive annotated bibliography is present, pp. 297–323.

National Science Foundation, *Current Projects on Economic and Social Implications of Science and Technology* [Publication NSF 66–21] (Washington, 1966; 187 pp.). Over 300 projects in 14 subject categories are described briefly, including Automation and Impacts on Labor, History and Philosophy of Science and Technology, International and Foreign Studies, and Sociology and Psychology. The projects are being pursued in universities in the United States.

A. M. Rosenthal and Arthur Gelb, *The Night the Lights Went Out* (New York: New American Library, 1965; 158 pp., illustrated). A paperback collection of *New York Times* reports on the massive power failure of November 9, 1965, in northeastern United States. A number of published official reports to state and U.S. governments followed this event.

Andrew G. Van Melsen, *Science and Technology* (Pittsburgh: Duquesne University Press, 1961, 373 pp.). A philosopher quietly thinks through the fundamental meanings of science and technology for human existence. Clearly and sensibly written. The impossibility of rejecting the technological culture is matched by the need for the technician to understand what he is about. "Possessing concepts proper to physical science is not the same as having a concept of what physical science is, because this concept cannot be expressed in terms of the notions that are proper to physical science."

Charles R. Walker, *Modern Technology and Civilization. An Introduction to Human Problems in the Machine Age* (New York: McGraw-Hill, 1962, 469 pp.). A book of readings. Bibliographies will be found at the beginning of each of the eight sections. The book is one of a series that has grown out of "the Yale Technology Project, organized in 1946 and devoted to a full-time study of the effects of modern technology on organizations and human relations."

Graham Wallas, *The Great Society; A Psychological Analysis* (New York: Macmillan, 1915, 383 pp.). The Great Society is mainly due to Great Industry. The author applies some psychological insights in dealing with the problems she perceives in the Great Society.

Aaron W. Warner, Dean Morse, and Alfred Richner, Eds., *The Impact of Science on Technology* (New York: Columbia University Press, 1965, 221 pp.). This book is the record of a conference on Technology and Social Change, in which important people participated. In a review of the book in *Technology and Culture*, 7 (Spring 1966), 255–257, J. R. Ravetz observes that in the conference a "genuine understanding of the subject given in the title was conspicuous by its absence."

See also
White, *Medieval Technology and Social Change* (p. 11)
ReQua and Statham, *The Developing Nations* (p. 37)
Clark, *Science and Social Welfare* (p. 13)

B. Philosophy of Technology

The philosophy of technology has been explored in *Technology and Culture*. A theme issue, "Toward a Philosophy of Technology," 7 (Summer 1966), 301–390, includes the contributions of seven authors; and a session in the 1966 Annual Meeting on "Social Aspects of the Philosophy of Technology: Technological Determinism" was reported in 8 (Apr. 1967), 319–320.

E. Herlitzius, "Technik und Philosophie," *Informationsdienst Geschichte der Technik* (Dresden), 5, No. 5 (1965), 1–36. The entire issue is devoted to a list, without annotations, of more than 500 titles on the philosophy of technology. Books and monographs before 1945, pp. 7–14; 1945–1964, pp. 15–25; recent periodical literature, pp. 25–36. Nearly all titles are German. The author of the list is interested in the Marx-Lenin approach to the question.

Donald Brinkmann, *Mensch und Technik. Grundzüge der Philosophie der Technik* (Berne: A. Franke AG, 1946, 167 pp.). A critical review of philosophers' and technologists' assumptions regarding technology. Bibliographical notes, pp. 145–161.

C. Human Ecology and Natural Resources

1. Ecologic Problems

Sylvia Crowe, "Civilization and the Landscape" (Smithsonian Institution, *Annual Report . . . for the Year Ended June 30, 1962* [1963], 537–544). The attention of the community is belatedly being drawn to the ecological balance necessary between destructive "development" of landscape and preservation and restoration of our heritage of decently habitable surroundings.

Gordon T. Goodman, Ed., *Ecology and the Industrial Society. A Symposium of the British Ecological Society* (New York: Wiley, 1965, 395 pp). There are chapters on air and water pollution, radioactive wastes, insecticides, and other matters of current concern.

Oscar Handlin and John Burchard, Eds., *The Historian and the City* (Cambridge, Mass.: The M.I.T. Press and Harvard University Press, 1963, 299 pp.). A publication of the Joint Center for Urban Studies, a cooperative venture of the Massachusetts Institute of Technology and Harvard University. Of particular value in this volume is Philip Dawson and Sam B. Warner, Jr., "A Selection of Works Relating to the History of Cities," pp. 270–290, a wide-ranging bibliography.

Ralph Sanders, *Project Plowshare. The Development of the Peaceful Uses of Nuclear Explosions* (Washington: Public Affairs Press, 1962; 206 pp., illustrated). The unstated assumption of this book is that change is progress, and the bigger the change the greater the progress. Utterly absent is an understanding of ecologic changes to be wrought by turning rivers around and making one-shot harbors. If "Plowshare" proceeds as this suggests, we can expect more damaging effects from peaceful uses of nuclear explosions than from military uses. Incredibly arrogant in a mindless way.

Stewart L. Udall, *The Quiet Crisis* (New York: Holt, Rinehart & Winston, 1963; 209 pp., illustrated). A review of our ecologic crisis by the U.S. Secretary of the Interior.

Sam B. Warner, Jr., *Streetcar Suburbs: the Process of Growth in Boston, 1870–1900* (Cambridge, Mass.: The M.I.T. Press and Harvard University Press, 1962, 208 pp.). A publication of the Joint Center for Urban Studies. Bibliography and notes, pp. 187–203.

Lynn White, Jr., "The Historical Roots of Our Ecologic Crisis," *Science,* 155 (Mar. 10, 1967), 1,203–1,207. The author argues that "we shall continue to have a worsening ecologic crisis until we reject the Christian axiom that nature has no reason for existence save to serve man."

Lynn White, Jr., "The Legacy of the Middle Ages in the American Wild West," *Speculum, 40* (Apr. 1965), 191–202. The author remarks, "By far the larger part of a man's ecology is what is inside his skull: a new external problem rarely begets an authentically novel solution."

2. *Use and Conservation of Natural Resources*

Archibald and Nan L. Clow, "The Timber Famine and the Development of Technology," *Annals of Science, 12,* No. 2 (June 1956), 85–102. Relationships between depletion of timber in Great Britain and innovations in metals and chemical industries are explored. Very good.

Energy and Man. A Symposium (New York: Appleton-Century-Crofts, 1960, 113 pp.). Columbia University's Graduate School of Business and the American Petroleum Institute sponsored a series of talks by Allan Nevins, Robert Dunlop, Edward Teller, Edward Mason, and Herbert Hoover, Jr. The occasion was the centenary of the beginning of full-scale exploitation of Pennsylvania oil. The speakers were Men of Affairs.

A. William Hoglund, "Forest Conservation and Stove Inventors, 1789–1850," *Forest History, 5,* No. 4 (Winter 1962), 2–8. The technology of wood-burning stoves is examined in relation to early fears of fuelwood shortage and forest devastation.

Thomas LeDuc, "The Historiography of Conservation," *Forest History, 9* (Oct. 1965), 23–28. A penetrating review of resource mismanagement in the United States, largely a result of Congressional pressure, from 1800 to the present. Very good; very disturbing.

Hermann H. B. Meyer, *Select List of References on the Conservation of Natural Resources in the United States* (Washington: Library of Congress, 1912, 110 pp.). About 600 references are listed on mineral, water, forest, soil, and human resources.

Resources for the Future, Inc., 1755 Massachusetts Ave., Washington, D.C. 20036, has held several symposia, beginning in 1958, on some aspect of resource use and development. The proceedings of each symposium have been published in successive books: *Perspectives on Conservation: Essays on America's National Resources* (1959); *Science and Resources: Prospects and Implications of Technological Advance* (1960): *Comparisons in Resource Management* (1962); *Cities and Space: The Future Use of Urban Land* (1963); *Natural Resources and International Development* (1964).

Nathan S. Shaler, *Man and the Earth* (New York: Fox, Duffield, 1905, 240 pp.). An early assessment of future resources by an eminent geologist.

Fremont P. Worth, *Discovery and Exploitation of the Minnesota Iron Lands* (Cedar Rapids, Iowa: Torch, 1937, 247 pp.). While concentrating on legal history, the author demonstrates the nature and magnitude of exploitation. Bibliography, pp. 223–230.

See also
Ayres and Scarlott, *Energy Sources* (p. 243)

XIV. SUBJECT FIELDS (monographs, articles, and bibliographies)

"Consider what it is we look for in a normal bibliography of a special subject. Reflection will show, I think, that we look, above all, for completeness . . . . We desire completeness even more than accuracy (painfully uncongenial though it is for me to make such a statement); for in most cases a bibliography is intended to give us particulars of publications to which we wish to refer; thus we can always judge for ourselves (waiving gross errors) whether the bibliographer has correctly described these publications. On the other hand, anything that is omitted is lost until rediscovered."

—Theodore Besterman
World Bibliography of Bibliographies (4th ed., Lausanne, 1965–1966), cols. 33–34.

"'Etre complet' est un idée enfantine et pédantesque qu'il vaut mieux abandonner. Le but n'est pas d'être complet mais d'être utile et ce but ne peut être rempli sans explications et sans critiques. Il faut donc absolument que la bibliographie soit 'raisonée.'"

—George Sarton
Archives Internationales d'Historie des Sciences, 6 (Oct.–Dec. 1953), 418.

The guiding principle in compiling this subject-matter part has been to list one or more standard works, when known, and specialized bibliographies. I have not aimed for completeness, but I hope that each part will be just complete enough so that a serious and reasonably diligent reader will not miss a major work in the subject field that he enters. When I have failed in this respect, I should appreciate learning the details.

Because subject matter is frequently mentioned in other parts of this book, the index should be checked to exhaust entries pertaining to a particular subject. The reader is also reminded that **Sotheran**, *Bibliotheca* (p. 24), generally has from a few to many dozens of annotated entries covering the subject fields treated here.

Singer, *History of Technology* (p. 2), is particularly good in providing brief reviews of subject fields when time does not permit more extensive reading.

A. Food Production, Preservation, and Preparation

1. Agricultural Machinery and Grain Milling

Everett E. Edwards, *A Bibliography of the History of Agriculture in the United States* (Washington: U.S. Department of Agriculture, 1930 [Misc. Pub. No. 84], 307 pp.). A convenient starting point in locating histories of various farm implements. See the part, "Farm Implements and Machinery," pp. 194–204. This title is to be superseded by "a new comprehensive bibliography," of which **Pursell and Rogers,** *Preliminary List* (following) is the second part.

The F. Hal Higgins Library of Agricultural Technology at the University of California Library in Davis, Calif., has a good collection of trade catalogues, ephemera, pictures, and manuscripts. The library was described by Wayne D. Rasmussen in *Technology and Culture*, 5 (Fall 1964), 575–577.

Carroll W. Pursell, Jr. and Earl M. Rogers, *A Preliminary List of References for the History of Agricultural Science and Technology in the United States* (Davis, Calif.: Agricultural History Center, University of California, 1966, 46 pp.). This bibliography is the second section of a series intended to supersede **Edwards,** *Bibliography* (preceding). It contains about 500 titles and "represents the results of a search of the obvious sources." Nevertheless, the sources would be obvious only to knowledgeable searchers, and the result is a first-rate bibliography of books, articles, and government publications, arranged by subjects. About three-fourths of the list is devoted to technology, including implements and machinery, fences, ditches, barns, silos, and fertilizers. The basic books or articles, many little-known, have been selected with a sure knowledge. Very good.

Rowman & Littlefield, Inc., 84 Fifth Ave., New York, N.Y. 10011, expects to publish the complete card catalogue of the **National Agricultural Library.** The U.S. Department of Agriculture's interest in agricultural machinery and in farm buildings of all kinds has resulted in important holdings in these fields. [Note by Jack Goodwin.]

John C. Morton, *A Cyclopaedia of Agriculture, Practical and Scientific* (2 vols., Glasgow, 1855). Perhaps the thickest of many handbooks of agriculture that were published in the United Kingdom during the middle third of the nineteenth century. J. C. Loudon and Cuthbert

Johnson were authors of others; Gouverneur Emerson in 1868 adapted one of Johnson's handbooks to American usage.

Percy W. Bidwell and John I. Falconer, *History of Agriculture in the Northern United States, 1620–1860* (Washington: The Carnegie Institution of Washington, 1925; 512 pp., illustrated). Although emphasis upon farm mechanization is relatively slight, Chapters 16, pp. 204–216, and 23, pp. 281–305, deal with laborsaving machinery. Classified and critical bibliography, pp. 454–473; alphabetical list of titles, pp. 474–492.

E. Cecil Curwen and Gudmund Hatt, *Plough and Pasture. The Early History of Farming* (New York: Schuman, 1953; 329 pp., illustrated). The emphasis is upon the period before 1500 A.D. The authors' interests are world-wide. Bibliographies, pp. 143–147 and pp. 309–320.

Paul W. Gates, *The Farmer's Age: Agriculture, 1815–1860* (New York: Holt, Rinehart & Winston, 1960 [The Economic History of the United States, Vol. 3]; 460 pp., illustrated). Bibliography, pp. 421–439. In the section of bibliography on agricultural machinery, the author notes, quite accurately, that "All works on agricultural history deal with this topic in some degree."

Lewis C. Gray, *History of Agriculture in the Southern United States to 1860* (2 vols., Washington: The Carnegie Institution of Washington, 1933; illustrated). The emphasis, as stated in the first sentence of the preface, is "economic rather than technological." Nevertheless, helpful, well-documented comments on techniques occur throughout the work. Bibliography, Vol. 2, pp. 945–1,016.

G. E. Mingay, "The 'Agricultural Revolution' in English History: A Reconsideration," *Agricultural History,* 37 (July 1963), 123–133. The author holds that the term should be retained; significant footnote references.

Wayne D. Rasmussen, *Readings in the History of American Agriculture* (Urbana: University of Illinois Press, 1960; 340 pp., illustrated). The improvements in farming techniques are of particular interest to the editor, a leading agricultural historian, who is particularly sensitive to the significance of the mechanization of farming. "Selected Readings in the History of American Agriculture," pp. 312–320, is intended to provide an entry into the literature of the subject. A section on machinery has seven titles, one on technology has thirteen titles, and one on transportation has seven titles.

Fred A. Shannon, *The Farmer's Last Frontier, Agriculture, 1860–1897* (New York: Farrar and Rinehart, 1945 [The Economic History of the United States, Vol. 5], 434 pp.). "Literature of the Subject," pp. 379–414, includes a section on farm machinery.

Russell H. Anderson, "Grain Drills through Thirty-Nine Centuries," *Agricultural History*, 10 (Oct. 1936), 157–205; with references.

Allan G. Bogus, *From Prairie to Corn Belt* (Chicago: University of Chicago Press, 1963). Chapter 8, pp. 148–168, summarizes innovations in farm machinery. Footnote references.

Clark, *The English Traction Engine* (p. 234), is a profusely illustrated album.

Gilbert C. Fite, "Recent Progress in the Mechanization of Cotton Production in the United States," *Agricultural History*, 24 (Jan. 1950), 19–28. This article refers to
Frank J. Welch and D. G. Miley, "Mechanization of Cotton Harvest," *Journal of Farm Economics*, 27 (Nov. 1945), 928–946; which in turn refers to
H. P. Smith *et al.*, "Mechanical Harvesting of Cotton," Texas Agricultural Experiment Station, *Bulletin 452* (Aug. 1932), which has a list of patents for cotton pickers from 1850.

G. E. Fussell, *The Farmer's Tools 1500–1900* (London: Melrose, 1952) is a standard work, nicely illustrated. Chronology of farmer's tools, pp. 218–222; bibliography, pp. 225–230, which may be supplemented by references at the end of appropriate articles in **Singer**, *A History of Technology* (p. 2).

R. B. Gray, *Development of the Agricultural Tractor in the United States, Part I: Up to 1919 inclusive* (Beltsville, Md.: U.S. Department of Agriculture, Agriculture Research Service, 1954). Unpaged, about 100 pp.; many illustrations; no references whatsoever. Reynold M. Wik expects to publish a book on tractor development in the United States. See also his *Steam Power*, following.

Leo Rogin, *The Introduction of Farm Machinery in its Relation to the Productivity of Labor in the Agriculture of the United States during the Nineteenth Century* (Berkeley: University of California Press, 1931 [University of California Publications in Economics, Vol. 9]; 260 pp., 84 illustrations). A classic work. The illustrations of machines are nearly all contemporaneous representations, mostly wood engravings. Extensive bibliographic footnotes.

Clark C. Spence, *God Speed the Plow: the Coming of Steam Cultivation to Great Britain* (Urbana: University of Illinois Press, 1960, 183 pp.) is adequate. Bibliography, pp. 167–172. Nice illustrations, not washed out by halftone screens.

Reynold M. Wik, *Steam Power on the American Farm* (Philadelphia: University of Pennsylvania Press, 1953) provides a sound survey of the subject. Bibliography, pp. 259–278.

Russell H. Anderson, "Technical Ancestry of Grain-Milling Devices," appeared in both *Agricultural History*, 12 (July 1938), 256–270, and *Mechanical Engineering*, 57 (1935), 611–617.

R. Bennett and J. Elton, *History of Corn Milling* (4 vols., London, 1898–1904) is, according to Moritz, still a standard work, but Storck and Teague "has in some respects superseded it." It will be remembered that in the United Kingdom "corn" is any grain.

Harold W. Brace, *History of Seed Crushing in Great Britain* (London: Land, 1960, 172 pp.) is sound, concerned largely with post-1850. Has a very short bibliography.

L. A. Moritz, *Grain-Mills and Flour in Classical Antiquity* (Oxford: Clarendon, 1958, 230 pp.) is authoritative. Notes, pp. 218–223, suggest sources.

Herman Steen, *Flour Milling in America* (Minneapolis: Denison, 1963, 455 pp.) has neither footnotes nor bibliography. Not recommended.

John Storck and W. D. Teague, *Flour for Man's Bread: a History of Milling* (Minneapolis: University of Minnesota Press, 1952, 382 pp.). A book that resulted from a projected Museum of Milling History, which was not built. Half of the book is devoted to the eighteenth century and earlier. An extensive bibliography, carefully classified, includes periodical literature, pp. 343–372.

2. *Food Preservation and Preparation* (See also Refrigeration, Part XIV.D.7.)

A. W. Bitting, *Appertizing or the Art of Canning; Its History and Development* (San Francisco: The Trade Pressroom, 1937, 852 pp.). A long, unsystematic, and diffuse work on many aspects of canning. Among the historical bits spread throughout the book, there is a short chapter on Nicolas Appert, whose book *L'Art de Conserver* was published in 1810; the English patent of 1810 of Peter Durand; a piece on Appert's work from *Edinburgh Review* of 1814 (20 pp.); and a few papers and patents of *ca.* 1900. In the chapter on "Appertizing" in **Clow and Clow,** *Chemical Revolution* (p. 262), pp. 569–581, the sources include Bitting and *Encyclopaedia Britannica* (7th ed., 1841).

Clarence Francis, *A History of Food and Its Preservation, A Contribution to Civilization* (Princeton: The Guild of Brackett Lecturers, 1937, 45 pp.). Although frequently cited, this pamphlet merely records a lecture on "the latest development in food preservation — quick-frozen Birdseye Foods." The absence of any solid historical work on the preservation of food is quite striking.

International Tin Research and Development Council, *Historic Tinned Foods* (2nd ed., Greenford, 1939 [Pub. No. 85]; 70 pp. illustrated). Descriptions and analyses of cans of meat and vegetables reopened after periods varying from 14 to 58 years. Short summary of developments in canning is to be found *passim*.

T. P. Christensen, "The First Cream Separator in the United States," *Annals of Iowa* (Series 3), 34 (July 1957), 56–58. Brought to Iowa in 1882 by a Danish immigrant. Listed in **Pursell and Rogers,** *Preliminary List* (p. 214).

Joe B. Frantz, *Gail Borden, Dairyman to a Nation* (Norman: University of Oklahoma Press, 1951; 310 pp., illustrated). The book is based on original Borden papers, private and corporate, but the author was not particularly interested in the techniques of milk preservation. Sources can be located through the bibliography, pp. 277–294.

G. E. Fussell, *The English Dairy Farmer 1500–1900* (London: Frank Cass, 1966; 357 pp., illustrated). The cow and its feeding come first, even today. Chapters on these essentials are followed by others on buildings, equipment, and on butter and cheese making and marketing. Bibliography, pp. 340–350.

Eric E. Lampard, *The Rise of the Dairy Industry in Wisconsin: a Study in Agricultural Change, 1820–1920* (Madison: State Historical Society of Wisconsin, 1963, 466 pp.). Material on technology. Listed in **Pursell and Rogers,** *Preliminary List* (p. 214).

T. R. Pirtle, *History of the Dairy Industry* (Chicago: Majonnier, 1926, 645 pp.). Listed in **Pursell and Rogers,** *Preliminary List* (p. 214).

Stanley Baron, *Brewed in America. A History of Beer and Ale in the United States* (Boston: Little, Brown, 1962, 424 pp.). A serious work by an enthusiast. Bibliography, unpublished materials, pp. 379–381; books and articles on brewing in the United States and England, pp. 381–384.

Peter Mathias, *The Brewing Industry in England 1700–1830* (Cambridge: University Press, 1959, 596 pp.). A thorough work, concerned with brewing, raw materials, markets, financing, and so forth. Chapter 3, pp. 63–98, reviews technical innovations, including process control and installation of steam engines. Bibliography, pp. 559–574.

B. Civil Engineering

1. *General*

William H. Burr, *Ancient and Modern Engineering and the Isthmian Canal* (New York: Wiley, 1902; 473 pp., illustrated). A series of lectures on Egyptian and Roman civil engineering works, modern bridges, water supplies, railroads, and the Nicaragua and Panama Canal routes, to which have been added chapters on modern bridge analysis and arch theory. The historical chapters exhibit no sources, but they do contain standard information. The chapters on analysis unfortunately are not connected by reference to the structures described in the historical chapters.

Daniel H. Calhoun, *The American Civil Engineer. Origins and Conflict* (Cambridge, Mass.: The M.I.T. Press, distrib. by Harvard University Press, 1960, 295 pp.). Considers sources of supply of engineers, their principal work, and the difficulties of divided loyalties, in the period 1816–1846. List of (87) chief engineers of projects in 1837; bibliographical note, pp. 221–237; text notes, pp. 241–283.

Carl Condit, *American Building Art: the Nineteenth Century* (New York: Oxford University Press, 1960, 4to; 371 pp., illustrated).

Carl Condit, *American Building Art: the Twentieth Century* (New York: Oxford University Press, 1961, 4to; 427 pp., illustrated).

Concerned with techniques, materials, and the architectural integrity of buildings, train sheds, bridges, dams, and highways, these two volumes are without peers. Each volume has spacious, explanatory, bibliographic notes and an adequate bibliography.

James Kip Finch, *Engineering and Western Civilization* (New York: McGraw-Hill, 1951) is particularly valuable for its extensive and well-selected bibliography, pp. 331–374. The bibliography is strongest in civil-engineering subjects, although it is useful in everything it touches.

John Gloag and Derek Bridgewater, *A History of Cast Iron in Architecture* (London: Allen & Unwin, 1948, 4to, 395 pp.). A profusely illustrated work, showing applications in bridges, buildings, water pipes, tunnels, and decoration. Bibliographic reference notes. Very good.

Rolt Hammond, *Engineering Structural Failures. The Causes and Results of Failure in Modern Structures of Various Types* (New York: Philosophical Library, 1956). Chapters on twentieth-century failures, most quite recent, of earthworks, dams, buildings, bridges, tunnels, and so forth. A few references are given at the end of each chapter. A good work would be welcomed on engineering failures in the nineteenth century and the lessons learned or ignored.

Richard S. Kirby and F. G. Laurson, *The Early Years of Modern Civil Engineering* (New Haven: Yale University Press, 1932, 324 pp.). A sound book, principally concerned with the period 1750–1850. Divided by subject — surveying, canals, waterworks, railroads, and so forth. Short biographical sketches of 114 men, many impossibly obscure. Bibliography, pp. 313–318.

G. W. MacGeorge, *Ways and Works in India; being an Acount of the Public Works in that Country from the Earliest Times up to the Present Day* (London, 1894). Describes railways, roads, water supply, irrigation, and so forth.

Charles J. Merdinger, *Civil Enigneering through the Ages* (Washington, D.C.: Society of American Military Engineers, 1963; 159 pp., illustrated). Thirty chapters covering construction materials, surveying, roads, bridges, canals, tunnels, water supply, sewerage, hydraulic and waterfront structures, and railroads. Reprinted from the *Military Engineer*, 1952–1962. Reviewed by J. K. Finch in *Technology and Culture*, 5 (Summer 1964), 435–438.

J. P. M. Pannell, *An Illustrated History of Civil Engineering* (London: Thames and Hudson, 1964; 370 pp., 228 illustrations). Essentially "a series of sketches of civil-engineering development in Great Britain in the eighteenth and nineteenth centuries," according to a review by Charles J. Merdinger in *Technology and Culture*, 6 (Summer 1965), 447–448. Selected bibliography, twenty-seven titles, pp. 369–370.

Royal Institute of British Architects, *Catalogue of the . . . Library* (2 vols., London, 1937–1938). Vol. 1, Author Catalogue of Books and Manuscripts, 1,138 pp.; Vol. 2, Classified Index and Alphabetical Subject Index, 541 pp. Sections on building and bridge construction, structural engineering, masonry, sewage disposal, exhibition buildings, cathedrals, and so forth. This impressive catalogue may well be, as claimed, the "most complete architectural bibliography that has yet been published."

Hans Straub, *A History of Civil Engineering*, trans. by E. Rockwell (Cambridge, Mass.: The M.I.T. Press, 1964). Translated from *Die Geschichte der Bauingenieurkunst* (Basle: Birkhäuser, 1949); published in hardback (London: Hill; Newton, Mass.: Branford, 1952) and in paper by The M.I.T. Press. A sensible survey from ancient to modern times. Has a short "selected bibliography," pp. 248–250.

Vitruvius, *The Ten Books on Architecture*, Morris Hicky Morgan, Trans. (Cambridge, Mass.: Harvard University Press, 1914; New York: Dover, 1960; 331 pp., illustrated). This is a standard translation of the first-century, B.C., Roman engineer. Passages on masonry, foundations, pilings, cranes, water mills, and various machines.

See also
Parsons, *Engineers . . . in the Renaissance* (p. 11)
Gille, *Renaissance Engineers* (p. 10)

2. *Buildings and Similar Structures*

R. J. C. Atkinson, *Stonehenge* (London: Hamish Hamilton, 1956). A very good illustrated summary of what is known of the building of this ancient monument, which is located in southern England.

Gerald S. Hawkins, *Stonehenge Decoded* (Garden City: Doubleday, 1965). An astronomer has resurrected a species of Pyramidology that will appeal to some readers.

Turpin Bannister, "The First Iron-Framed Buildings" (*Architectural Review, 107* [Apr. 1950], 231–246) is a pioneering article on the first structural use of iron in buildings at the end of the eighteenth century in England. Copious footnotes, bibliographic in scope.

Marion Elizabeth Blake, *Ancient Roman Construction in Italy from the Prehistoric Period to Augustus* (Washington: Carnegie Institution of Washington, 1947; 421 pp. + 57 pls.). Based upon careful, firsthand study. There is a good chapter on Roman cement, pp. 309–352.

Marion Elizabeth Blake, *Roman Construction in Italy from Tiberius through the Flavians* (Washington: Carnegie Institution of Washington, 1959; 195 pp. + 31 pls.). See also preceding title.

Carl Condit, *The Chicago School of Architecture: A History of Commercial and Public Building in the Chicago Area, 1875–1925* (Chicago: University of Chicago Press, 1964; 238 pp., 196 pls.). An outgrowth of the author's earlier *Rise of the Skyscraper* (1952), but an essentially new, thoroughly matured work. Reviewed in *The New York*

Times, Aug. 30, 1964; in *Technology and Culture,* 6 (Summer 1965), 472–475. See also **Condit's** *American Building Art* (p. 219).

Carl Condit, "The First Reinforced-Concrete Skyscraper," *Technology and Culture,* 9 (Jan. 1968), 1–33. The building of the title, completed in 1903, is in Cincinnati. The first third of the article is a history of reinforced concrete construction in Europe and America. The footnote references lead to basic works on the subject.

I. E. S. Edwards, *The Pyramids of Egypt* (rev. ed., Baltimore: Penguin, 1961; 320 pp., illustrated). A knowledgeable review. Chapter 7, "Construction and Purpose," brings together much scattered information. Bibliography, pp. 299–311.

Ahmed Fakhry, *The Pyramids* (Chicago: University of Chicago Press, 1961; 262 pp., illustrated). A well-illustrated survey of the many Egyptian pyramids by the Director of Pyramid Studies in Egypt. A thoroughly sound book.

Jacob Feld, *Lessons from Failures of Concrete Structures* (Ames: Iowa State University Press, 1964, 179 pp.). "Historical" chapter, pp. 9–16. Entire book is by its nature essentially historical. References, pp. 161–170, from *ca.* 1900 to the present.

Herbert W. Ferris, *Rolled Shapes: Historical Record Dimensions and Properties Steel and Wrought Iron Beams & Columns as Rolled in U.S.A., Period 1873 to 1952* (New York: American Institute of Steel Construction, 1953, 141 pp.). Gives detailed dimensions of rolled shapes, derived from makers' contemporary catalogues. Apparently a careful and thorough tabulation.

Banister Fletcher, *A History of Architecture on the Comparative Method* (17th ed., New York: Scribner, 1961, 1,366 pp.) is not only a monumental compendium of graphic and pictorial information, but each chapter is followed by an important bibliography.

John Fitchen, *The Construction of Gothic Cathedrals* (Oxford: Clarendon, 1961, 4to; 344 pp., 73 figs.). A first-rate monograph on a nearly undocumented subject. Most of the figures have been beautifully drawn by the author to illustrate the details of his thesis. Bibliography, pp. 317–336.

Henry Russell Hitchcock, Jr., *American Architectural Books: a List of Books, Portfolios, and Pamphlets published in America before 1895 on Architecture and Related Subjects* (2nd ed., Middletown, Conn.: the Author, 1939). Mimeographed; over 1,000 entries.

Antoine Moles, *Histoire des Charpentiers, leurs Travaux* (Paris, 1949, 405 pp.). Many illustrations of arch- and vault-centering, of scaffolding, framing, rigging, and so forth. Suggestive of eighteenth-century sources.

Lonnie R. Shelby, *The Technical Supervision of Masonry Construction in Medieval England* (unpub. dissertation, University of North Carolina, 1962) (Ann Arbor: University Microfilms, 1963, 273 pp.). Technical problems of construction are explored. Bibliography, pp. 250–273.

Arthur Vierendeel, *La Construction Architecturale en Fonte, Fer et Acier* (Louvain & Paris, 1903). Text vol., 4to, 879 pp., atlas of 135 pls., double fo. Illustrations showing structural details are in the grand manner. Includes railroad stations, international exhibition buildings, Sydenham Crystal Palace, lighthouses, winter gardens, reinforced concrete schemes, metal ceilings, and so forth in the United Kingdom, low countries, France, and Germany.

Robert M. Vogel, "Elevator Systems of the Eiffel Tower, 1889" (U.S. N.M., *Bulletin 228*, Paper 19 [1961], 1–40). The first part of this paper is a well-illustrated history of elevator development from about 1845. Very good.

See also
Hamilton on "Structural Theory" (p. 304)
Timoshenko, *Strength of Materials* (p. 275)

3. *Bridges and Tunnels*

An informative symposium on the proposed rebuilding of the Brooklyn Bridge appeared in **American Society of Civil Engineers**, *Proceedings*, 72 (1946), 5–68, 534–536, 725–726, 864–866, 1,437.

On American bridges, the best review and the starting point for serious study is **Condit**, *American Building Art* (p. 219).

James K. Finch, "Wind Failures of Suspension Bridges, or Evolution and Decay of the Stiffening Truss" (*Engineering News Record*, 126 [Mar. 13, 1941], 402–407), which puts the failure of the Tacoma Narrows Bridge into historical perspective, is an excellent case study in the utility of historical knowledge in one's own field. In spite of the author's disclaimer (*ibid.*, Mar. 27, 1941, p. 43), the article suggests that a designer who was aware of earlier failures *might* (not *would*) have avoided this one.

Arne A. Jakkula, "A History of Suspension Bridges in Bibliographical Form" (College Station: Texas Engineering Experiment Station, *Bulletin*, 4th ser., Vol. 12, No. 7, [July 1941], 564 pp.). Impressively full for all times and all places. Includes periodical literature. For example, Brooklyn Bridge has more than 500 entries, annotated. First rate in every way.

"List of works in the New York Public Library relating to Bridges and Viaducts" (N.Y.P.L., *Bulletin*, 9 [Aug. 1905], 295–329). Arranged by type and location. This list was not published separately.

H. Shirley Smith, *The World's Great Bridges* (rev. ed., New York: Harper, 1965; 250 pp., illustrated). A helpful though brief bibliography is on pp. 238–239.

David B. Steinman, *The Builders of the Bridge* (New York: Harcourt, Brace, 1945, 457 pp.) is the story by a well-known bridge designer of building the Brooklyn Bridge. While I find the early biographical portion tedious, the account of building is accurate and intensely interesting.

On bridges of the world, **Henry G. Tyrell**, *Bridge Engineering* (Evanston, Ill., 1911) is still useful, though devoid of scholarly apparatus.

Robert M. Vogel, "The Engineering Contributions of Wendel Bollman" (U.S.N.M., *Bulletin 240*, Paper 36 [1964], 77–104). A remarkable group of photographs of Bollman truss bridges, including those at Harper's Ferry (1862), supplements the description of Bollman's work in introducing iron bridges. Bibliography, twenty-two entries.

W. J. Watson, *Bridge Architecture* (New York: Helburn, 1927), is notable for its excellent illustrations, as is the following title.

Charles S. Whitney, *Bridges* (New York: Rudge, 1929).

Squire Whipple, *Bridge-Building: being the Author's Original Work, published in 1847, with an Appendix, containing Corrections, Additions* [etc.] (Albany, N.Y., 1869). This important pioneering work has not been, to my knowledge, analyzed, criticized, and compared with Haupt's book, which follows.

Herman Haupt, *General Theory of Bridge Construction* (New York: Appleton, 1853).

Many published reports of major bridges are full and informative.

For example, the Eads Bridge of 1874 is described in **Calvin M. Woodward**, *A History of the St. Louis Bridge* (St. Louis, 1881). Profusely illustrated. Subsequent reports, such as the following, if they can be located, are valuable for their comments on adequacy of the original design: **Shortridge Hardesty**, *Report on Inspection of the Eads Bridge* (1940, 35 pp. + 17 photos + 15 dwgs.). May be catalogued under Terminal Railroad Association of St. Louis.

William C. Copperthwaite, *Tunnel Shields* (London, 1906, 390 pp.), describes the Greathead shield system of tunneling.

Henry S. Drinker, *Tunneling* (New York: Wiley, 1882; 1,143 pp., over 1,000 illustrations). A monumental work on all historical and practical aspects of tunnels.

Gilbert H. Gilbert, *The Subways and Tunnels of New York: Methods and Costs* (New York: Wiley, 1912), is informative. Illustrated.

Gosta E. Sandström, *The History of Tunneling* (London: Barrie & Rockliff, 1963, 427 pp.) is very good. Published also in the United States as *Tunnels* (New York: Holt, Rinehart & Winston, 1963).

See also
Hamilton on "Structural Theory" (p. 304)
Timoshenko, *Strength of Materials* (p. 275)

4. *Hydraulic Engineering, Water Supply, and Dams*

Thomas J. Bell, *History of the Water Supply of the World* (Cincinnati, 1882, 134 pp.). Very slight.

Nelson M. Blake, *Water for the Cities: a History of the Urban Water-Supply Problem in the United States* (Syracuse: Syracuse University Press, 1956, 341 pp.). A social history of the supplies of, principally, Philadelphia, New York, Boston, and Baltimore. Includes adequate references to sources of technical details.

Arthur H. Frazier, "William Gunn Price and the Price Current Meters" (U.S.N.M., *Bulletin* 252, Paper 70 [1967], 37–68). The author, in this and in the following article, relates the development of stream flow measuring devices and the activities of the U.S. Geological Survey. Both articles are nicely illustrated.

Arthur H. Frazier, "Daniel Farrand Henry's Cup Type 'Telegraphic' River Current Meter," *Technology and Culture*, 5 (Fall 1964), 541–565.

Clemens Herschel, *The Two Books on the Water-Supply of the City of Rome by Sextus Julius Frontinus* (2nd ed., London: Longmans, Green, 1913; 4to, 296 pp., illustrated). A well-known work, translated by an American waterworks engineer. The translation, pp. 1–100, is followed by "Explanatory Chapters," pp. 103–296.

A. G. Keller, "Renaissance Waterworks and Hydromechanics," *Endeavour,* 25 (Sept. 1966), 141–145; 6 illustrations, 12 refs. A brief but significant interpretive article. Intense interest during the sixteenth and seventeenth centuries in hydraulic machines and devices led to a wider understanding of pneumatic and hydraulic phenomena.

Stephen Kolupaila, *Bibliography of Hydrometry* (South Bend, Ind.: Notre Dame University Press, 1961, 975 pp.). A monumental, helpfully annotated bibliography, not a descriptive work, of the measurement of water flow, very broadly interpreted. Works in more than thirty languages are listed. Chapters on water stage, runoff determination, current meters, hydrometry of tidal waters, sediment measurements, and hydrometeorological measurements. There are historical sections in several of the chapters.

"List of Works in the New York Public Library relating to Hydraulic Engineering" (N.Y.P.L., *Bulletin,* 11 [Nov., Dec. 1907], 512–552, 565–626) (also separately reprinted) includes titles on hydromechanics, canals, aqueducts, cement, dredges, meters, and so forth.

F. W. Robins, *The Story of Water Supply* (London: Oxford University Press, 1946, 207 pp.). Nearly half of this competent survey is concerned with ancient and classical times; the other half brings the story to the nineteenth century. A welcome book in a scarcely worked field.

Esther B. Van Deman, *The Building of the Roman Aqueducts* (Washington: Carnegie Institution of Washington, 1934). A standard work.

Detailed information on many individual water supply systems is available:

Henry W. Dickinson, *The Water-Supply of Greater London* (London: Newcomen Society, 1954) is the republication of a series of twenty-three articles in *The Engineer,* Vol. 186 (July 9–Dec. 10, 1948).

Edward Wegmann, *The Water-Supply of the City of New York 1658–1895* (New York: Wiley, 1896, 316 pp.) includes the first and second Croton projects.

Complete Report on Construction of the Los Angeles Aqueduct (Los Angeles: Department of Public Service, 1916, 330 pp.) is heavily illustrated.

Sewerage (systems of sewers) of London and Paris are described in several papers in **Institution of Civil Engineers,** *Minutes of the Proceedings, ca.* 1870. Papers may be located through **Finch,** *Engineering* (p. 4), p. 342.

Edward Wegmann, *The Design and Construction of Dams* (8th ed., New York: Wiley, 1927) is a proper starting point for a detailed inquiry into dams. **Condit,** *American Building Art, Twentieth Century* (p. 219), pp. 219–273, is concerned with dams.

See also
Rowe, *Rivers and Harbors* (p. 232)
Rouse and Ince, *History of Hydraulics* (p. 305)

5. Surveying and Mapping

Leo Bagrow, *History of Cartography.* Revised and enlarged by R. A. Skelton (Cambridge, Mass.: Harvard University Press, 1964, 321 pp.). Sumptuous and very important volume on maps and mapping to about 1750. List of cartographers, pp. 227–280; select bibliography, pp. 283–300.

Lloyd A. Brown, *The Story of Maps* (Boston: Little, Brown, 1949; 397 pp., illustrated). Bibliography, pp. 341–373.

A. W. Richeson, *English Land Measuring to 1800: Instruments and Practices* (Cambridge, Mass.: The Society for the History of Technology and The M.I.T. Press, 1966; 214 pp., illustrated). Bibliography, pp. 189–207.

Walter W. Ristow and Clara E. LeGear, Comps., *A Guide to Historical Cartography: a Selected, Annotated List of References on the History of Maps and Map Making* (Washington: Library of Congress, Map Division, 1960). A comprehensive list of sixty-seven titles.

R. V. Tooley, *Maps and Map-Makers* (New York: Bonanza, 1949, 1962; 140 pp., 104 illustrations, some in color). A basic and authoritative work. Bibliography.

See also
Keily, *Surveying Instruments* (p. 41)
Bedini, *American Instruments* (p. 287)
Scientific Instruments, Part XIV.G.5

C. Transportation

1. General

A large number of canal and railroad reports may be located through notes in **Calhoun**, *American Civil Engineer* (p. 219).

G. P. de T. Glazebrook, *A History of Transportation in Canada* (Toronto: Ryerson, 1938, 475 pp.), surveys the field and suggests available primary sources.

W. T. Jackman, *The Development of Transportation in Modern England* (London: Cass, 1962, 820 pp.). This standard work, covering the period *ca.* 1500–1850, first published in 1916, has been reprinted with the addition of an "Introductory Guide" (15 pp.) by W. H. Chaloner. The new matter is bibliographic, supplementing the original sixty-two-page bibliography. The coverage of roads and canals is comprehensive.

B. H. Meyer, Ed., *History of Transportation in the United States before 1860* (Washington: Carnegie Institution, 1917). A mass of detailed information, to be used with caution. Footnote references are numerous and suggestive, however, and the bibliography, pp. 609–649, is useful in any extended study.

The **Parsons Collection** of early U.S. railroad and canal materials is described in the "Catalogue" (p. 42).

Although written as an economic history, the best guide for its period and place (U.S.A.), is **Taylor**, *The Transportation Revolution 1815–1860* (p. 13). An excellent critical bibliography is on pp. 399–438.

2. Ships and Boats, Navigation and Charting

Robert G. Albion, *Naval & Maritime History. An Annotated Bibliography* (3rd ed., Mystic., Conn.: The Marine Historical Assn., 1963, 230 pp.). *Third Edition, First Supplement, 1963–65* (1966, 62 pp.). This comprehensive bibliography is indispensable to any serious work in maritime history, including navigation and cartography. Further supplements may be expected.

Edwin T. Adney and Howard I. Chapelle, *The Bark Canoes and Skin Boats of North America* (Washington: Smithsonian Institution, 1964 [U.S.N.M., *Bulletin 230*]; 242 pp., 224 illustrations). Descriptions of techniques of building are based upon personal observations by Adney over a forty-year period. Informatively reviewed by Olaf H. Prufer in *Technology and Culture*, 6 (Summer 1965), 460–462.

Romola and R. C. Anderson, *The Sailing-Ship: Six Thousand Years of History* (New York: W. W. Norton, 1963; 212 pp., illustrated). A standard work, first published in 1926. Unfortunately, there are neither notes nor bibliography.

For early books on shipbuilding, rigging, and seamanship, see Anderson's bibliographic articles (p. 40).

K. C. Barnaby, *The Institution of Naval Architects 1860–1960* (London, 1960, 4to, 645 pp.). A review of the Institution's *Transactions*, year by year; the table of contents of each volume is given since 1860.

Henry T. Bernstein, *Steamboats on the Ganges. An Exploration in the History of India's Modernization through Science and Technology* (Bombay: Orient Longmans, 1960; 239 pp., illustrated). The period covered by the book is 1819–1840; however, the author probes the larger issues as indicated by the subtitle.

Howard I. Chapelle, *The History of American Sailing Ships* (New York: W. W. Norton, 1935; 4to, 400 pp., illustrated). The author, an authority on ship design and shipbuilding, insists upon having the evidence of drawings and models. His several books are very good.

Carl C. Cutler, *Greyhounds of the Sea. The Story of the American Clipper Ship* (New York: Putnam, 1930; 4to, 592 pp., illustrated). A massive, authoritative work, in which the writing reflects the stirring nature of the subject. Very good.

James T. Flexner, *Steamboats Come True. American Inventors in Action* (New York: Viking, 1944; 406 pp., illustrated). The development is traced through *ca.* 1815. Bibliography and notes, pp. 379–401.

Charles Green, *Sutton Hoo: The Excavation of a Royal Ship-Burial* (New York: Barnes & Noble, 1963, 168 pp.). A narrative history of the Viking ship unearthed in 1939 and a review of the Viking ship in general. Beautifully illustrated; many excellent drawings; two color plates; and a fine job of English bookmaking. Bibliography, pp. 148–154.

Codman Hislop, "The S.S. *Novelty*" (New-York Historical Society, *Quarterly*, 49 [Oct. 1965], 327–340). This anthracite-burning steamboat of 1836 gave its name to the "Novelty Works," one of New York's leading machine and boiler works.

Louis C. Hunter, *Steamboats on the Western Rivers. An Economic and Technological History* (Cambridge, Mass.: Harvard University Press, 1949; 684 pp., illustrated). A model of technological history: painstaking, thoroughly reliable. There is no bibliography, but footnotes form a rich mine, well worth the time required to extract specific titles. The period covered is 1810–1880, approximately. The western rivers of the title include the Mississippi and tributaries.

L'Illustration, *Histoire de la marine* (Paris, 1937; 575 pp., illustrated). A collaborative work on shipbuilding, navigation, and so forth.

Bjorn Landstrom, *The Ship* (Garden City, N.Y.: Doubleday, 1961). An expertly illustrated story of development of a wide variety of watercraft from ancient times to the present. The artist author has not given as much documentation in his restorations as I should like, but many illustrations reveal careful consideration of alternatives and a generally well-informed judgment. Very attractive.

Frederic C. Lane, *Venetian Ships and Shipbuilders of the Renaissance* (Baltimore: The Johns Hopkins Press, 1934; 285 pp., illustrated). A standard work.

Mariner's Museum Library, Newport News, Va., *Dictionary Catalog* (9 vols.); *Catalog of Maps, Ships' Papers and Logbooks* (1 vol.); *Catalog of Marine Photographs* (5 vols.); *Catalog of Marine Prints and Paintings* (3 vols.). All of these catalogues are photoprints, available from G. K. Hall & Co. See note on p. 26.

F. Moll, "History of the Anchor," *Mariner's Mirror,* 13 (1927), 293–332. I have not seen this article.

Edgar C. Smith, *A Short History of Naval and Marine Engineering* (Cambridge: University Press, 1938, 376 pp.). A general survey by an English authority.

Society of Naval Architects and Marine Engineers [U.S.], *Historical Transactions, 1893–1943* (1945). A collection of papers on historical topics; list of articles in the Society's *Transactions,* pp. 455–466.

David B. Tyler, *The American Clyde* (Newark: University of Delaware Press, 1958). A history of iron and steel shipbuilding on the Delaware River from 1840 up to World War I by a careful student of modern maritime affairs.

David B. Tyler, *Steam Conquers the Atlantic* (New York: Appleton, 1939, 425 pp.). A scholarly monograph covering the period before 1880. Bibliography, pp. 393–406.

A number of books have been written about particular boat lines. Two examples only are listed here.

Roger W. McAdam, *The Old Fall River Line* (New York: Daye, 1955, 288 pp.). First published in 1937. Popular; no references; no bibliography.

Donald C. Ringwald, *Hudson River Day Line: the Story of a Great American Steamboat Company* (Berkeley, Calif.: Howell-North, 1965; 228 pp., illustrated). "A distinguished addition to the published history of New York City and State," according to C. Bradford Mitchell in New-York Historical Society, *Quarterly*, 50 (July 1966), 320–321.

Edmund Guyot, *Histoire de la détermination des longitudes* (La Chaux-des-Fonds: La Chambre Suisse de l'horlogerie, 1955; 254 pp., illustrated). A review by the director of the Neuchâtel Observatory of modern methods of determining longitude, starting with lunar distances, which were used at sea in the eighteenth century, and ending with radio methods. There is a long chapter on methods using chronometers, pp. 91–171. I know of no work in English that reviews the eighteenth- and nineteenth-century methods of navigation.

Frederic C. Lane, "The Economic Meaning of the Invention of the Compass," *American Historical Review*, 68 (Apr. 1963), 605–617. A major interpretive essay.

Humphrey Quill, *John Harrison the Man who Found Longitude* (London: Baker, 1966; 255 pp., illustrated). This volume was written as complementary to R. T. Gould, *The Marine Chronometer, Its History and Development* (London: Holland, 1960), first published in 1923, which describes in detail the mechanical aspects of the Harrison chronometers.

E. G. R. Taylor and M. W. Richey, *The Geometrical Seaman. A Book of Early Nautical Instruments* (London: Institute of Navigation, 1962). An illustrated introduction to navigational instruments.

E. G. R. Taylor, *The Haven-Finding Art. A History of Navigation from Odysseus to Captain Cook* (New York: Abelard-Schuman, 1957; 295 pp., illustrated). The best general survey; bibliography, pp. 264–266.

D. W. Waters, *The Art of Navigation in England in Elizabethan and Early Stuart Times* (New Haven: Yale University Press, 1958; 696 pp., illustrated). Bibliography, pp. 597–628.

See also
Anderson, *Oared Fighting Ships* (p. 295)

3. *Canals, Rivers, and Harbors*

Hugh G. J. Aitken, *The Welland Canal. A Study in Canadian Enterprise* (Cambridge, Mass.: Harvard University Press, 1954; 178 pp., 2 maps). A balanced narrative of promotion and building of the Welland Canal from Lake Erie to Lake Ontario, commencing in 1824.

Robert G. Albion, *The Rise of New York Port, 1815–1860* (New York: Scribner, 1939, 1961; 485 pp., illustrated). A comprehensive and classic work, in which relationships between New York and the inland empire and New York and rival ports are thoroughly explored. Bibliography, pp. 425–470.

"American Interoceanic Canals; a List of Works in the New York Public Library" (N.Y.P.L., *Bulletin*, 20 [Jan. 1916] 11–81). Also reprinted separately.

Calhoun (p. 219), p. 220, notes: "On early American canals, the subject catalogue of the New York Public Library far surpasses in bibliographic usefulness any published compilation."

Hubertis Cummings, *Pennsylvania Board of Canal Commissioners' Records . . . Descriptive Index* (Harrisburg: Pennsylvania Department of Internal Affairs, 1959, 235 pp.). A carefully made survey of an important but as yet largely untouched body of original records of canals and allied works in Pennsylvania.

Carter Goodrich, Ed., *Canals and American Economic Development* (New York: Columbia University Press, 1961; 303 pp., 1 map). A collection of papers on the New York, Pennsylvania, and New Jersey Canals. The notes, pp. 256–291, are bibliographic in scope.

Charles Hadfield, *British Canals* (2nd ed., London: Phoenix, 1959; 291 pp., illustrated). A general survey. Bibliography, pp. 277–279.

John Marlowe, *World Ditch. The Making of the Suez Canal* (New York: Macmillan, 1964, 294 pp.). The bibliography, pp. 281–283, is adequate to permit a student to enter the serious works on the subject. The same book was published in London under the title *Making of the Suez Canal*.

Robert S. Rowe, *Bibliography of Rivers and Harbors and Related Fields in Hydraulic Engineering* (Princeton: Department of Civil Engineering, Princeton University, 1953, 407 pp.). More than 6,000 references, a few briefly annotated, to books, bulletins, congresses, dictionaries, encyclopaedias, reports, and treatises. Periodical titles are included, but no references are given to journal articles. Rivers and harbors (135 pp.); canals, including interoceanic (74 pp.); reclamation (45 pp.); dams, reservoirs, water power and water supply

(86 pp.). Detailed author and subject index. Most titles are nineteenth- and twentieth-century items, but earlier items are included also. An extremely useful work. [Note by Jack Goodwin.]

Walter S. Sanderlin, *The Great National Project. A History of the Chesapeake and Ohio Canal* (Baltimore, 1946 [Johns Hopkins University Studies in Historical and Political Science, Series 46, No. 1], 331 pp.). A careful and thorough work. Bibliography, pp. 295–302. The author deposited copies of his original dissertation, which included an augmented bibliography, in the Library of Congress and University of Maryland Library.

Ronald E. Shaw, *Erie Water West. A History of the Erie Canal 1792–1854* (Lexington: University of Kentucky Press, 1966; 449 pp., illustrated). The author has attempted to give a balanced account of the Erie, including political, technical, economic, and social aspects. Essay on bibliography, pp. 419–432.

For an introduction to the canal era in the United States, see **Taylor,** *Transportation Revolution* (p. 13), Chapter 3 and bibliography, pp. 405–406.

L. F. Vernon-Harcourt, *Rivers and Canals* (2 vols., Oxford, 1882). Not historical in intent. Works of the 1870's are described in full technical detail, including the Amsterdam Ship Canal, the Suez Canal, and many lesser works.

The annual reports of the **U.S. Army, Corps of Engineers,** are rich sources on river and harbor works. See the Index (p. 99) for the period 1866–1912; there is no index for the earlier reports, 1851–1865.

4. Roads and Vehicles

American Association of State Highway Officials, *Public Roads of the Past* (2 vols., Washington, 1952–1953). Vol. 1, 3500 B.C. to 1800 A.D.; Vol. 2, Historic American Highways. Written by Albert C. Rose. The first volume has an indifferent bibliography, pp. 97–101; the second volume is much better. Many standard and obscure works are listed, pp. 165–183.

Joseph A. Durrenberger, *Turnpikes, A Study of the Toll Road Movement in the Middle Atlantic States and Maryland* (Valdosta, Ga.; Southern Stationery and Printing, 1931, 188 pp.). A standard work. Bibliography, pp. 166–181.

R. J. Forbes, *Notes on the History of Ancient Roads and Their Construction* (2nd ed., Amsterdam: Halkkert, 1964, 182 pp.). See also **Forbes,** *Studies* (p. 9), Vol. 2.

Walter W. Ristow, Ed., *A Survey of the Roads of the United States of America, 1789, by Christopher Colles* (Cambridge, Mass.: Harvard University Press, 1961, 250 pp.) includes eighty-three contemporary strip maps and much explanatory matter.

Frederic J. Wood, *The Turnpikes of New England and the Evolution of the Same through* [sic] *England, Virginia, and Maryland* (Boston: Marshall Jones, 1919; 4to, 461 pp., heavily illustrated). A labor of love.

Don H. Berkebile, "Conestoga Wagons in Braddock's Campaign, 1755" (Washington: Smithsonian Institution, 1959 [U.S.N.M., *Bulletin 218*], 142–155). An authoritative article, illustrated with photographs and drawings.

William L. Gannon, *Carriage, Coach, and Wagon: the Design and Decoration of American Horse-Drawn Vehicles* (Unpub. dissertation, University of Iowa, 1960 [University Microfilms, No. 60–5656]; 404 pp., 240 illustrations). Bibliography, pp. 366–404. An abstract is in *Dissertations Abstracts*, 22 (Aug. 1961), 531.

Jack D. Rittenhouse, *Carriage Hundred: A Bibliography of Horse-Drawn Transportation* (Houston, Tex.: Stagecoach, 1961, 49 pp.). A sensible selection of 100 titles from 1671 to the present; annotated.

George Shumway, Edward Durrell, and Howard C. Frey, *Conestoga Wagon 1750–1850* (York, Pa.: George Shumway, 1964; 206 pp., 73 illustrations). Reviewed in *Technology and Culture*, 6 (Summer 1965), 469–471.

C. F. Cauntner, *The History and Development of Cycles* (London: Her Majesty's Stationery Office, 1955; 70 pp., illustrated), is a Science Museum monograph. The emphasis is upon British development. A short bibliography of "the more important books" is on p. 66.

Ronald H. Clark, *The Development of the English Steam Wagon* (Norwich: Goose, 1963; 237 pp., illustrated). A history, 1895–1915, of the steam truck and omnibus.

Ronald H. Clark, *The Development of the English Traction Engine* (Norwich: Goose, 1960; 390 pp., 582 illustrations). Agricultural engines, plows, digging machines, and tram, road traction engines, and road rollers are all considered from the technical viewpoint. The period is very nearly the latter half of the nineteenth century.

C. St. C. B. Davison, *History of Steam Road Vehicles Mainly for Passenger Transport* (London: Her Majesty's Stationery Office, 1953;

60 pp., illustrated) is a Science Museum monograph. A short bibliography is on p. 58.

Detroit Public Library, *The Automotive History Collection* (2 vols., Boston: G. K. Hall). This photoprinted catalogue is in four parts: Dictionary catalogue of books, periodical shelflist, checklist of automobile catalogues, and descriptions of special collections. The collection includes 11,000 volumes of books and serials, 200,000 photographs, and over 55,000 advertising brochures, catalogues, and technical manuals. See note on p. 26.

L'Illustration, *Histoire de la Locomotion Terrestre* (Paris, 1936). Although I have not seen this book, I have seen many favorable comments about it.

Allan Nevins and Frank E. Hill, *Ford* (3 vols., New York: Scribner, 1957–1963). A comprehensive study of one automobile company; unfortunately weak on technical subjects.

John B. Rae, *The American Automobile* (Chicago: University of Chicago Press, 1965; 265 pp., illustrated). The jacket says, quite accurately, that this is "the first complete, authoritative treatment of the whole span of the automobile industry." Critical bibliography, pp. 249–253.

John B. Rae, *American Automobile Manufacturers: The First Forty Years* (Philadelphia: Chilton, 1959). A thoroughly sound work in technical, economic, and business history. Bibliography, pp. 209–212.

L. T. C. Rolt, *Horseless Carriage. The Motor-Car in England* (London, 1950) is a title I have not seen. Rolt's other books are technically sound.

"Automobile Tires; a List of References in the New York Public Library" (N.Y.P.L., *Bulletin,* 27 [Feb. 1923], 124–146). Also reprinted separately.

Arthur Du Cros, *Wheels of Fortune* (London: Chapman & Hall, 1938, 316 pp.). A history of rubber tires, with emphasis upon Dunlop.

On origins of the automobile in continental Europe, Lynwood Bryant has kindly selected for me a list of standard works on the subject. All of these books are indifferent so far as systematic bibliographies and careful documentation of sources are concerned, but all are based upon careful study of basic materials.

Herbert O. Duncan, *The World on Wheels* (Paris: the Author, 1926). A popular book on bicycles and automobiles, based, however, upon extensive study.

Jacques Ickx, *Ainsi naquit l'automobile* (2 vols., Lausanne: Edita S.A., 1961; 371, 376 pp.). A popular but generally reliable work. An exceptionally informative review by Lynwood Bryant of this book is in *Technology and Culture*, 9 (Apr. 1968), 237–239.

John Nerén, *Automobilens Historia* (2nd ed., Stockholm: Motor Byråns Förlag, 1949; 874 pp., illustrated). Proportionate coverage is given to the various nations, although Swedish contributions are forgivably emphasized.

Gerhard Schultz-Wittuhn, *Von Archimedes bis Mercedes; eine Geschichte des Kraftfahrzeuges bis 1900* (Frankfurt a.M.: Internationale Motor-Edition, 1952). Quite good on the automobile industry and associated developments in fuel and rubber.

Die Welt im Zeichen des Motors (Zürich: Metz Verlag, 1962). This collaborative work is the first volume of an illustrated series, "Forum der Teknik: Eine Rundschau über die wichtigsten Zweige der Technik als Beitrag zum Kulturgeschehen unserer Zeit."

5. *Railroads and Vehicles*

Bureau of Railway Economics (American Association of Railroads), *Railway Economics: a Collective Catalogue of Books in Fourteen American Libraries* (Chicago, 1912, 446 pp.), gives an idea of the secondary holdings of the remarkable library of the Bureau of Railway Economics in Washington, D.C.

Bureau of Railway Economics Library, *Railroad Bibliographies — A Trial Check-List* (Washington, 1938). A 71-page mimeographed list of about 500 titles. The bureau's library maintains a union catalogue of railroad items in other libraries and has done a great deal with the indexing, on cards, of periodical literature. See entry for this library in Reeves, *Resources for Economic History* (p. 32), pp. 2–3; also in *D.C. Library Reference Facilities* (p. 36), entry 38.

Daniel C. Haskell, *A Tentative Check-List of Early European Railway Literature 1831–1848* (Boston: Baker Library, 1955). Over 4,000 entries; locations of books are given when known.

Elisabeth C. Jackson and Carolyn Curtis, *Guide to the Burlington Archives* (Chicago: Newberry Library, 1949, 374 pp.). A model calendar of a very large body of primary materials in the library.

Carolyn Curtis Mohr, *Guide to the Illinois Central Archives* (Chicago: Newberry Library, 1951, 210 pp.). By one of the authors of the preceding title.

George Ottley, *A Bibliography of British Railway History* (London: Allen & Unwin, 1965, 683 pp.). A painstakingly constructed work, comprising nearly 8,000 entries, genealogical charts of systems and consolidations, and an index of over 200 pages. A companion volume, which will cover the periodical literature, is promised (p. 17 and item 7,895). Short and helpful annotations are given. Mr. Ottley is a librarian in the British Museum Library. A work of this kind brings into sharp focus a whole field of study; the absence of a similar work for U.S. railroads is thus more keenly felt.

The Railway and Locomotive Historical Society (Baker Library, Soldier's Field Rd., Boston, Mass.) had issued, through 1961, 105 numbers of its *Bulletin,* in addition to at least nine unnumbered publications. Most of the bulletins that I have seen have been the work of enthusiasts. A great deal of valuable material has thus been assembled. I have seen two indexes: *Bulletins 1–81* (1951, 16 pp.) and *Bulletins 82–101* (1960, 8 pp.), and a list of contents of *Bulletins 69–105* (1962, 6 pp.). Information will be found on railroad and locomotive builders, depots, locomotives, valve gear, railroad companies, and so forth. For example, *Bulletin 79* is by P. C. Dewhurst, on *The Norris Locomotives in the United States and Europe* (80 pp.).

Adequate general bibliographies for U.S. railroads are given in **Taylor,** *Transportation Revolution* (p. 13), pp. 407–410, and **Kirkland,** *Industry Comes of Age* (p. 13), pp. 411–416.

Thomas R. Thomson, *Check List of Publications on American Railroads before 1841* (New York: New York Public Library, 1942, 250 pp.). A union list of printed books and pamphlets, including state and federal documents, dealing with charters, legislation, speeches, debates, land grants, officers' and engineers' reports, travel guides, maps, and so forth.

Pierre-C. L. de Villedeuill, *Bibliographie des Chemins de Fer 1771–1837* (Paris, 1903, 240 pp.). I have not seen this bibliography, which I think is annotated.

John F. Due, *The Intercity Electric Railway Industry in Canada* (Toronto: University of Toronto Press, 1966; 118 pp., illustrated). Extends the earlier work by Hilton and Due (see following).

Cuthbert Hamilton Ellis, *British Railway History* (2 vols., London: Allen & Unwin, 1954–1959). Vol. 1, 1830–1876; Vol. 2, 1877–1947.

A comprehensive account of the entire British network of railways. While it is surpassed in detail by the many histories of individual companies, it is noteworthy for its breadth, balance, and insight. It is definitely the starting point for anyone who has not read in the subject. [Note by Hugh R. Gibb.]

Georges le Franc, "The French Railroads 1823–1842," *Journal of Economic and Business History,* 2 (1929–1930), 299–331, relates the experience in France, under central state planning, to that in England and the United States. Most of the footnotes refer to French archival materials.

George W. Hilton and John F. Due, *The Electric Interurban Railways in America* (Stanford: Stanford University Press, 1960, 463 pp.). A serious work in technical and economic history. Individual interurban lines are reviewed, pp. 254–423; there is also a list of carbuilders, pp. 424–425.

G. O. Holt, *A Short History of the Liverpool & Manchester Railway* (2nd ed., Caterham, Surrey: Railway & Canal Historical Society, 1965, 30 pp.). A substantial and original contribution, making use of material unavailable to earlier historians, according to a note by C. E. Lee in *The Newcomen Bulletin,* March, 1967.

Charles E. Lee, "Some Railway Facts and Fallacies," Newcomen Society, *Transactions,* 33 (1960–1961), 1–16. A summary statement on the history of railroads before 1800.

C. F. Dendy Marshall, *Centenary History of the Liverpool and Manchester Railway* (London: Locomotive Pub. Co., 1930; 4to, 192 pp., illustrated). The emphasis is upon origins and early history of this pioneer line. Included are reproductions, in color, of fifteen Ackermann lithographs issued upon opening of the railroad. This outstandingly attractive book is representative of several others written by Dendy Marshall. For his other titles, see **Ottley** (p. 237).

Richard C. Overton, *Burlington Route: A History of the Burlington Lines* (New York: Knopf, 1965; 663 pp., illustrated). There are many books on individual railroad companies in the United States, but this is probably the best of all. Bibliography, pp. 589–623. The author was instrumental in having important railroad papers preserved in the Newberry Library. See **Jackson and Curtis,** preceding.

Michael Robbins, *The Railway Age* (London: Routledge & Paul, 1962; Harmondsworth: Penguin, 1965; 191 pp.). A cultural history of railways, principally but not entirely in nineteenth-century Britain. "Notes on Sources" (pp. 164–176) contain a number of brief critical comments on a wide variety of printed works, including periodical articles.

Jack Simmons, *The Railways of Britain. An Historical Introduction* (London: Routledge & Paul, 1962; 264 pp., illustrated). The author, who is a historian and editor of *Journal of Transport History*, supplies in one chapter of this volume a critical guide to British railway literature (pp. 233–252), under headings of general works, individual companies, permanent way, rolling stock, biography, periodicals, management and organization, government records, maps, and relics. Very good.

John F. Stover, *American Railroads* (Chicago: University of Chicago Press, 1961; 302 pp., illustrated). The broad outlines are covered in this volume in The Chicago History of American Civilization series. Bibliography, pp. 272–281.

George R. Taylor and Irene D. Neu, *The American Railroad Network 1861–1890* (Cambridge, Mass.: Harvard University Press, 1956). The book is concerned principally with the development of a standard gauge and opposition to its establishment. Bibliographic notes, pp. 85–99.

J. Elfreth Watkins, "The Development of the American Rail and Track" (U.S.N.M., *Annual Report, 1888–1889*, pp. 651–708), is by far the best study that has appeared on the subject; but many questions remain unanswered.

J. N. Westwood, *A History of Russian Railways* (London: Allen & Unwin, 1964; 326 pp., illustrated). An account by a Canadian who has ridden on Russian railways. The nineteenth century, including Siberian railway building, occupies 128 pages; the rest of the book is devoted to the twentieth century. Footnotes and a short bibliography, pp. 318–319, refer in general to Russian works.

Ernest L. Ahrons, *The British Steam Railway Locomotive 1825–1925* (London: Locomotive Pub. Co., 1927; 391 pp., 473 illustrations). Reprint (London: Ian Allen, 1961). Rather technical but unquestionably the most accurate and complete locomotive history ever written in the English language. [Note by Hugh R. Gibb.]

E. P. Alexander, *Iron Horses: American Locomotives 1829–1900* (New York: W. W. Norton, 1941, 239 pp.). A book of pictures, which must be used critically. A wood engraving, published in 1880, of an 1830–1840 locomotive, is likely to be wildly in error. A scholarly history of locomotives is long overdue.

William H. Brown, *The History of the First Locomotive in America* (New York: Appleton, 1871, 242 pp.) is useful, though slanted.

Alfred W. Bruce, *The Steam Locomotive in America* (New York: Bonanza, 1952; 443 pp., illustrated). A mechanical history of locomotive development from 1900 to 1950. Written by an official of the American Locomotive Company, it is thorough yet not overly technical. [Note by Hugh R. Gibb.]

Cuthbert Hamilton Ellis, *Railway Carriages in the British Isles from 1830 to 1914* (London: Allen & Unwin, 1965; 279 pp., illustrated). Detailed descriptions and footnote references are given throughout the book.

August Mencken, *The Railroad Passenger Car: an Illustrated History of the First Hundred Years with Accounts by Contemporary Passengers* (Baltimore: The Johns Hopkins Press, 1957; 209 pp., illustrated) is disappointing but represents a first effort on its subject.

Angus Sinclair, *Development of the Locomotive Engine* (New York, 1907), perhaps the best history of early locomotives in the United States, is woefully inadequate.

J. G. H. Warren, *A Century of Locomotive Building by Robert Stephenson & Co. 1823–1923* (Newcastle upon Tyne: Reid, 1931; 4to, 461 pp.). Profusely illustrated. Despite its sponsorship, this is a nearly definitive work on the English locomotive before 1840. Many original drawings of locomotives, including some exported to the United States.

John H. White, *Cincinnati Locomotive Builders 1845–1868* (Washington: Smithsonian Institution, 1965 [U.S.N.M., *Bulletin 245*]; 167 pp., illustrated). A sound regional study which contains much original material. Reviewed by Carl Condit in *Technology and Culture,* 7 (Summer 1966), 428–429.

See also
Cassier's Magazine has a large number of descriptive articles on railways in Brazil, the Philippines, China, Egypt, and so forth. See the *Subject Index* (p. 157), pp. 29–30.

6. Aircraft and Spacecraft

Paul Brockett, *Bibliography of Aeronautics* (Washington: Smithsonian Institution, 1910, Misc. Coll., Vol. 55). A compilation of 13,500 titles from perhaps 150 journals from 1860 down to June 30, 1909. There are no annotations and the classified index is weak, but the work is indispensable.

National Advisory Committee on Aeronautics, *Bibliography of Aeronautics* (Washington: Government Printing Office, 1909/16, 1917/19,

1920/21, annually 1922–1932). This is a continuation of Brockett's bibliography.

William H. Gamble, Comp., *History of Aeronautics: a Selected List of References to Material in the New York Public Library* (New York: New York Public Library, 1938). Includes 5,574 annotated entries; also author and subject indexes. First published serially in New York Public Library, *Bulletin*, Vols. 40–41 (Jan. 1936–Sept. 1937).

For sources and information on repositories, see **James F. Sunderman,** "Documentary Collections related to the U.S. Air Force" (*Air University Quarterly Review*, Vol. 13, No. 1 [Summer 1961], about 10 pp.). Coverage is more extensive than title indicates. A review of location and nature of manuscript, pictorial, and published materials on U.S. aviation. Describes holdings of depositories in the Pentagon, Maxwell AFB, Air Force Museum, Wright-Patterson AFB, as well as resources of NASA, National Archives, Library of Congress, New York Public, and several other libraries. Addresses to which inquiries may be directed are also given. Quite worthwhile.

Griffith Brewer and Patrick Y. Alexander, *Aëronautics: An Abridgment of Aëronautical Specifications Filed at the Patent Office from A.D. 1815 to A.D. 1891* (Amsterdam: B. M. Israel N.V., 1965; 60 pp., illustrated). Reprint of the original London edition of 1893. A selected group of about 350 specifications, many with illustrations (sources not given). Published unofficially; not a number in the abridgment series of the Patent Office.

R. E. G. Davies, *A History of the World's Airlines* (London: Oxford University Press, 1964; 591 pp., illustrated). Widely accepted as the standard work of reference, according to Peter W. Brooks in *Journal of Transport History* (May 1965), pp. 60–61.

Eugene M. Emme, *Aeronautics and Astronautics. An American Chronology of Science and Technology in the Exploration of Space 1915–1960* (Washington: National Aeronautics and Space Administration, 1961, 240 pp.). A chronological list of events. Sources for the chronology are nowhere mentioned, nor is their nature indicated. Airplane records of various kinds are given as compiled by Fédération Aéronautique Internationale (pp. 154–160), and a list of balloon flights, 1927–1961, is given (pp. 162–165). Starting with 1961, the chronology of each year occupies a separate volume.

Spacecraft can be pursued in **Eugene M. Emme,** Ed., *The History of Rocket Technology* (Detroit: Wayne State University Press, 1964), which contains an extensive "bibliographic note" by Arthur G. Renstrom.

Charles H. Gibbs-Smith, *The Aeroplane: an Historical Survey of its Origins and Development* (London: Her Majesty's Stationery Office, 1960; 375 pp., illustrated). A thoroughly competent, well-reasoned work, truly remarkable in scope, which supersedes all other general histories that I have seen. Has a short bibliography, pp. 306–309.

Charles H. Gibbs-Smith, *Sir George Cayley's Aeronautics 1796–1855* (London: Her Majesty's Stationery Office, 1962; 269 pp., illustrated). A monographic sequel to J. Laurence Pritchard's *Sir George Cayley: the Inventor of the Aeroplane* (1961).

Leslie H. Hayward, *History of Air Cushion Vehicles* (London: Kalerghi-McLeavy, 1963, 45 pp.). On the genesis and development of "ground effect" machines. Based on patents and other sources. The principle can be recognized now in a 1716 design by Swedenborg.

Fred C. Kelly, *The Wright Brothers* (New York: Harcourt, Brace, 1943). An "authorized" biography, which shares the strengths and weaknesses of autobiography, for Wilbur Wright read the manuscript and suggested changes. However, the book is essentially accurate, is well written, reflects the point of view of the subjects, and suffers only from the absence of critical judgments.

Robert B. Meyer, Jr., "Three Famous Early Aero Engines" (Smithsonian Institution, *Annual Report for 1961* [1962], pp. 357–371), describes Clement (French), Manly-Balzer (U.S.), and Wright (U.S.) engines, all represented in the U.S. National Air Museum collections.

J. L. Nayler and E. Ower, *Aviation: Its Technical Development* (Philadelphia: Dufour Editions, 1965; 290 pp., illustrated). The book "records only events of technical significance, either in themselves or as representative of the state of progress at any given time" of the twentieth century. There are chapters on balloons, airships, light airplanes, fighters, large bombers, civil transport, seaplanes, piston engines, jet propulsion, vertical flight, guided missiles, rockets, navigation, and aeronautical science. Tables of chronology and a glossary are included.

L. T. C. Rolt, *The Aeronauts. A History of Ballooning 1783–1903* (London: Longmans, Green, 1966; 267 pp., illustrated). Rolt uses few footnotes, but further study may be pursued through the bibliography, pp. 255–257.

Robert Schlaifer and S. D. Heron, *Development of Aircraft Engines and Development of Aviation Fuels* (Boston: Graduate School of Business Administration, Harvard University, 1950, 754 pp.). Careful analysis of British, American, and German engine developments from about 1925.

Richard K. Smith, *The Airships Akron & Macon. Flying Aircraft Carriers of the United States Navy* (Annapolis: United States Naval Institute, 1965; 229 pp., illustrated). A detailed work on the last of the Navy's rigid airships, 1933–1935.

K. E. Tsiolkovskiy, *Collected Works* (Vols. 1–3, Washington: National Space and Aeronautics Administration, 1965). First published in Moscow, these works of Tsiolkovskiy, b. 1857, a Russian pioneer in jet aircraft and rocket development, are now available in English from the Clearinghouse for Federal Scientific and Technical Information. Vol. 1: Papers on aerodynamic theory, flight of birds and insects, and wind-tunnel tests, 1890–1902; Vol. 2: Reactive flying machines; Vol. 3: Dirigibles. Two more volumes may be expected.

Frank Whittle, "Early History of the Whittle Jet Propulsion Gas Turbine," Institution of Mechanical Engineers, *Proceedings*, 152 (1945), 419–435. A technical explanation, profusely illustrated. This supplements the authors popular, and very good, *Jet* (New York: Philosophical Library, 1954).

D. Energy Conversion

1. *General*

Eugene Ayres and Charles A. Scarlott, *Energy Sources — Wealth of the World* (New York: McGraw-Hill, 1952; 344 pp., illustrated). Undergirding this book is an historical sense that I fail to find in much of the work on this subject. The bibliography, pp. 328–338, includes sections on solar energy, photosynthesis, volcanic steam, atomic energy, heat pump, and so forth.

2. *Horsepower and Manpower*

The preoccupation with relationships between manpower and horsepower in the nineteenth century has not been adequately reported. A short and partial summary appears under "Animal Strength" in John Nicholson, *The Operative Mechanic, and British Machinist* (2nd American ed., Philadelphia, 1831), Vol. 1, pp. 52–61. I know of no summary article, even, on the establishment of "horsepower" as a power unit and the final acceptance of 33,000 ft.-lb. per minute (approximately 75 kg.-m. per sec.). **Farey**, *Steam Engine* (p. 247), pp. 437–440, relates a bit carelessly Watt's contribution.

Frank Atkinson, "The Horse as a Source of Rotary Power" (Newcomen Society, *Transactions*, 33 [1960–1961], 31–55), is a thorough survey of horse wheels, mills, and gins from classical times. There are nine text figures and fifteen photographs. Footnote references.

Thomas H. Brigg, "Haulage by Horses," American Society of Mechanical Engineers, *Transactions, 14* (1893), 1,014–1,066. A "scientific" approach is taken to horse-drawn vehicle management.

Albert C. Leighton, "The Mule as a Cultural Invention," *Technology and Culture, 8* (Jan. 1967), 45–52. The author throws a fresh light on the nature of technological innovation.

3. Waterwheels, Hydro-power, Windmills, and Pumps

Edward Dean Adams, *Niagara Power: History of the Niagara Falls Power Company 1886–1918* (2 vols., Niagara Falls, 1927). An exhaustive account of the exploitation of the Niagara Falls.

James B. Francis, *Lowell Hydraulic Experiments* (New York, 1855; 4th ed., 1883) reports the classic developmental work that led to the Francis turbine.

Louis C. Hunter, "Origines des Turbines Francis et Pelton," *Revue d'Histoire des Sciences et de leurs Applications, 17,* No. 3 (July–Sept. 1964), 209–242. A major review of the development of the hydraulic turbine in the United States, giving adequate attention to the transmission and diffusion of information on the French turbines of Fourneyron, Jonval, and so forth. Also treats development of the Pelton wheel in the Far West. Footnote references.

Anders Jesperson, *The Lady Isabella Waterwheel of the Great Laxey Mining Company, Isle of Man 1854–1954* (Virum, Denmark: the Author, 1954) is a thorough study of one great wheel.

A. Stowers, "Observations on the History of Water Power," Newcomen Society, *Transactions, 30* (1955–1957), 239–256. This brief introduction to water mills has useful footnote references and a bibliography of twenty-nine titles.

Rex Wailes, "Some Windmill Fallacies" (Newcomen Society, *Transactions, 32* [1959–1960], 93–109), lists twenty-four books on windmills, including two of his own. Many other papers on windmills in the *Transactions* can be located through the general index, 1920–1960.

Thomas Ewbank, *A Descriptive and Historical Account of Hydraulic and Other Machines for Raising Water* (2nd ed., New York: Greeley

& McElrath, 1847; 608 pp., 290 illustrations). A discursive, sometimes tedious, account of all kinds of devices. Ewbank had read everything in his field and much else besides. Thus a reader ignores Ewbank at his own peril. Sources can be determined from text references.

Arthur M. Greene, Jr., *Pumping Machinery* (2nd ed., New York: Wiley, 1919). Chapter 1, pp. 1–55, reviews early history of pumping, to about 1840; Chapter 2, pp. 56–129, brings the story to about 1909. References to recent articles in the technical press are given on pp. 679–691.

See also
For dams, see Part XIV.B.4.

4. Internal-Combustion Engines

Although it is difficult to point to any other technical device that has so drastically changed the quality of life in the West, I know of no recent book in English on the general development of the internal-combustion engine.

Lynwood Bryant, "The Silent Otto," *Technology and Culture,* 7 (Spring 1966), 184–200. This paper and the one following are more informative and informed on the subject than anything else I know of in English. The footnotes are bibliographic in scope.

Lynwood Bryant, "The Origin of the Four-Stroke Cycle," *Technology and Culture,* 8 (Apr. 1967), 178–198.

Dugald Clerk, *The Gas, Petrol and Oil Engine* (rev. ed., New York: Wiley, 1909; 380 pp., illustrated) contains a historical sketch, pp. 1–50, colored but not spoiled by the author's deep involvement in engine development.

Eugen Diesel, Gustav Goldbeck, and Friedrich Schildberger, *From Engines to Autos: Five Pioneers in Engine Development and their Contributions to the Automotive Industry* (Chicago: Regnery, 1960; 302 pp., illustrated). Translated from the German by Peter White. A popularized series of historical sketches of Otto, Daimler, Benz, Diesel, and Bosch; perhaps useful, but to be used with caution. No references; no bibliography; no index.

Bryan Donkin, Jr., *A Text-Book on Gas, Oil, and Air Engines* (London: Charles Griffin, 1894; 419 pp., illustrated). More wide ranging and better illustrated than Clerk, just listed; on the other hand, Donkin's technical judgments are less reliable than Clerk's.

Gustav Goldbeck, *Siegfried Marcus, Ein Erfinderleben* (Düsseldorf: VDI-Verlag, 1961; 88 pp., illustrated). The only reliable biography of an important pioneer in the development of a satisfactory engine for automobiles.

Hugo Güldner, *Das Entwerfen und Berechnen der Verbrennungsmotoren. Handbuch für Konstrukteure und Erbauer von Gas- und Ölkraftmaschinen* (2nd ed., Berlin: Springer Verlag, 1905). The American edition, translated by H. Diedrichs, *The Design and Construction of Internal-Combustion Engines* (New York: Van Nostrand, 1910; 4to, 672 pp., heavily illustrated), omits "the entire first part treating of the history of the gas engine." Information on current American engines has been added in its place.

Friedrich Sass, *Geschichte des deutschen Verbrennungsmotorenbaues* (Berlin: Springer Verlag, 1962; 4to, 667 pp., illustrated). This book contains detailed technical explanations, good figures, photographs of engines, and bibliographic notes. The work is fully and informatively reviewed in *Technology and Culture,* 5 (Winter 1964), 82–86, by Lynwood Bryant, who calls it an "ideal monograph" and a definitive account of the development of the internal-combustion engine in Germany from 1860 to 1920.

American Society of Mechanical Engineers, Gas Turbine Division, *Bibliography on Gas Turbines, 1896–1948* (New York, 1962). Nearly 2,000 titles, mostly articles, are described briefly. Subject index is arranged chronologically within each classification.

R. Tom Sawyer, *The Modern Gas Turbine* (New York: Prentice-Hall, 1945). The short review (pp. 15–41) of gas turbine history, 1791–1926, while far from definitive, will furnish most readers with information they have not encountered before. Bibliography, pp. 199–202.

Henry H. Suplee, *The Gas Turbine* (Philadelphia: Lippincott, 1910; 262 pp., illustrated). The author has brought together "such theoretical and practical data as are now available in the solution of the problem of the gas turbine." An early technical work on a promising but impractical engine.

C. Gordon Peattie, "A Summary of Practical Fuel Cell Technology to 1963," Institute of Electrical and Electronics Engineers, *Proceedings,* 52 (May 1963), 795–806. A brief survey of fuel cell systems and a bibliography of seventy-four articles, 1842–1963.

See also
Some aspects of the internal-combustion engine are treated in works on the automobile, Part XIV.C.4, and on the airplane, Part XIV.C.6.

5. Steam Engines, Turbines, and Boilers

Curiously, there is no comprehensive bibliography of works on the steam engine. Considering the continuing intense interest in the subject, I should expect that such a bibliography would become a standard work immediately, much appreciated, with an enviable life expectancy.

George H. Babcock, "Substitutes for Steam," American Society of Mechanical Engineers, *Transactions*, 7 (1885–1886), 680–741. This is an excellent review of attempts to use naphtha, ammonia, and other fluids in a vapor cycle in order to improve its economy. Babcock was a founder of the present firm of Babcock & Wilcox.

D. B. Barton, *The Cornish Beam Engine* (Truro, Cornwall: Bradford Barton, 1966; 286 pp., illustrated). A survey of the history and development, in Cornwall and Devon from 1800 to the present, of a notable type of medium-pressure, condensing engine.

D. S. L. Cardwell, *Steam Power in the Eighteenth Century. A Case Study in the Application of Science* (London and New York: Sheed and Ward, 1963, 102 pp.). A critical study of the elements of innovation in the Newcomen and Watt engines. A number of points ordinarily ignored or glossed over are examined in some detail, as for example the eighteenth-century confusion of heat conduction and heat capacity.

Henry W. Dickinson, *A Short History of the Steam Engine* (London: Cass, 1963; 255 pp., illustrated). This work is still the best-informed general history in English. The present edition is a reprint of the 1938 (first) edition, to which has been added an eighteen-page introduction by A. E. Musson. Works since 1938 are noted in the introduction.

Henry W. Dickinson and Rhys Jenkins, *James Watt and the Steam Engine* (Oxford: Clarendon, 1927; 4to, 415 pp., 39 text illustrations, 104 pls.), is a massive and exhaustive study of the Watt engine, based on Boulton and Watt papers. Bibliography, pp. 359–372.

Many illustrated histories of the steam engine were published around 1820–1850; one of the best is **John Farey,** *Treatise on the Steam Engine* (London, 1827; 4to, 728 pp. + 25 pls.). Other titles are listed in **Williams,** *Guide* (p. 25), Vol. 2, pp. 1–35, *passim*.

Arthur M. Greene, Jr., *History of the ASME Boiler Code* (New York: American Society of Mechanical Engineers, 1953) describes one of the standards-setting activities of U.S. engineering societies. Most of the book appeared in *Mechanical Engineering* in 1952–1953.

House Document 21, *Report on the Steam Engines in the United States* (25 Cong., 3 sess., dated Dec. 12, 1838 [Document series No. 345]). This 472-page compilation brings together much quantitative information on engines and boilers, reports of explosions, and so forth.

Milton Kerker, "Science and the Steam Engine," *Technology and Culture,* 2 (Fall 1961), 381–390. The author shows that the steam engine was not built by merely practical men, ignorant of science. It was this article that brought home to me the way in which seventeenth-century science was an indispensable prerequisite to the steam engine.

Conrad Matschoss, *Die Entwicklung der Dampfmaschine* (2 vols., Berlin: Springer, 1908) is a detailed history of engines and boilers and their application. Profusely and clearly illustrated.

A series of papers on U.S. turbine development, by Westinghouse, Allis-Chalmers, and General Electric authors, appeared in **Mechanical Engineering,** 58 (1936), 683–696, and 59 (1937), 71–82, 239–256.

A. E. Musson and E. Robinson, "Early Growth of Steam Power," *Economic History Review,* 2nd ser., *11* (1959), 418–439. This is an important example of a number of careful articles re-evaluating traditional beliefs about the development of the steam engine.

Joseph Needham, "The Pre-Natal History of the Steam-Engine," Newcomen Society, *Transactions,* 35 (1962–1963), 3–58. In another of his tours de force, Needham has pointed to origins of steam-engine elements in the Near East and China. Extensive bibliography, pp. 51–55.

Charles A. Parsons, *The Development of the Parsons Steam Turbine* (London, 1936, 420 pp.) first appeared as a series of articles in *The Engineer,* 1934–1935.

L. T. C. Rolt, *Thomas Newcomen* (Dawlish, Devon: David & Charles, 1963; 158 pp., illustrated) has brought together essentially all the pertinent researches on the Newcomen engine. A short bibliography, pp. 143–145, is thoroughly discriminating.

Robert H. Thurston, *Steam-Boiler Explosions in Theory and in Practice* (New York: Wiley, 1887, 173 pp.). This account of the hazards of boiler explosions and ways to prevent them is only incidentally historical in view. The account appears also as Chapter 15 of the author's *A Manual of Steam-Boilers* (New York: Wiley, 1888).

See also
Ladislao Reti on Leonardo's contributions to the steam engine (p. 50).

6. Atomic Energy

L. J. Anthony, *Sources of Information on Atomic Energy* (Oxford: Pergamon, 1966, 245 pp.). This is a highly literate introduction to sources of information on atomic energy in general and on special aspects of it. Summary statements on the various branches of the field are followed by rich supporting bibliographies. For example, a chapter on "Nuclear Power and Engineering," pp. 157–176, considers energy resources in general, economic aspects of nuclear power, technical and scientific information on reactors, reactor safety, and nuclear fuels and the attendant problems of radioactive wastes.

7. Refrigeration and Heating

Oscar E. Anderson, *Refrigeration in America: a History of a New Technology and its Impact* (Princeton: University Press, 1953, 344 pp.). The bibliographical note, pp. 321–325, is supplemented by full footnotes throughout the text. Author's approach is cultural, but references to technical materials are adequate.

Richard O. Cummings, *The American Ice Harvests. A Historical Study in Technology, 1800–1918* (Berkeley: University of California Press, 1949; 184 pp., illustrated). Although this book is disappointingly brief, the various aspects of the ice trade — Frederic Tudor's overseas shipment of American ice, ice supplies for cities, and the frozen-meat trade — all can be entered through this book's reference notes and short bibliography.

A. F. Dufton, "Early Application of Engineering to the Warming of Buildings" (Newcomen Society, *Transactions*, 21 [1940–1941], 99–107), lists a dozen book titles, including the principal English works of the nineteenth century.

Margaret Ingels, *Willis Haviland Carrier, Father of Air Conditioning* (Garden City: Country Life, 1952, 170 pp.). A brief (100 pp.) life of Carrier (1876–1950) is followed by a "Chronological Table of Events Which Led to Modern Air Conditioning, 1500–1952," pp. 110–162. References are given for the events listed; there is also a list of Carrier's publications, pp. 162–170.

A review of English work is in **Edgar C. Smith**, "Some Pioneers of Refrigeration" (Newcomen Society, *Transactions*, 23 [1942–1943], 99–107). Bibliographic in scope.

8. *Lighting*

Walter Hough, *Collection of Heating and Lighting Utensils in the United States National Museum* (Washington: Smithsonian Institution, 1928 [U.S.N.M., *Bulletin 141*]; 113 pp. + 99 pls.). A catalogue of torches, candles, lamps of all ages, candlesticks, chandeliers, lanterns, snuffers, Argand lamps, flat-wick lamps, gas lamps, footwarmers, stoves, blowers, bellows, spits, sadirons, ovens, and so forth.

"List of Works in the New York Public Library Relating to Illumination" (N.Y.P.L., *Bulletin, 12* [Dec. 1908], 686–734). Also separately reprinted. Very rich on all kinds of lighting.

William T. O'Dea, *A Social History of Lighting* (London: Routledge & Paul, 1958; New York: Macmillan, 1959, 254 pp.). A meticulous, well-illustrated summary of lighting for the home, for travel, for work, for great occasions, and for lighthouses. References in text. The author is a keeper in Science Museum, London.

F. W. Robins, *The Story of the Lamp (and the Candle)* (London: Oxford University Press, 1939; 155 pp., 27 pls.). Treats all times and places. References are made to collections in some seventy-five museums. Very short bibliography.

Leroy Thwing, *Flickering Flames: A History of Domestic Lighting through the Ages* (Rutland, Vt.: Tuttle, 1958, 138 pp.). A useful work despite absence of scholarly apparatus.

See also
For electric lighting, see Part XIV.E.2.
For fuels, see Part XIV.F.6, Coal, Oil, and Gas.

E. Electrical and Electronic Arts

1. *General*

Thomas J. Higgins, *A Classified Bibliography of Publications on the History and Development of Electrical Engineering and Electrophysics,* (University of Wisconsin, Engineering Experiment Station, Reprint No. 198, 26 pp.). Reprinted from *The Bulletin of Bibliography, 20,* Nos. 3–7 (Sept.–Dec. 1950 to Jan.–Apr. 1962). A total of over 1,000 titles; not annotated.

W. James King, "The Development of Electrical Technology in the 19th Century" (U.S.N.M., *Bulletin 228,* Papers 28–30 [1961–1962], 231–407). Paper 28: The Electrochemical Cell and the Electromagnet;

Paper 29: The Telegraph and the Telephone; Paper 30: The Early Arc Light and Generator. These well-illustrated papers trace the "hardware history" of successive steps in the development of equipment.

"List of Works on Electricity in the New York Public Library" (N.Y.P.L., *Bulletin*, Vols. 6–7 [Nov. 1902–Jan. 1903]). Also separately reprinted. There are 125 "history" references.

Katharine Maynard, Ed., *A Bibliography of Bibliographies in Electrical Engineering 1918–1929* (Providence, R.I.: Special Libraries Association, 1931, 56 pp.). The book is valuable enough, but the title is misleading. More than 2,500 journal articles, not historical, in English, German, and French, appearing in the years noted, were selected for their bibliographic interest. Individual articles yield from five to several hundred references, either as separate or footnote citations. Classification is by subject, covering the whole field of generation, distribution, and electrical and electronic applications of all kinds. Useful for state-of-the-art articles that also provide further references.

Paul F. Mottelay, *Bibliographic History of Electricity and Magnetism, Chronologically Arranged* (London, 1922, 673 pp.). A systematic analysis of work in the field through 1821.

National Electrical Manufacturers Association, *A Chronological History of Electrical Development from 600 B.C.* (New York: National Electrical Manufacturers Assn., 1946, 150 pp.). In common with many other chronologies, this one is slight but not entirely negligible. Appendix (23 pp.) lists member companies of N.E.M.A. with founding dates, names of founders, and original names of companies.

Francis Ronalds, *Catalogue of Books and Papers Relating to Electricity, Magnetism, the Electric Telegraph, Etc., Including the Ronalds Library*, Alfred J. Frost, Ed. (London, 1880, 564 pp.). The emphasis is upon works on electrical science, before 1873.

Harold Sharlin, *Making of the Electrical Age: From the Telegraph to Automation* (New York: Abelard-Schuman, 1963; 248 pp., illustrated). A general view of electrical technology in the nineteenth century, including chapters on telegraph, telephone, central power station, electric motor, and "automation." Footnote references.

William D. Weaver, *Catalogue of the Wheeler Gift of Books, Pamphlets, and Periodicals to the Library of the American Institute of Electrical Engineers* (2 vols., New York: American Institute of Electrical Engineers, 1909). About 6,000 items, heavily technical and industrial, including trade catalogues and pamphlets. Carefully annotated. Very good.

2. *Power Generation and Transmission, Motors, and Lighting*

On long-distance power transmission, see **Adams**, *Niagara Power* (p. 244). See also a series of papers in *Cassier's Magazine:* 8 (1895), 333–362; 9 (1896), 219–228, 359–374; 20 (1901), 3–20.

"Bibliography of Rotating Electric Machinery for 1948–1961," Institute of Electrical and Electronic Engineers, *Transactions on Power Apparatus and Systems*, 83, No. 6 (June 1964), 589–606. Based upon eight U.S. serials, one Russian, and one British, this is a list of 920 titles. It is a supplement to *Bibliography of Rotating Electric Machinery, 1886–1947* (American Institute of Electrical Engineers, Publication S–32), which I have not seen.

Arthur A. Bright, Jr., *The Electric-Lamp Industry: Technological Change and Economic Development from 1800 to 1947* (New York: Macmillan, 1949, 526 pp.). This comprehensive work is accurately described by its title. Bibliography, pp. 507–512.

Bern Dibner, *Early Electrical Machines* (Norwalk, Conn.: Burndy Library, 1959; 4to, 57 pp., illustrated). On machines of the seventeenth and eighteenth centuries. A short but discriminating bibliography is annotated.

Mel Gorman, "Charles F. Brush and the First Public Electric Street Lighting System in America," *The Ohio Historical Quarterly*, 70, No. 2 (Apr. 1961), 128–144. Brush (1849–1929) lighted Cleveland's public square in 1879.

Thomas P. Hughes, "British Electrical Industry Lag: 1882–1888" (*Technology and Culture*, 3 [Winter 1962], 27–44), is a comparative study whose notes are bibliographic in scope.

Malcolm MacLaren, *Rise of the Electrical Industry during the Nineteenth Century* (Princeton: Princeton University Press, 1943). Bibliographic notes, pp. 199–218. A factual, not very critical, survey by an electrical engineer who was active from the 1890's.

Joseph C. Michalowicz, "Origin of the Electric Motor," *Electrical Engineering*, 67 (Nov. 1948), 1,035–1,040. The author refers to an earlier series of articles by Franklin L. Pope in *Electrical Engineer*, 11 (1891), 1–5, 33–39, 65–71, 93–98, 125–130.

SUBJECT FIELDS 253

R. H. Parsons, *The Early Days of the Power Station Industry* (Cambridge, 1939, 217 pp.). Traces the development of central stations in London. There is a good chapter on the battle of the systems (alternating vs. direct current).

Harold C. Passer, *The Electrical Manufacturers 1875–1900* (Cambridge, Mass.: Harvard University Press, 1953; 412 pp., illustrated). A first-rate study of the rise of U.S. firms from about 1880 and of their response to innovations and needs in light, power, and railway traction. Bibliography, pp. 391–400.

Harry J. White, *Industrial Electrostatic Precipitation* (Reading, Mass.: Addison-Wesley, 1963). The introductory chapter, pp. 1–32, traces development of precipitators from about 1885. List of twenty-odd U.S. patents, 1886–1954, and forty-three references.

See also
For competing systems of lighting, see Part XIV.D.8, Lighting.

3. *Communications and Electronic Arts, Including Electronic Computers*

C. K. Moore and K. J. Spencer, *Electronics: A Bibliographical Guide* (2 vols., New York: [Vol. 1] Macmillan and [Vol. 2] Plenum Press, 1961–1965, 411, 369 pp.). The first volume covers literature of the years 1949–1959 and the second, 1959–1964, plus a number of historical works and bibliographies of earlier dates. The work is intended for scientists and engineers who require a technical review of or entry into specialized aspects of the field of electronics (radio, computer, automatic control, scientific instrumentation, biomedical, and many, many other applications of electronic tubes and solid-state devices). Although marginal for anyone but a historian of electronic devices, the entries are fully, intelligently, and helpfully annotated.

S. J. Begun, *Magnetic Recording* (New York: Murray Hill, 1949; 242 pp., illustrated). Chapter 1 is historical; includes forty-six references and ten patents.

G. G. Blake, *History of Radio Telegraphy and Telephony* (London, 1928, 425 pp.), has in the bibliography 1,125 references, many of which are of U.S., U.K., and German patents.

Bern Dibner, *The Atlantic Cable* (Norwalk, Conn.: Burndy Library, 1959; 4to, 95 pp., illustrated). A handsomely illustrated account of the laying of the cable, 1857–1866. List of sixty-one books on the subject in the Burndy Library, pp. 94–95.

J. J. Fahie, *A History of Electric Telegraphy to 1837* (London, 1884). See next entry also.

J. J. Fahie, *A History of Wireless Telegraphy* (New York, 1899, 325 pp.). This book and the preceding one are chronological accounts, heavy with quotations from papers and books. The author was an English electrical engineer.

Frederick V. Hunt, *Electroacoustics: the Analysis of Transduction, and its Historical Background* (Cambridge, Mass., 1954, 91 pp.). A history of the conversion of sound to electrical energy and back again. Footnote references.

Lawrence Lessing, *Man of High Fidelity: Edwin Howard Armstrong* (Philadelphia: Lippincott, 1956). A popular but sound biography of a pioneer in FM radio.

William R. Maclaurin, *Invention and Innovation in the Radio Industry* (New York: Macmillan, 1949) is the standard work.

E. A. Marland, *Early Electrical Communications* (London and New York: Abelard-Schuman, 1964; 220 pp., illustrated). Telegraphy, pp. 11–152; submarine telegraphy, pp. 153–181; telephone, pp. 182–200; bibliography, pp. 209–214. The author was interested primarily in equipment and techniques.

J. Savin, "L'effet Laser dans les solides," *Revue d'histoire des sciences*, 16 (Oct.–Dec. 1963), 359–372. A review of laser development since 1954. Appended is an annotated list of fourteen key papers, 1914–1961.

R. Serrel and others, "The Evolution of Computing Machines and Systems," Institute of Radio Engineers, *Proceedings*, 50 (May 1962), 1,039–1,058. A review of the Harvard Mark I (1944) and later computers, with a bibliography, 1946–1961. No history of computers, starting with Pascal and bringing the story to electronic analog and digital machines, has yet been published.

Thomas Shaw, "The Conquest of Distance by Wire Telephony," *Bell System Technical Journal*, 23, No. 4 (Oct. 1944), 377–421. On the development of long-distance telephone service in the United States, 1904–1940.

Robert L. Thompson, *Wiring a Continent: the History of the Telegraph Industry in the United States 1832–1866* (Princeton: Princeton University Press, 1947; 544 pp., illustrated). Based upon the study of very extensive sources, which are detailed in footnotes and in the bibliography, pp. 518–526.

Carmen Wilson, *Magnetic Recording 1900–1949* (Chicago: The John Crerar Library, 1950). There are 339 fully annotated references in this "Bibliography No. 1" of The John Crerar Library. Unfortunately, No. 2 (1953) on abstracting and indexing sources on metals fabrication was the last of the series.

F. Materials and Processes

1. *General*

Henry Hodges, *Artifacts: An Introduction to Primitive Technology* (New York: Praeger, 1964; 248 pp., illustrated). A review of many materials — pottery, glass, metals, textiles, leather, and so forth — used by early man and methods of working them. This is a knowledgeable handbook on how to look at objects in order to extract from them a maximum of information and insights. Reviewed in *Antiquity*, 38 (Dec. 1964), 315–316.

A. Lucas and J. R. Harris, *Ancient Egyptian Materials and Industries* (4th ed., London: Arnold, 1962, 523 pp.). This work has been a standard one since its first modest appearance in 1926. Detailed discussion of topics; extensive bibliographic footnotes; adequate indexes. Indispensable to a scholar's bibliographic control of early technology.

L. F. Salzmann, *English Industries of the Middle Ages* (London: Constable, 1913, 260 pp.). Chapters on coal, iron, lead and silver, tin, quarrying, metalworking, pottery, tile, bricks, textiles, leatherworking, brewing.

George Sarton, "Charles Fremont, Historien de la technologie (1855–1930)," *Isis*, 27 (Nov. 1937), 475–484. The article includes a bibliography of eighty-two titles, concerned with the history of tools, materials, processes, testing and failure of materials, and so forth. Fremont wrote on copper founding (360 pp.), the nail (143 pp.), evolution of tools (157 pp., 41 pp.), the file (176 pp.), the pulley, windlass (72 pp.), the bellows (98 pp.), the lock (51 pp.). Many titles, generally short, on such subjects as the testing and failures of metals, rails, boilers. There is also a history of testing materials of construction (104 pp.).

Cyril Stanley Smith, "Materials and the Development of Civilization and Science," *Science*, 148 (May 14, 1965), 908–917, 21 refs. The author recognizes that materials have fundamentally determined what man could do at every stage of history, but he goes a big step further: he also recognizes that most of the decisions as to what materials could or should be used at any period have rested upon aesthetic qualities as well as empirical utility.

Mary E. Weeks, *Discovery of the Elements* (6th ed., Easton, Pa.: Journal of Chemical Education, 1956; 910 pp., illustrated). A well-constructed, in some ways elegant, handbook that gives essential information for each element. Extensive references. A new edition, extensively revised by Henry M. Leicester, was published in 1968.

2. Manufacturing in General

J. Leander Bishop, *A History of American Manufactures from 1608 to 1860* (3rd ed., 3 vols., Philadelphia, 1868). A valuable compendium of organizational and biographical information; little on techniques.

Albert S. Bolles, *The Industrial History of the United States* (3rd ed., Norwich, Conn., 1878; 936 pp., illustrated). Sections on agriculture, manufacturing, shipping and railroads, mining and oil, banking and commerce, and trade unions.

Victor S. Clark, *History of Manufactures in the United States* (3 vols., New York: McGraw-Hill, 1929). Reprint (3 vols., New York: Peter Smith, 1949). Vol. 1, 1607–1860; Vol. 2, 1860–1893; Vol. 3, 1893–1928. A standard economic history of great value. The bibliographic essay, Vol. 3, pp. 400–407, and the bibliography, pp. 407–442, are thorough and informative.

M. de Moléon, Ed., *Musée Industrielle; . . . produits de l'Industrie française fait en 1834* (4 vols., Paris, 1838). I have not seen this work.

See also
Turgan, *Les grandes usines de France* (p. 126)
Eighty Years' Progress of the United States (p. 64)
Greeley, *Great Industries* (p. 64)

3. Metals: Mining, Metallurgy, and Metallography

Leslie Aitchison, *A History of Metals* (2 vols., London: Macdonald & Evans, 1960, 4to). This is the best general work on the winning and working of metals. All metals, all ages. First rate in every way, spaciously literate, judicious in reaching conclusions. Illustrated, with reference notes, and an extensive bibliography, pp. 622–627.

Cyril Stanley Smith, *A History of Metallography* (Chicago: University of Chicago Press, 1960; 291 pp., illustrated). An authoritative and handsome book. Bibliographic notes, pp. 261–280.

Cyril Stanley Smith, Ed., *The Sorby Centennial Symposium on the History of Metallurgy* (New York: Gordon and Breach, 1965, 580 pp.). The papers comprising this volume were prepared for the 1963 symposium. Reviewed in detail by John G. Burke in *Technology and Culture*, 8 (Jan. 1967), 82–86.

D. C. Barton, *A Historical Survey of the Mines and Mineral Railways of East Cornwall and West Devon* (Truro, Cornwall: Truro Bookshop, 1964; 102 pp., illustrated). A detailed, nicely made history of this little-known mining area whose peak of exploitation was reached in the latter half of the nineteenth century. Illustrated with attractive drawings and maps. This is one of a fresh and interesting group of books being published by Barton in his Truro Bookshop. One who is interested in this region should obtain this publisher's list.

Walter Cline, *Mining and Metallurgy in Negro Africa* (Menasha, Wis.: Banta, 1937; 155 pp., illustrated). This work, which consists of descriptive and comparative studies of actual techniques and associated ritual and beliefs, will introduce a reader to the literature of the subject through some three hundred references.

A note on recent work in Africa is in **Hamo Sassoon**, "Iron-Smelting in the Hill Village of Sukur, North-Eastern Nigeria," *Man*, 64 (Nov.–Dec. 1964), 174–178.

Walter R. Crane, *Index of Mining Engineering Literature, Comprising an Index of Mining, Metallurgical, Civil, Mechanical, Electrical and Chemical Engineering Subjects as Related to Mining Engineering* (2 vols., New York: Wiley, 1909–1912). I have not seen this title, which was primarily concerned with periodical literature from about 1880. See entry EI168 in **Winchell**, *Guide to Reference Books* (p. 33).

Olivier Davies, *Roman Mines in Europe* (Oxford: Clarendon, 1935, 291 pp.). A standard work which I have often seen cited, but I have not seen the book.

E. W. Davis, *Pioneering with Taconite* (St. Paul: Minnesota Historical Society, 1964; 246 pp., illustrated). The author, who conceived, promoted, and developed the present taconite (medium-grade iron ore) industry, has been obsessed with taconite for fifty years. Much quantitative information; good illustrations.

Ernst Darmstaedter, *Berg-, Probier-, und Kunstbüchlein* (München: Münchner Drucke, 1926; 109 pp., illustrated). A bibliography of the early literature of the mineral industries.

A. K. Hamilton Jenkin, *Mines and Miners of Cornwall* (10 pts., Truro, Cornwall: Truro Bookshop, 1961–1965). Each part, 50–75 pages, supplies detailed information, including illustrations and maps, on the mining of the several metals found in Cornwall.

For a bibliography of mining in the sixteenth through nineteenth centuries, see Koch, *Bergmännischen Schrifttums* (p. 42).

David Lavender, *The Story of the Cyprus Mines Corporation* (San Marino, Calif.: Huntington Library, 1962; 387 pp., illustrated) tells of modern reopening of ancient copper mines in Cyprus. Interesting.

Thomas A. Rickard, *Man and Metals: a History of Mining in Relation to the Development of Civilization* (2 vols., New York: McGraw-Hill, 1932). A long work by an intelligent journalist in the metals field, sometime president of A.I.M.E. A list of references is at the end of each chapter.

Robert E. Stewart, Jr., *Adolph Sutro* (Berkeley: Howell-North, 1962, 243 pp.) tells of the building of the four-mile Sutro drainage tunnel of the Comstock Lode, during the 1870's.

Theodore Wertime, "Man's First Encounters with Metallurgy," *Science, 146* (Dec. 4, 1964); 1,257–1,267, 41 refs. The latest and I think best opinions on the origins of metals in the Near East.

T. S. Ashton, *Iron and Steel in the Industrial Revolution* (3rd ed., Manchester: Manchester University Press, 1963, 265 pp.). Bibliographic footnotes. This edition is a reprint of the second edition (1951), to which has been added a chapter by W. H. Chaloner, "Bibliographical Note to the 1963 Edition."

Ludwig Beck, *Geschichte des Eisens in seiner technischen und kulturgeschichtlichen Beziehung* (5 vols., Brunswick: Vieweg & Sohn, 1891–1903). A monumental work; contains valuable biographical and bibliographical sketches.

Arthur C. Bining, *Pennsylvania Iron Manufacture in the Eighteenth Century* (Harrisburg, Pa.: The Pennsylvania Historical Commission, 1938; 227 pp., illustrated), is thoroughly sound. The author was an ironmaker before he became a distinguished historian. Bibliography, pp. 197–214.

Harry Brearley, *Steel-Makers* (London: Longmans, Green, 1933, 156 pp.). A finely wrought collective portrait of the crucible steelmakers of Sheffield of around 1900, written by the English inventor of stainless steel. The author writes of the men in the works; managers, if they are mentioned, are upstage. The processes are described in detail, and pride of craftsmanship and intuitive sense of fitness are brought to life.

[Carnegie Library, Pittsburgh]. Esther Cheshire, *Review of Iron and Steel Literature for 1964* (Pittsburgh: Carnegie Library, 1965, 74 pp.). A reprint, in pamphlet form, of the forty-eighth annual review compiled for and appearing in *Blast Furnace and Steel Plant*. About ten historical works were listed in the 1964 review.

J. C. Carr and W. Taplin, *History of the British Steel Industry* (Cambridge, Mass.: Harvard University Press, 1952, 632 pp.). The industry before Bessemer is reviewed on pp. 1–16; the first major innovations, 1856–1875, on pp. 19–90. Most of the book is concerned with the twentieth century.

Charles B. Dew, *Ironmaker to the Confederacy: Joseph R. Anderson and the Tredegar Iron Works* (New Haven: Yale University Press, 1966; 345 pp., illustrated). The Tredegar works, founded in 1837, was the principal iron works available to the South in the U.S. Civil War. This book, which is concerned with the period 1859–1867, is based upon manuscript records of the company, now in the Virginia State Library. "Bibliographical Essay," pp. 327–334.

E. N. Hartley, *Ironworks on the Saugus* (Norman: University of Oklahoma Press, 1957; 328 pp., illustrated). Tells of the reconstruction of a seventeenth-century Massachusetts furnace, forge, and mill.

Helen Hughes, *The Australian Iron and Steel Industry 1848–1962* (Melbourne: Melbourne University Press, 1964, 213 pp.). The period before 1911 is treated in about fifty pages.

Louis C. Hunter, "Influence of the Market upon Technique in the Iron Industry in Western Pennsylvania up to 1860," *Journal of Economic and Business History*, 1 (1928–1929), 241–281. An absence of demand, not deficient techniques, delayed the establishment of Pittsburgh's coke-based iron industry until 1860.

Otto Johannsen, *Die Geschichte des Eisens* (3rd ed., Düsseldorf: Verlag Stahleisen, 1953; 621 pp., illustrated). A standard, authoritative work. According to **Wertime** (p. 261), p. xi, "the English version of such a history still needs to be written, though it has a very promising precursor in Hans Schubert's *History*" (following).

Jeanne McHugh, *Index to the Year Books and Regional Papers of the American Iron and Steel Institute 1910–1953* (Norman: University of Oklahoma Press, 1955, 593 pp.). A *Supplement, 1954–1956*, was published in 1962. This index provides entry into the corporate view of the iron and steel industry in the United States.

Joseph Needham, *The Development of Iron and Steel Technology in China* (Cambridge: Heffer, for The Newcomen Society, 1964; 76 pp. + 31 pls.). A review of three thousand years. Bibliography, pp. 52–63.

Earl R. Parker, *Brittle Behavior of Engineering Structures* (New York: Wiley, 1957; 323 pp., illustrated). A chapter on failures of steel structures other than ships, pp. 253–272, will bring the reader to recognize that brittle failure has been known since Bessemer steel was introduced just before 1860.

Carroll W. Pursell, Jr., "Tarrif and Technology: The Foundation and Development of the American Tin-Plate Industry, 1872–1900," *Technology and Culture*, 3 (Summer 1962), 267–284.

Franz M. Ress, *Bauten, Denkmäler und Stiftungen deutscher Eisenhüttenleute* (Düsseldorf: Verlag Stahleisen, 1960; 4to, 319 pp.). A beautifully and lavishly illustrated volume on the iconography and cultural milieu of the early (fifteenth–nineteenth centuries) iron industry in Germany. Many details of furnaces, forges, tools, and statues, paintings, and so forth.

H. R. Schubert, *History of the British Iron and Steel Industry from ca. 450 B.C. to A.D. 1775* (London: Routledge & Paul, 1957; 445 pp., illustrated). A thorough, scholarly work, in which processes are carefully explained and equipment is described in words and pictures.

Cyril Stanley Smith, "The Discovery of Carbon in Steel," *Technology and Culture*, 5 (Spring 1964), 149–175. Developments that occurred largely in the eighteenth century led to a recognition around 1800 of the role of carbon in iron technology.

James M. Swank, *History of the Manufacture of Iron in All Ages, and Particularly in the United States . . . from 1585 to 1885* (Philadelphia, 1884, 428 pp.), contains much detailed and fairly reliable information, having been written by the secretary of the American Iron and Steel Association.

Peter Temin, *Iron and Steel in Nineteenth-Century America: An Economic Inquiry* (Cambridge, Mass.: The M.I.T. Press, 1964, 304 pp.). Many statistics of production, from about 1810 onward, are assembled and analyzed. Information on technical developments is included.

Constance F. Tipper, *The Brittle Fracture Story* (Cambridge: Cambridge University Press, 1962; 197 pp., illustrated). This book, like Parker's, preceding, is technical in approach. Both books are clear and informative.

Theodore A. Wertime, *The Coming of the Age of Steel* (Chicago: University of Chicago Press, 1962, 330 pp.) is very, very good. Deals with the period *ca.* 1400–1860. Bibliography, pp. 299–320. Reviewed by Cyril Stanley Smith in *Technology and Culture*, 5 (Winter 1964), 87–88.

W. B. Gates, *Michigan Copper and Boston Dollars* (Cambridge, Mass.: Harvard University Press, 1951) is an economic history of the upper peninsula copper industry. Bibliography, pp. 273–283.

Daniel R. Hull, *Casting of Brass and Bronze* (Cleveland: American Society of Metals, 1950, 186 pp.). This contains an excellent technical discussion, by an active and perceptive participant, of developments from about 1900 onward.

F. B. Howard-White, *Nickel: an Historical Review* (London: Methuen, 1963, 350 pp.). The history is traced from the discovery of nickel (about 1750) to the present.

James Lewis Howe, *Bibliography of the Metals of the Platinum Group 1748–1896* (Washington: Smithsonian Institution, 1897 [*Misc. Colls.*, 38]). The metals included in the group are Pt, Pd, Ir, Rh, Os, and Ru. A revision to 1917 of Howe's work was published under the same title as U.S. Geological Survey, *Bulletin 694* (1919), 558 pp. Further revision to 1950 was published, also by the Geological Survey, in three volumes, 1947–1956.

Franz Kirchheimer, *Das Uran und seine Geschichte* (Stuttgart: Schweitzerbart'sche Verlagsbuchhandlung, 1963, 372 pp.). A history of uranium to 1898, including mining, geology, and chemistry. Bibliography, 1789–1898, pp. 304–330.

Donald McDonald, *A History of Platinum from the Earliest Times to the Eighteen-Eighties* (London: Johnson Matthey, 1960, 254 pp.). A careful work, published by a firm of platinum fabricators.

Joseph W. Richards, *Aluminum: its History . . . Metallurgy* [etc.] (3rd ed., Philadelphia, 1896). The first edition was published in 1887. A bibliography is included.

Elva Tooker, *Nathan Trotter, Philadelphia Merchant, 1787–1853* (Cambridge, Mass.: Harvard University Press, 1955, 276 pp). A distinguished study of a metal merchant, based upon the voluminous papers of Nathan Trotter and Co. in the Baker Library. The metals include brazier's copper, bar copper, brass wire, spelter, terneplate, Phelps tin plate (1822), Babbitt metal (1850), and so forth.

H. A. Wilhelm, "Development of Uranium Metal Production in America," *Journal of Chemical Education,* 37 (Feb. 1960); 56–68, 22 refs. The author was intimately involved in the production of uranium for the first U.S. atomic bomb.

The bibliographies of manganese, columbium, uranium, thallium, and zirconium, mentioned on p. 98, can be located through **Rhees** (p. 97).

4. *Chemical Industries, Including Photography*

Histories of the science of chemistry are not listed here. However, H. C. Bolton's massive bibliographies and other works published before 1953 are listed in **Sarton,** *Guide* (p. 17), pp. 163–167. Later titles can be located through the annual *Isis* "Critical Bibliography."

John J. Beer, *The Emergence of the German Dye Industry* (Urbana: University of Illinois Press, 1959 [Illinois Studies in the Social Sciences: Vol. 44], 168 pp.). A wide-ranging but well-disciplined study of German sources, in Germany, by an American historian. Bibliography, pp. 153–157.

Ernest Child, *The Tools of the Chemist* (New York: Reinhold, 1940; 220 pp., illustrated). Emphasis is upon American instruments and apparatus and their European origins. Evidently based upon extensive research but disappointingly free of scholarly apparatus.

Archibald Clow and Nan L. Clow, *The Chemical Revolution* (London: Batchworth, 1952; 680 pp., illustrated). A thorough and detailed study of industrial chemistry from 1750 to 1830. At the latter date "the sheer mass of material to be handled enforced a stop." Chronology, 1610–1856, pp. 623–632; bibliography, pp. 633–661. Very good.

L. F. Haber, *The Chemical Industry during the Nineteenth Century: A Study of the Economic Aspects of Applied Chemistry in Europe and North America* (London: Oxford University Press, 1958, 292 pp.). Emphasis is upon four inorganic chemicals — sulphuric acid, soda ash, caustic soda, and bleaching powder — which governed the rate of expansion of the industry as a whole. Critical bibliographical note, pp. 261–264; bibliography, pp. 265–280.

Williams Haynes, *American Chemical Industry* (6 vols., New York: 1945–1954). A monumental and encyclopaedic work. Vol. 1, background and beginnings, 1608–1910; Vols. 2–3, World War I era; Vol. 4, merger era; Vol. 5, new products; Vol. 6, company histories of 219 firms.

M. Kauffmann, *The First Century of Plastics: Celluloid and its Sequel* (London: Plastics Institute, 1963; 130 pp., illustrated) is good and stands alone on the subject of plastics, so far as I know. Celluloid development, pp. 13–54; synthetic resins, pp. 55–73; thermoplastics, pp. 74 ff. Contributions outside Great Britain are included.

Martin Levey, *Chemistry and Chemical Technology in Ancient Mesopotamia* (Amsterdam: Elsevier; Princeton: Van Nostrand, 1959, 242 pp.). A standard work. Good bibliography and chronology.

C. M. Mellor and D. S. L. Cardwell, "Dyes and Dyeing 1775–1860," *British Journal for the History of Science*, 1, No. 3 (June 1963), 265–275. A review of natural dyes, their supply and use. "Before 1856 the solution to the problem of supplying the great textile industries with dye stuffs was a commercial, not a scientific or technical one."

Charles Singer, *The Earliest Chemical Industry* (London: Folio Society, 1948; fo., 337 pp., illustrated). A spacious, beautiful book on the alum trade and industry. Very full, bibliographic notes, pp. 308–322.

F. Sherwood Taylor, *A History of Industrial Chemistry* (New York: Abelard-Schuman, 1957; 467 pp., illustrated). A sensible, well-illustrated book, starting with the earliest times and coming to the present but preserving a balanced emphasis. Bibliography, pp. 436–447. The author was a chemist, museum curator, and director of the Science Museum, London.

Peter C. Welsh, *Tanning in the United States to 1850* (Washington: Smithsonian Institution, 1964 [U.S.N.M., *Bulletin 242*]; 99 pp., illustrated). An attractive review of a traditional industry. Bibliography, pp. 91–95.

Clarence J. West and D. D. Berolzheimer, "Bibliography of Bibliographies on Chemistry and Chemical Technology 1900–1924" (National Research Council, Washington, *Bulletin*, 50 [1925]; supplement, 1924–1928, *Bulletin*, 71 [1929]; supplement, 1929–1931, *Bulletin*, 86 [1932]; are essentially current bibliographies but include a "history" rubric and historical studies under subject headings.

H. Brunswig, *Explosives: A Synoptic and Critical Treatment of the Literature of the Subject*, Charles E. Munroe and A. L. Kibler, Trans. (New York: Wiley, 1912, 350 pp.). This is a technical treatise on the behavior of explosives, translated from the German. Not historical in view.

Henry S. Drinker, *A Treatise on Explosive Compounds, Rock Drills and Blasting* (New York: Wiley, 1883; 406 pp., illustrated). A reissue of parts of the author's *Tunneling* (p. 225).

The library of Oscar Guttman, a collection of several hundred rare books and pamphlets on explosives since the fifteenth century, is in the **Eleutherian Mills Historical Library,** Greenville, Delaware.

Oscar Guttman, *The Manufacture of Explosives: A Theoretical and Practical Treatise on the History, the Physical and Chemical Properties, and the Manufacture of Explosives* (2 vols., London, 1895, illustrated). Also published in the same year in German.

Charles E. Munroe, *Index to the Literature of Explosives* (Baltimore: 1886), and *Part 2* (Baltimore, 1893), give titles of articles "relating in any way to explosives" in a remarkable group of periodicals: *American Journal of Science,* from 1819; *Philosophical Transactions,* from 1665; *Dingler's,* from 1820; *Nicholson's,* 1797–1813; *Edinburgh Journal of Science,* 1824–1832; *Brande's Journal of Science,* 1816–1830; American Chemical Society, *Proceedings,* from 1879; *Popular Science Monthly,* from 1872; Royal United Service Institution, *Journal,* from 1857; U.S. Naval Institute, *Proceedings,* from 1874; *Revue d' Artillerie,* from 1871; H.M. Inspector of Explosives, *Reports,* from 1873. I have seen only part 1.

Arthur P. Van Gelder and Hugo Schlatter, *History of the Explosives Industry in America* (New York: Columbia University Press, 1927, 1,132 pp.), is a voluminous and generally sound history prepared for the Institute of Makers of Explosives. No bibliography.

Wolfgang Baier, *A Source Book of Photographic History* (New York: Focal Press, 1964; 703 pp., illustrated). Text in German. In a review of this book in *Technology and Culture,* 7 (Spring 1966), 237–240, Beaumont Newhall, who is author of histories of photography, writes that he knows "of no other book in any language that covers the triple fields of the science, technology, and art of photography so thoroughly."

Albert Boni, *Photographic Literature* (New York: Bowker, 1962, 335 pp.). Concerned with historical development from fifteenth-century camera obscura through present-day electronic cameras. About 12,000 citations arranged by subject, with an author index. "Additional References," given after each general subject, refer the user to works listing more than a million additional titles. The literature of photography is too extensive for comprehensive treatment, but this selective bibliography is very useful. [Note by Jack Goodwin.]

SUBJECT FIELDS 265

Robert Taft, *Photography and the American Scene: A Social History, 1839–1889* (New York: Macmillan, 1938; Dover, 1964; 546 pp., 300 illustrations). Bibliographic notes, pp. 453–516. This book was written by a chemist.

5. *Glass, Ceramics, and Cement*

Pearce Davis, *The Development of the American Glass Industry* (Cambridge, Mass.: Harvard University Press, 1949, 316 pp.). Treats technical, economic, labor, and tariff questions, ca. 1800–1929. Bibliography, pp. 295–305.

George S. Duncan, *Bibliography of Glass* (London: Dawsons of Pall Mall for The Society of Glass Technology, Sheffield, 1960; 4to, 544 pp.). A major contribution, containing over 16,000 entries covering all periods. No titles later than 1940 are listed. Subject index, pp. 505–544. A supplementary volume, 1941–1960, is being prepared.

Warren G. Scoville, "Growth of the American Glass Industry to 1880," *Journal of Political Economy*, 52 (1944), 193–216, 340–355. A solid article which emphasizes techniques. Many references. The later period has been treated in more detail by the same author in the following title.

Warren G. Scoville, *Revolution in Glassmaking: Entrepreneurship and Technology 1880–1920* (Cambridge, Mass.: Harvard University Press, 1948, 398 pp.). Bibliography, pp. 349–361.

Paul N. Perrot, "Special Collections Somewhat off the Beaten Track" (*Wilson Library Bulletin*, 41 [Feb. 1967], 593–597). Introduces the library of the Corning Museum of Glass, which is concerned with the technical and artistic aspects of glass in all ages. This already outstanding collection is being further augmented as a service to scholars.

S. Tolansky, *The History and Use of Diamond* (London: Methuen, 1962, 166 pp.). An enlightening work by an author who has studied diamond and has written on its microstructure.

Edward G. Acheson, *A Pathfinder . . . How the World Came to Have . . . Carborundum, Artificial Graphite and Other Valuable Products of the Electric Furnace* (New York, 1910; 145 pp., portrait). The pathfinder is the author, who found and exploited carborundum.

R. M. Barton, *A History of the China-Clay Industry* (Truro: D. Bradford Barton Ltd., 1966; 212 pp., illustrated). Since the eighteenth century Cornwall quarries have supplied most of the clay for porcelain pottery made in England. The author deals with equipment and processes as well as with organization of the industry.

John C. Branner, *A Bibliography of Clays and the Ceramic Arts* (Columbus, Ohio: American Ceramic Society, 1906, 451 pp.). An alphabetical list by authors and a list of pertinent periodicals.

P. Gooding and P. E. Halstead, "The Early History of Cement in England" (International Symposium on the Chemistry of Cement [Third, London, 1952], London [1954], pp. 1–29).

P. E. Halstead, "The Early History of Portland Cement," Newcomen Society, *Transactions*, 34 (1961–1962), 37–54. A discussion of the chemistry of various cements is included in this account of the period 1820–1850.

A. W. Skempton, "Portland Cements, 1843–1887," Newcomen Society, *Transactions*, 35 (1962–1963), 117–152. An impressively thorough quantitative and analytical review of improvements in concrete and mortar.

Charles Spackman, *Some Writers on Lime and Cement from Cato to Present Time* (Cambridge, 1929, 287 pp.), has brief reviews of books and articles, about twenty before 1800. Pages 21–165 are devoted to the nineteenth century.

On ceramics in the modern period (from 1500), see **Clow and Clow** (p. 262), and by the same authors, "Ceramics from the Fifteenth Century to the Rise of the Staffordshire Potteries," in **Singer,** *History of Technology* (p. 2).

Roman cement, mortar, and concrete are described in modern terms in **Blake,** *Ancient Roman Construction* (p. 221), pp. 309–352. Voluminous reference notes.

6. Coal, Oil, and Gas

Thomas S. Ashton, *The Coal Industry of the Eighteenth Century* (Manchester, 1929, 268 pp.), has three chapters devoted to techniques, including methods of working, combating water and fire, and winding and drawing. Bibliography, pp. 255–261.

Howard N. Eavenson, *The First Century and a Quarter of American Coal Industry* (Pittsburgh: Privately Printed, 1942; 701 pp., tables, maps). A compendious work in which a great quantity of original material has been reprinted. Much quantitative information on coal production and consumption appears in the numerous tables. A bibliography of over 1,000 titles is arranged alphabetically, and titles are often incomplete and cryptic. Nevertheless, the work done by the author can be of great help to a patient scholar.

Robert F. Munn, *The Coal Industry in America. A Bibliography and Guide to Studies* (Morgantown: West Virginia University Library, 1965, 230 pp.). A list of 1,928 items, briefly annotated, are divided under roughly a dozen rubrics, such as history, labor relations, economics, strip mining, racial and minority groups. Includes citations of a wide range of periodicals, books, government documents, theses, and so forth. List of 450 men prominent in the industry (with emphasis on period before 1930) refers reader to a published biographical sketch of each. A sound work, indispensable to any inquiry in its field.

John U. Nef, *Rise of the British Coal Industry* (2 vols., London: Routledge, 1932). Very good; very thorough. Covers the period to about 1700; has full discussions of mining techniques, including drainage.

Kendall Beaton, "Dr. Gesner's Kerosene: the Start of American Oil Refining," *Business History Review,* 29 (Mar. 1955), 28–53. This article on the kerosene industry of the 1850's should be tested against the following article.

John Butt, "Legends of the Coal-Oil Industry (1847–1864)" (*Explorations in Entrepreneurial History,* 2nd series, 2, No. 1 [Fall 1964], 16–30). New evidence, which questions Gesner's priority, is introduced.

R. J. Forbes, *Studies in Early Petroleum History* (Leiden: Brill, 1958; 199 pp., illustrated) and by the same author, *More Studies in Early Petroleum History 1860–1880* (Leiden: Brill, 1959; 199 pp., illustrated) are wide ranging. The second volume contains at least two chapters that do not fall within the chronological limits of the title. Reference notes but no bibliography.

P. H. Giddens, *Beginnings of the Petroleum Industry: Sources and Bibliography* (Harrisburg: Pennsylvania Historical Commission, 1941, 206 pp.). I have not seen this.

John L. Loos, *Oil on Stream! A History of the Interstate Oil Pipeline Company, 1909–1959* (Baton Rouge: Louisiana State University Press, 1959; 411 pp., illustrated). Written for the company by an independent historian, based on company records and many interviews. Attention is paid to construction and operating details; there are some good pictures of early pipeline construction. About half of the book is concerned with the period before 1945. Bibliography, pp. 387–398.

Boverton Redwood, *Petroleum: a Treatise* (3rd ed., 3 vols., London, 1913). Treats geology, properties, production, refining, transport, and storage of petroleum and gas; also shale oil and allied industries. History is reviewed in Vol. 1, pp. 1–109. The bibliography, Vol. 3, pp. 187–349, which comprises 8,804 entries, is arranged alphabetically by author and is inadequately indexed by subject. Valuable, but not designed for a reader who demands rapid retrieval of bits of information.

Edward B. Swanson, *A Century of Oil and Gas in Books: a Descriptive Bibliography* (New York: Appleton, 1960, 214 pp.). A brief summary, not critical, is given of the contents of each work cited. Reviewed in *Technology and Culture*, 5 (Winter 1964), 134–136.

Harold F. Williamson and others, *The American Petroleum Industry* (2 vols., Evanston: Northwestern University Press, 1959–1963). Vol. 1, "The Age of Illumination, 1859–99," 864 pp.; Vol. 2, "The Age of Energy, 1900–59," 928 pp. The history to 1859 is reviewed in Vol. 1, pp. 3–60. Bibliographic notes in both volumes. The authors are economic historians, but techniques are nowhere overlooked. The "others" are, for Vol. 1, Arnold R. Daum, for Vol. 2, Ralph L. Andreano, Arnold R. Daum, and Gilbert C. Klose.

Dean Chandler and A. Douglas Lacey, *The Rise of the Gas Industry in Britain* (London: The Gas Council, 1949; 156 pp., illustrated). From *Technical Book Review Index* (p. 32), 1949, I learn that this book is "authoritative and interesting," deals with the first fifty years of the industry, and that several early articles and pamphlets are reprinted in appendixes.

Johannes Körting, *Geschichte der deutschen Gasindustrie* (Essen: Vulkan Verlag, 1965; 558 pp., 144 illustrations). A systematic survey of the gas industry in Germany and elsewhere from about 1825 to the present. Bibliographic footnotes. Reviewed in *Blätter für Technikgeschichte*, 27 (1965), 206–207.

Alfred Lief, *Metering for America; 125 Years of the Gas Industry and American Meter Company* (New York: Appleton-Century-Crofts, 1961; 154 pp., illustrated). While there are neither notes nor bibliography, the student can extract from text and pictures many suggestions that point to source materials in the measuring of gas.

Louis Stotz, *History of the Gas Industry* (New York: Privately Printed, 1938, 534 pp.). A "human interest non-technical story of the gas business," based upon public-relations information supplied by gas companies. Negligible.

7. Paper and Printing, Including Inks

On papermaking in the United States, several recent pioneering articles should be consulted. **Harold B. Hancock and Norman B. Wilkinson,** "The Gilpins and Their Endless Paper Machine" (*Pennsylvania Magazine of History and Biography*, 81 [Oct. 1957], 391–405), has an accurate drawing of the original Gilpin machine; see also by the same authors, "Thomas and Joshua Gilpin, Papermakers" (*The Paper Maker*, 27, No. 2 [1958], 1–10). Based principally upon the rich manuscript collections of the American Philosophical Society are: **John W. Maxson, Jr.,** "Nathan Sellers, America's First Large Scale Maker of Paper Moulds" (*The Paper Maker*, 29, No. 1 [1960], 1–16); and by the same author, "Coleman Sellers, Machine Maker to America's First Mechanized Paper Mills" (*The Paper Maker*, 30, No. 1 [1961], 13–27).

W. Turner Berry and H. Edmund Poole, *Annals of Printing; A Chronological Encyclopaedia from the Earliest Times to 1950* (London: Blandford, 1966; 315 pp., illustrated). An enormously impressive, painstaking work. References to further information are given at the ends of articles when published works exist. Some of the works used by the authors are listed on pp. x–xi and 287–294. This of itself constitutes a respectable specialized bibliography of printing. Primary references for each entry would have been very desirable, but the authors' decision to omit these is understandable. As one reads it, the work develops its own authority.

Edward C. Bigmore and C. W. H. Wyman, *A Bibliography of Printing* (3 vols., London, 1880–1886). This richly comprehensive, extensively annotated work was published serially from 1876 in *Printing Times and Lithographer*.

David Bland, *History of Book Illustration* (Cleveland, 1958), is long on aesthetics, very short on technology. The bibliography, pp. 429–432, however, is quite useful. The same author's very similar work, *The Illustration of Books* (3rd ed., London, 1962), pays more attention to technical aspects. Bibliography, pp. 185–187. [Note by Jack Goodwin.]

André Blum, *On the Origin of Paper*, H. M. Lydenberg, Trans. (New York: Bowker, 1934, 79 pp.) tells of arrival of paper in the West and resistance of parchment guilds to its introduction.

André Blum, *The Origins of Printing and Engraving*, H. M. Lydenberg, Trans. (New York: Scribner, 1940, 226 pp.). Origins of type, pp. 20–37; of engraving, pp. 41–192. Bibliographic notes, pp. 193–211.

Thomas F. Carter, *The Invention of Printing and Its Spread Westward,* revised by L. Carrington Goodrich (2nd ed., New York: Ronald, 1955; 293 pp., illustrated). A classic work, corrected and brought up to date. A chronological table summarizes the development of paper and printing from 200 B.C. to 1500 A.D. Bibliography, pp. 255–278.

D. C. Coleman, *The British Paper Industry 1495–1860* (Oxford: Clarendon, 1958; 367 pp., illustrated). Chapter 7, pp. 179–199, on mechanization after 1800 is quite good. There is no bibliography, but the footnote references are adequate.

P. M. Handover, *Printing in London from 1476 to Modern Times* (Cambridge, Mass.: Harvard University Press, 1960; 224 pp., illustrated). Equipment and techniques are described. Other titles can be located through the bibliographic notes at the end of each chapter.

Thomas C. Hansard, *Typographia: An Historical Sketch of the Origin and Progress of the Art of Printing; with Practical Directions for Conducting Every Department in an Office: with a Description of Stereotype and Lithography* (London, 1825; 939 pp.). A standard work, frequently cited. Another edition, issued in 1869, contained only the "practical" portions of the book.

Dard Hunter, *Papermaking* (2nd ed., New York: Knopf, 1947; 421 pp., illustrated). This is the standard work. Well over half the book is on papermaking by hand. There is as yet no satisfactory history of machine-made paper in the United States. Chronology, pp. 311–374; a good bibliography, pp. 374–389.

Jacob Kainen, *George Clymer and the Columbian Press* (San Francisco: Book Club of California, 1950; 60 pp., illustrated) is a very good study of a Philadelphia innovation that was taken to England around 1820 by its maker and was generally forgotten in the United States.

Hellmut Lehmann-Haupt, *Gutenberg and the Master of the Playing Cards* (New Haven: Yale University Press, 1966; 4to, 83 pp., illustrated). The author suggests that Gutenberg envisaged the mechanical reproduction in color of the miniatures and borders of manuscript originals and that in so doing he contributed to the development of copper-plate engraving. The copper plate was to be used as the mold for a raised cast printing plate. In the words of the author, Gutenberg "started at the top" in his conception of the way a printed book should be made. The magnificence of the dream must be evident to those who have contemplated at firsthand the Gutenberg Bible.

Douglas C. McMurtrie, *Invention of Printing: a Bibliography* (Chicago, 1942; 413 pp., mimeographed), is concerned chiefly with the fifteenth century, as indicated in the title. However, a list of "Bibliographies," pp. 388–392, covers a much wider field and should be consulted. Full bibliographic information but not annotated.

The very extensive bibliographic activities of Douglas C. McMurtrie, covering bibliographies of early imprints, classified geographically, both U.S. and foreign, can be followed in **Charles F. Heattman,** *McMurtrie Imprints* (Hattiesburg, Miss., 1942, 55 pp.), and supplement (Biloxi, Miss., 1946, 16 pp.). Iowa State University Library has copies of Heattman's works. McMurtrie's primary interest is in completed books and typography; only incidentally does he treat techniques and apparatus.

Joseph A. Miller, *Pulp and Paper History: a Selected List of Publications on the History of the Industry in North America* (St. Paul, Minn., 1963). This forty-one page mimeographed work was published by the Forest History Society (now located in Marsh Hall, 360 Prospect St., Yale University, New Haven, Conn. 06520), which is actively preparing extensive bibliographic lists and guides. There are 350 entries on technical and trade aspects; good coverage of standard works and periodical literature.

James Moran, "An Assessment of Mackie's Steam Type-composing Machine," *Journal of the Printing History Society,* 1 (1965), 57–67. Alexander Mackie (1825–1894) was the first to use a previously perforated tape to actuate a composing machine.

Joseph Moxon, *Mechanick Exercises on the Whole Art of Printing,* Herbert Davis and Harry Carter, Eds. (2nd ed., London: Oxford University Press, 1962; 487 pp., illustrated). A careful presentation of this 1683–1684 work, which is an invaluable source for the middle period of the art.

Newberry Library, Chicago, *Dictionary Catalogue, History of Printing* (6 vols., Boston: G. K. Hall). See note on p. 26.

St. Bride Institute, *Catalogue of the Technical Reference Library of Works on Printing and the Allied Arts of the St. Bride Institute* (London, 1919). I have not seen this title. This library of printing and allied trades, founded in 1895 and located in Bride Lane, Fleet Street, London, E.C. 4, has a collection of about 30,000 volumes on publishing, typography, bookbinding, and reproduction processes, including machines.

Isaiah Thomas, *The History of Printing in America* (2 vols., Albany, 1874). Reprinted by Burt Franklin, New York. Biographical and anecdotal information on printers, newspapers, and booksellers, arranged geographically. Contained in 2, 309–666, is a catalogue of American publications prior to 1775.

John S. Thompson, *History of Composing Machines* (Chicago: The Inland Printer Co., 1904; 200 pp., illustrated). This was published serially in *Inland Printer*.

Oriol Valls i Subira, "A Lively Look at Papermaking," *The Paper Maker*, 35, No. 1 (1966), 33–40. A series of (24) eighteenth-century Spanish woodcut prints illustrating the art of making paper are reproduced and described. The prints are on cards that were included in packets of cigarette papers.

Clarence J. West, Comp., *Bibliography of Pulp and Paper Making 1900–1928* (New York: Lockwood, 1929, 982 pp.), is clogged with trivial articles in the trade press but should be consulted for any extended study of the industry. "History" items, pp. 362–379. Several supplementary volumes were issued in succeeding years.

Frank B. Wiborg, *Printing Ink: a History with a Treatise on Modern Methods of Manufacture and Use* (New York: Harper, 1926, 299 pp.). A valuable if idiosyncratic and uncritical work. Much miscellaneous historical information, pp. 1–113; modern inks, pp. 114–252; bibliography, pp. 253–275.

Lawrence C. Wroth, Ed., *A History of the Printed Book, being the Third Number of The Dolphin* (New York: Limited Editions Club, 1938; fo., 507 pp., illustrated). Sound and interesting chapters on manuscripts, invention of printing, spread in the fifteenth century; separate chapters for sixteenth through nineteenth centuries; chapters on type, the printing press (the best connected account I have seen), and papermaking. Short on bibliography, however.

8. Textiles and Allied Industries

Larson, *Guide to Business History* (p. 23) has annotated lists of titles on textile history on pp. 133–135, 538–549, and elsewhere. See the detailed index. A selected list of titles is in **Singer,** *History of Technology* (p. 2), Vol. 4, pp. 306–307, 326–327.

C. Aspin and S. D. Chapman, *James Hargreaves and the Spinning Jenny* (Helmshore, Lancs.: Helmshore Local History Society, 1964; 79 pp., illustrated). A fresh work, based upon meticulous research. A jenny was constructed for the authors from patent drawings and was successfully operated.

Arthur H. Cole, *The American Wool Manufacture* (2 vols., Cambridge, Mass.: Harvard University Press, 1926). A major work in economic and business history which pays balanced attention to technical problems and machines. Bibliography, Vol. 2, pp. 303–314.

Melvin T. Copeland, *The Cotton Manufacturing Industry of the United States* (Cambridge, Mass.: Harvard University Press, 1923), has a good chapter on "Technical Development," pp. 54–111, and a bibliography, pp. 398–405.

Herman Freudenberger, *Waldstein Woolen Mill* (Boston: Baker Library, Graduate School of Business Administration, Harvard University, 1963; 68 pp. + 20 pls.). Technical description of an eighteenth-century textile complex in Bohemia. Plates are from a fresh work of 1728. Very good.

George S. Gibb, *Saco-Lowell Shops* (Cambridge, Mass.: Harvard University Press, 1950; 835 pp., illustrated), follows the Saco [Maine] Manufacturing Company from 1825 and the Lowell, Mass., machine shops from 1845. Bibliographic notes, pp. 737–807.

F. O. Howitt, *Bibliography of the Technical Literature on Silk* (London: Hutchinson's Scientific and Technical Publications, 1946, 248 pp.). A series of (15) critical essays on the literature of the various aspects of silk are supported by voluminous lists of titles. The author, as "Head of the Silk Section of the British Cotton Industry Research Association," reviewed everything he could lay his hands on. His searches appear to have been very thorough from about 1880 onward.

L. G. Lawrie, *A Bibliography of Dyeing and Textile Printing Comprising a List of Books from the Sixteenth Century to the Present Time (1946)* (London: Chapman & Hall, 1949, 143 pp.). The first part of the book consists of an alphabetical list of 804 books; the second part is a chronological list of the same books. A large number of reference works and library catalogues (detailed on pp. 10–12) were combed by the author.

Thomas R. Navin, *The Whitin Machine Works since 1831* (Cambridge, Mass.: Harvard University Press, 1950; 654 pp., illustrated), has bibliographic notes, pp. 561–625. The author's comment that there "is virtually no published information on the textile machinery industry" remains substantially correct, so far as I know, except for this book and **Gibb**, *Saco-Lowell Shops* (preceding).

Emil Ernst Ploss, *Ein Buch von alten Farben. Technologie der Textilfarben im Mittelalter mit einem Ausblick auf die festen Farben* (Heidelberg and Berlin: Impuls Verlag Heinz Moos, 1962, 168 pp.).

A beautifully illustrated history of dyes, dyeing, and textile printing in the Middle Ages and later. Bibliography, pp. 158–161.

Samuel E. Morison, *The Ropemakers of Plymouth: a History of the Plymouth Cordage Company 1824–1949* (Boston: Houghton Mifflin, 1949, 177 pp.) is a disappointing book. No reference notes; no bibliography.

C. J. H. Woodbury, *Bibliography of the Cotton Manufacture* (Waltham, Mass., 1909). Five thousand titles are listed, 1,361 on history and economics, 728 on engineering and machinery; many nineteenth-century titles. Not annotated; alphabetical by author.

The following three titles are useful for the firsthand information they contain and for the attitudes they reflect.

Edward Baines, Jr., *History of the Cotton Manufacture in Great Britain* (2nd ed., New York: Kelley, 1966; 544 pp., illustrated). The second edition consists of a reprint of the first (London, 1835) and a comprehensive and highly informative ten-page "Bibliographical Introduction" by W. H. Chaloner. In the original work, processes and factories are described and illustrated.

James Montgomery, *A Practical Detail of the Cotton Manufacture of the United States . . . Contrasted and Compared with That of Great Britain* (Glasgow, 1840, 219 pp.). The author supervised textile operations in the United States and the United Kingdom.

George S. White, *Memoir of Samuel Slater, Connected with a History of the Rise and Progress of the Cotton Manufacture in England and America* (Philadelphia, 1836; 448 pp., illustrated). Information is given on technical, economic, and moral aspects of the industry.

9. Timber and Wood Industries

A major bibliography of books and articles in forest history, including the techniques of logging, sawmills, and subsequent processing, as well as economic and business aspects of the various industries, is being prepared by Joseph A. Miller for Forest History Society at Yale University, New Haven.

Clodaugh M. Neiderheiser, *Forest History Sources of the United States and Canada* (St. Paul: Forest History Foundation, 1956, 140 pp.). Lists 972 groups of manuscripts in 108 institutions, from the seventeenth to the twentieth centuries. Analytical index, pp. 125–140.

An addendum to the preceding title is in **Richard C. Berner,** "Sources for Research in Forest History: the University of Washington

Manuscripts Collection" (*Business History Review*, 35 [Autumn 1961]).

Ralph W. Hidy, Frank E. Hill, and Allan Nevins, *Timber and Men: The Weyerhauser Story* (New York: Macmillan, 1963, 704 pp.). Information on forest, lumber, and wood technology is an integral part of this important company history. Reviewed in *Journal of Forest History*, 7 (Fall 1963) and in *Journal of Forestry*, 62 (Feb. 1964).

The following two works are useful for contemporary evidence.

John Richards, *A Treatise on the Construction and Operation of Wood-working Machines* (London: Spon, 1872, 283 pp.).

Manfred P. Bale, *Woodworking Machinery: Its Rise, Progress, and Conclusion* (3rd ed., London, 1914). The first edition was published in 1880.

W. S. Worsam, *History of the Bandsaw* (Manchester, 1892). No further information is at hand.

10. *Testing of Materials*

Chester H. Gibbons, *Materials Testing Machines* (Pittsburgh: Instruments Publishing Co., 1935; 90 pp., illustrated). The author made one of the first surveys of mechanical testing machines from the time of Galileo; in his book he illustrates the machines of Musschenbroek, Gauthey, Soufflet, Perronet, Rondelet, Barlow, Hodgkinson, Fairbairn, Kirkaldy, and, in the United States, Riehle, Eads, Olsen, Emery, and others. The bare outline is here, but when used with **Timoshenko**, following, the significant sources can be readily located.

Charles Fremont, "Évolution des méthodes et des appareils employés pour l'essai des matériaux de construction" (*Congrès international des méthodes d'essai des matériaux de construction* [held in Paris, July 9–16, 1900], [1900]; 104 pp., 166 figs.). More than twenty papers by Fremont on the testing of metals, published after 1900, are listed in **Sarton**, "Charles Fremont" (p. 255).

Stephen P. Timoshenko, *History of Strength of Materials with a Brief Account of the History of Theory of Elasticity and Theory of Structures* (New York: McGraw-Hill, 1953; 452 pp., illustrated). Because much of the theoretical understanding of the way materials behave has been derived from physical tests, the history of materials testing is woven into this book. Tests are described for ultimate and yield strengths, bending, shear, buckling, and fatigue. The photoelastic approach (from 1815), resistance strain gages, and brittle coatings are also included. The author expects that the reader will have "had a course in" the strength of materials, but the reader not so endowed will find a great deal of solid information and a mine of bibliographic reference, located in footnotes and in the text.

See also
   The series of volumes published between 1880 and 1920 by the **U.S. Board for Testing** (p. 101) and **U.S. Ordnance Bureau** (p. 101). Metals and other materials were tested.

G. Mechanical Technology

1. Hand Tools and Machine Tools

Walter Bernt, *Altes Werkzeug* (Munich: Georg D. W. Callwey, 1939; 197 pp., 228 illustrations). The lavishly decorated tools of 1600–1800 A.D. are handsomely displayed. Bibliography, pp. 28–30.

Otto Dick, *Die Feile und ihre Entwicklungsgeschichte* (Berlin: Julius Springer, 1925; 4to, 278 illustrations). This is the most exhaustive and thorough treatise on the file. Bibliography contains 108 references. [Note by Warren G. Ogden, Jr.]

F. M. Feldhaus, *Die Säge* (Berlin, 1921, 72 pp.), on the saw, is heavily illustrated.

   The following titles by Charles Fremont are representative of his work on tools. For other titles, see **Sarton**, "Charles Fremont" (p. 255).

Charles Fremont, *Origine et évolution des outils* (Paris: Société d'encouragement pour l'industrie nationale, 1913; 157 pp., 322 figs.).

Charles Fremont, *La vis* (Paris, 1928; 61 pp., 96 figs.).

Charles Fremont, "La scie" (Société d'encouragement pour l'industrie nationale, *Bulletin* [1928], pp. 643–721, 142 figs.).

Charles Fremont, *Files and Filing* (London: Pitman, 1920; 148 pp., illustrated). A translation of the author's *La lime*, first published in 1916, and issued in a second edition in 1930.

W. L. Goodman, *The History of Woodworking Tools* (London: Bell, 1964; 208 pp., illustrated). This admirable work supersedes Mercer (see following) as the standard work on hand tools. While the early chapters should be supplemented by the works of Petrie and others, the treatment from Roman times onward is definitive. Very good.

Reference list, p. 204 (40 titles); list of (31) British museums displaying tools, p. 205 ("by no means complete"). The author, in collaboration with R.A. Salaman, is preparing a *Dictionary of Hand Tools.* This will be particularly useful in identification and comparison of tools.

W. L. Goodman, *Woodwork* (Oxford: Blackwell, 1962; 72 pp., 38 illustrations). A spare but graceful summary of woodworking tools and techniques through the ages. This work complements the preceding title.

Rudolf Kellermann and Wilhelm Treue, *Die Kulturgeschichte der Schraube* (Munich: Bruckmann, 1962; 4to, 308 pp., illustrated). A history of screws from Archimedes to international standardization and an attractive example of bookmaking. Bibliography, pp. 303–308. Reviewed by Edwin A. Battison in *Technology and Culture,* 7 (Winter 1966), 76.

Henry C. Mercer, *Ancient Carpenters' Tools* (Doylestown, Pa.: Bucks County Historical Society, 1929, 1950; 339 pp., illustrated). For long the only general work in English on hand tools (and I know of none in other languages), this book is based upon Mercer's remarkable collections in the Bucks County Historical Museum and (apparently) extensive correspondence with European museums.

Kenneth P. Oakley, *Man the Tool-Maker* (3rd ed., London: British Museum, 1957; Chicago: University of Chicago Press, 1957; 159 pp., illustrated). An introduction to early toolmaking, to *ca.* 5000 B.C. Selected references, pp. 144–146.

S. A. Semenov, *Prehistoric Technology: An Experimental Study of the Oldest Tools and Artefacts from Traces of Manufacture and Wear,* M. W. Thompson, Trans. (London: Adams & Mackay; New York: Barnes & Noble, 1964; 211 pp., illustrated). The author shows that an observer may learn much about how an object was made if he will study traces on the object. The translator supplied also a critical preface.

Analogous methods were used on recent artifacts by **Edwin A. Battison,** "Eli Whitney and the Milling Machine" (*The Smithsonian Journal of History,* 1, No. 2 [Summer 1966], 9–34). The author used a Whitney musket of *ca.* 1805 to show that a "true milling machine" was not used in its manufacture.

Eric Sloane, *A Museum of Early American Tools* (New York: Funk, 1964, 108 pp.). Profusely illustrated. My reaction to this book is the same as to that of **Tunis,** following.

H. R. Bradley Smith, *Blacksmiths' and Farriers' Tools at Shelburne Museum* [Museum Pamphlet Series, Number 7] (Shelburne, Vt.: The Shelburne Museum, Inc., 1966; 272 pp.). An exhaustive, profusely illustrated compendium of information on the tools and skills of the smith. Pages 164–203 are devoted to horseshoeing and other arts of the farrier.

Edwin Tunis, *Colonial Craftsmen and the Beginnings of American Industry* (Cleveland: World Publishing, 1965, 160 pp.). Profusely illustrated. The original drawings by the author are attractive and provide in conjunction with the text a great deal of interesting information on tools, techniques, and products. Not definitive, but often very good.

Peter C. Welsh, "Woodworking Tools 1600–1900" (Washington, 1966 [U.S.N.M., *Bulletin 241*], paper 51, pp. 178–228). The gradual evolution of the tools is illustrated in sixty-seven figures, which include pages from books by Comenius, Moxon, Roubo, Martin, Nicholson, Jost Amman, and many other sources.

Frank H. Wildung, *Woodworking Tools at Shelburne Museum* (Shelburne, Vt.: Shelburne Museum, 1957; 79 pp., illustrated). A descriptive catalogue of the Wildung collection of tools in the Shelburne Museum. Good illustrations. All three titles — Goodman, Mercer, and Wildung — should be on the shelf of a student of hand tools of any woodworking trade.

Sydney G. Abell, John Leggat, and Warren G. Ogden, Jr., *Bibliography of the Art of Turning and Lathe and Machine Tool History* (2nd ed., Isleworth, Msx.: Society of Ornamental Turners, 1956, 89 pp.). This meticulously annotated list of 338 titles from the seventeenth century to the present is basic to any extended study of machine tools.

Edwin A. Battison, "Screw-Thread Cutting by the Master-Screw Method since 1480" (Washington, 1964 [U.S.N.M., *Bulletin 240*], paper 37, pp. 105–120). A detailed, knowledgeable description of machines for producing screw threads using a master screw as a template, still the most accurate method.

Orlan W. Boston, *A Bibliography on Cutting of Metals, 1864–1943* (New York: American Society of Mechanical Engineers, 1945, 547 pp.). Arranged chronologically, with an analytical index. Bibliographies, pp. 1–2; there are twenty-one references before 1900; the titles from 1900 to 1926 occupy pp. 5–66. The author was one of the first to investigate the forces exerted by metal-cutting tools.

A. S. Britkin and S. S. Vidinov, *A. K. Nartov, an Outstanding Machine Builder of the 18th Century* (Jerusalem: Israel Program for Scientific Translations, 1964 [TT64-11111, U.S. Clearing House for Federal Scientific and Technical Information, Springfield, Va.]; 120 pp., illustrated). Nartov was a turner and builder of fine lathes. Reviewed by Warren G. Ogden, Jr., in *Technology and Culture*, 7 (Summer 1966), 426–428.

Charles and John Jacob Holtzapffel, *Turning and Mechanical Manipulation* (5 vols., London: Holtzapffel, 1843–1884). Extensive historical, philosphical, and practical notes on hand and machine tools and materials are in all volumes, but particularly Vols. 1–3. Index to Vols. 1–3 in Vol. 3. A standard work. **Abell, Leggat, and Ogden** (p. 278), No. 142.

Cameron Knight, *The Mechanician and Constructor for Engineers* (London: Spon, 1869). Heavily illustrated practical manual on laying out and executing machinists' work.

Joseph W. Roe, *English and American Tool Builders* (New York, 1916, 1926). This pioneering work has been largely superseded by **Rolt's** *Short History*, following.

L. T. C. Rolt, *A Short History of Machine Tools* (Cambridge, Mass.: The M.I.T. Press, 1965; 256 pp., illustrated). An authoritative review of machine-tool development in the United Kingdom and United States. Bibliography, pp. 245–247. I reviewed the book in *Science*, 150 (Oct. 8, 1965), 201–202; Robert Woodbury reviewed it in *Technology and Culture*, 7 (Spring 1966), 231–233. This book was first published in England under the title *Tools for the Job*.

The following series of monographs on the mechanical development of individual machine tools is based upon intensive original research.

Robert S. Woodbury, *History of the Gear-Cutting Machine* (Cambridge, Mass.: The M.I.T. Press, 1958; 135 pp., illustrated). Bibliography, pp. 127–130.

Robert S. Woodbury, *History of the Grinding Machine* (Cambridge, Mass.: The M.I.T. Press, 1959; 191 pp., illustrated). Bibliography, pp. 185–186.

Robert S. Woodbury, *History of the Milling Machine* (Cambridge, Mass.: The M.I.T. Press, 1960; 107 pp., illustrated). Bibliography, pp. 103–104.

Robert S. Woodbury, *History of the Lathe to 1850* (Cleveland, O.: Society for the History of Technology, 1961 [Monograph Series No. 1]; 124 pp., illustrated). Bibliography, pp. 120–122.

F. N. Zagorsky, *An Outline of the History of Metal-cutting Machine Tools up to the Middle of the Nineteenth Century* (Moscow–Leningrad Akademizdat, 1960). In Russian; a review of Russian contributions.

See also
For woodworking machinery, See Part XIV.F.9.

2. *Crafts and Craftsmen*

Martin S. Briggs, *A Short History of the Building Crafts* (Oxford: Clarendon, 1925; 296 pp., illustrated). Chapters on brickwork, masonry, concrete, carpentry, joinery, ironwork, roof coverings, plasterwork, external plumbing, and glazing from Egyptian times through the eighteenth century.

Somers Clarke and R. Engelbach, *Ancient Egyptian Masonry, The Building Craft* (London: Oxford University Press, 1930; 242 pp., 268 illustrations). A detailed work by men who lived for many years with the problems they attacked. No odd theories: "The more [the Egyptian] constructional methods are studied, the more one is convinced that if any detail in a piece of work has to be explained by an apparatus of any complication, then that explanation is certainly wrong." (p. 95). Very good.

R. Engelbach, *The Problem of the Obelisks* (London: Unwin, 1923; 134 pp., illustrated). A study of the unfinished obelisk in the Aswan quarry. Quarry methods and the problems of moving and erecting an obelisk are carefully considered. For further thoughts by the author, some twenty years later, see his chapter in S. R. K. Glanville, *The Legacy of Egypt* (Oxford: Clarendon, 1942), pp. 120–159.

Edgar B. Frank, *Old French Ironwork: The Craftsman and His Art* (Cambridge, Mass.: Harvard University Press, 1950; 221 pp., 96 pls.). Tools of the craftsman are illustrated as well as his products. Bibliography, pp. 217–218. A French edition was published in 1948.

Rudolf P. Hommel, *China at Work: An Illustrated Record of the Primitive Industries of China's Masses, Whose Life is Toil, and thus an Account of Chinese Civilization* (New York: Day, for Bucks County Historical Society, 1937; 4to, 336 pp., 535 illustrations). This book is the impressive record of an expedition financed by Henry C. Mercer.

J. Geraint Jenkins, *Traditional Country Craftsmen* (London: Routledge and Paul, 1965; 236 pp., 239 illustrations). A distinguished, craftsmanlike book of which Peter Welsh wrote, in *Technology and Culture,* 8 (Jan. 1967), 104–105, that in format and in the class of information included, this was the kind of "a book that one always hoped that an American might write before it was too late." The trades include chair bodger, hoopmaker, charcoal burner, bowl turner, rakemaker, cooper, wheelwright, farrier, thatcher, brickmaker, dry stonewaller, slatter, ropemaker, saddler, bootmaker, clogmaker, and many others.

L. A. Mayer, *Islamic Armourers and their Works* (Geneva: Albert Kundig, 1962; 128 pp., 20 pls). Contents include a list of swordsmiths, list of cannonmakers, chronological list of cannon, and location of arms by museum and by collection. Bibliography, pp. 97–125.

This title is the only number I have seen of a series that includes *Mamluk Costume* (1952), *Islamic Architects and their Works* (1956), *Islamic Astrolabists and their Works* (1956), *Islamic Woodcarvers and their Works* (1956), and *Islamic Metalworkers and their Works* (1959).

G. M. A. Richter, *The Furniture of the Greeks, Etruscans, and Romans* (London: Phaidon, 1967; 369 pp., 668 pls.). A revised and expanded version of a standard work, which first appeared in 1926 and for which every scrap of evidence concerning the subject has been sifted. *The Times Literary Supplement* (August 17, 1967, p. 738) calls it a "miracle of patience, order and judgment: well-organized, sumptuously produced, and easy to consult."

Hans E. Wulff, *The Traditional Crafts of Persia* (Cambridge, Mass.: The M.I.T. Press, 1967; 404 pp., illustrated). This handsome and learned book is by a German engineer who, as principal of the Technical College at Shiraz, was responsible for teaching traditional crafts as well as modern techniques. Over 400 photographs supplement the clear descriptions of apparatus and techniques. A technical glossary of fifty pages captures terms nowhere else recorded. Bibliography, pp. 305–314, and a critical "Review of Relevant Literature," pp. 315–329. A short review is in *Scientific American,* 217, No. 1 (July 1967), 129–130.

A valuable and delightful series of books, published by Cambridge University Press, tells of the tools of the several trades and how they were used, generally in the latter half of the nineteenth century. With the exception of Freese's book, each of the narratives was written by a craftsman.

George Sturt, *The Wheelwright's Shop* (1923, 1958).

Eric Benfield, *Purbeck Shop: A Stoneworker's Story of Stone* (1940).

Walter Rose, *The Village Carpenter* (1952).

Stanley Freese, *Windmills and Millwrighting* (1957).

Carl Graf von Klinckowstroem, "Alte Handwerkerdarstellungen," *Börsenblatt für den Deutschen Buchhandel* (Frankfurter Ausgabe), Nr. 78a, Sept. 30, 1959, pp. 1304–1305. An annotated bibliography of seventeen early works, in German, that illustrate craftsmen at their trades. Included are Jost Amman (1568), Chr. Weigel (1698), Joh. von Justi (1762–1795), the latter a German republication of the French *Description des Arts et Métiers* (see p. 58). Not included are the "technologischen Jugendschriften," which I suppose encompass the boy's books of trades. The juvenile works were listed by Klinckowstroem in *ibid.*, 1956, Nr. 13. The present article was kindly pointed out to me by Warren G. Ogden, Jr.

Illustrations of craftsmen of many countries are in the collections of the **Kress Library** (p. 29).

Paintings of seventeenth-century Japanese spinning and weaving, metalworking, and leatherworking are reproduced in **Mumford**, *Myth of the Machine* (p. 207), plates 22–23. The paintings are "in the Kitain, a Buddhist temple in Kawagoye."

The following list of illustrated books of trades and tools was selected from a much longer list supplied by Warren G. Ogden, Jr., whose library consists of a very outstanding collection of works on hand and machine tools. The titles that are described in **Abell, Leggat, and Ogden** (p. 278) are designated ALO.

Das Hausbuch der Mendelschen Zwölfbrüderstiftung (p. 51). This foundation for retired craftsmen produced a handsome record of the crafts of Nürnberg in the fifteenth and sixteenth centuries.

Hans Sachs, *Das Ständebuch* (p. 47). The illustrations of trades in 1570 were by Jost Amman.

Stradanus, *New Discoveries* (p. 47). Trades of the 1580's.

André Felibien, *Les principes de l'Architecture, de la Sculpture, de la Peinture, et des autres arts qui en dépendent* (Paris, 1676). Includes the building trades. ALO No. 99.

Joseph Moxon, *Mechanick Exercises* (p. 46). An English work issued in parts, beginning in 1677. ALO No. 208.

Comenius, *Orbis Sensualium Pictus* (p. 45). An English work of 1685.

Christoff Weigel, *Abbildung der gemein-nützlichen Haupt-Stände von denen Regenten* (Regenspurg, 1698, 677 pp.), The plates, engraved by Weigel (1654–1725) from designs by Caspar Luyken, have never been equaled. There are many engravings of helve hammers, worked by water power, and interiors of smiths' shops represented in minute detail. Only two copies recorded in America: Harvard College Library and Chemist's Club, New York City. [Note by Warren G. Ogden, Jr.] ALO No. 293.

Jan Luiken, *Spiegel van het menselyk bedryf* (Amsterdam, 1730, 100 engravings). The engravings show the activities of craftsmen and of the professions.

Martin Engelbrecht, *Assemblage nouveau des manouvries habilles* (Augsburg, ca. 1730, 168 pls.). Illustrations of the skilled trades and professions. Cited in Library of Congress, *Quarterly Journal*, 22 (Apr. 1965), inside front cover.

Abbé Noël Antoine Pluche, *Le spectacle de la nature* (8 vols. in 9, Paris: Veuve Estienne, 1735–1748). First of the encyclopaedias of arts and crafts. The illustrations in Vols. 6 and 7 provide an interesting view of the trades and industrial processes of the eighteenth century. [This note by Warren G. Ogden, Jr.]. **Brunet** (p. 40) wrote, in 1839, "this work loses from day to day its earlier reputation; only the first edition is sought, and that at a very ordinary price: two or three francs per volume." **Sotheran** (p. 24) notes that the work is "instructive and agreeable" and that an English version was dedicated to the youthful William Augustus, Duke of Cumberland, known after 1746 as "the Butcher."

Book of the Trades, or Library of the Useful Arts (3 vols., First American Edition, Philadelphia, 1807). A total of sixty-seven copper-plate engravings depicts craftsmen in their shops. There were many similar works published in England and the United States in the first half of the nineteenth century.

Thomas Martin, *The Circle of the Mechanic Arts; Containing Practical Treatises on the Various Manual Arts, Trades and Manufactures* (London: Richard Rees, 1813). ALO No. 199.

The Book of English Trades & Library of the Useful Arts (London, 1825, 454 pp.). There are seventy-four plates showing trades and twelve illustrations in the appendix on machinery.

Edward Hazen, *Popular Technology, or, Professions and Trades* (2 vols., New York, 1842). Another book of the mechanical trades.

See also
Higgins Armory, *Catalog* (p. 191), which contains an excellent reconstruction of an armorer's shop, pp. 28–30.

3. *Mechanisms and Automatic Control*

Alfred Chapuis and Edmond Droz, *Automata: a Historical and Technological Study,* Alec Reid, Trans. (Neuchâtel, 1958; 4to, 408 pp., illustrated) is a standard work, but has no bibliography. The reader is, however, referred to

Alfred Chapuis and Edmund Gelis, *Le monde des automates* (2 vols., Paris, 1928), which has an extensive bibliography. I have not seen this latter book.

H. G. Conway, "Some Notes on the Origins of Mechanical Servo Mechanisms," Newcomen Society, *Transactions,* 29 (1953–1955), 55–75. On control devices of the nineteenth century.

Ananda K. Coomaraswamy, *The Treatise of Al-Jazari on Automata* (Boston: Museum of Fine Arts, 1924). On automata of the thirteenth century. I have not seen this title.

A. G. Drachmann, *Ktesibios, Philon and Heron; A Study in Ancient Pneumatics* (Copenhagen: Ejnar Munksgaard, 1948; 197 pp., illustrated). The automata of Alexandrian Greek times are analyzed by the chief authority on the subject.

A. G. Drachmann, *The Mechanical Technology of Greek and Roman Antiquity; A Study of the Literary Sources* (Copenhagen: Munksgaard; Madison: University of Wisconsin Press, 1963; 220 pp., illustrated). Exhaustive study of the best-known early technical equipment: that of Hero, Vitruvius, and others. Engines of war, winch, toothed wheels, lever, pulley, wedge, screw, and so forth. Reviewed by R. J. Forbes in *Technology and Culture,* 5 (Winter 1964), 69–74.

Derek J. de S. Price and Silvio A. Bedini, "Automata in History" (*Technology and Culture,* 5 [Winter 1964], 9–42), consists of two papers that will set the student on the way to an understanding of mechanisms and machines. Bibliographic footnotes.

Bennet Woodcroft, Ed., *The Pneumatics of Hero of Alexandria* (London, 1851). I know of no modern edition of this important work.

Drachmann, in the analytical works that precede, does not reproduce the entire book.

See also
Bertrand Gille, "Machines," in **Singer**, *History of Technology* (p. 2), Vol. 2, pp. 629–658.

For the systematic academic treatment of mechanisms, see titles pertaining to kinematics in Part XIV.K.

4. *Timekeepers*

Granville Hicks Baillie, *Clocks and Watches, an Historical Bibliography* (London: N.A.G. Press, 1951; 427 pp., 120 illustrations). Covers published and manuscript works on the subject of devices for measuring time from 1344 (The Orloge of St. Paul's Cathedral) through 1799 (Thomas Mudge, Jr.). Works of historical value, whether manuscript, book, or pamphlet, are included with concise abstracts of contents and occasional personal notes by the author with references to other works. Illustrations of many of the works are included. Over 1,000 titles are listed over a period covering four centuries. Many works, not strictly horological, are compendia on mechanics, and so forth, in which horological examples are included. Name and subject indexes. [This note by Silvio A. Bedini.]

G. H. Baillie, C. Clutton, and C. A. Ilbert, *Britten's Old Clocks and Watches and Their Makers* (7th ed., New York: Bonanza, 1956; 4to, 518 pp., over 200 illustrations). A comprehensive work that includes a glossary, pp. 297–308, a list of 14,000 clock- and watchmakers, and an extensive bibliography, pp. 312–317.

Ernst von Bassermann-Jordan, *The Book of Old Clocks and Watches*, revised by Hans von Bertele (4th ed., New York: Crown, 1964; 522 pp., 70 illustrations, 20 color pls.). Meticulous revision of a standard work, covering evolution of time measurement from the sundial and astrolabe through clocks of the nineteenth century. Very comprehensive and valuable bibliography. [This note by Silvio A. Bedini.]

The water clock is fully introduced in **Silvio A. Bedini**, "The Compartmented Cylindrical Clepsydra" (*Technology and Culture*, 3 [Spring 1962], 115–141 + 31 figs.). Reference notes.

Silvio A. Bedini, "The Scent of Time," American Philosophical Society, *Transactions*, n.s., 52, pt. 5 (Aug. 1963), 1–51. An exhaustive study of the use of fire and incense for time measurement in oriental countries. Bibliography, about eighty titles.

Silvio A. Bedini and Francis R. Maddison, *Mechanical Universe. The Astrarium of Giovanni de' Dondi* (Philadelphia, 1966 [American Philosophical Society, *Transactions*, n.s., 56, pt. 5]; 69 pp., 53 illustrations). This impressively thorough account of the great astronomical clock completed by de' Dondi in 1364 is clearly illustrated and heavily documented. A hardback edition is available from the Smithsonian Institution.

George Eckhardt, *United States Clock and Watch Patents 1790–1890* (New York: Privately Printed, 1960, 231 pp.). Classified by subject headings, the patents can be readily scanned. I could find no statement of the author's method of locating patents, so comprehensiveness is in doubt.

Joseph Needham, Wang Ling, and Derek J. de S. Price, *Heavenly Clockwork: the Great Astronomical Clocks of Medieval China* (Cambridge: Cambridge University Press, 1960; 254 pp., illustrated). Review of a thousand years of instruments and timekeepers before Su Sung's great clock of 1090 A.D. and subsequent transmission to the West. Bibliography, pp. 206–215.

Walter A. R. Pertuch, Comp., *Horological Books and Pamphlets in the Library of the Franklin Institute* (Philadelphia: Franklin Institute, 1956, 50 pp.). A list, without annotations, of a collection of about 1,000 books, pamphlets, and some early advertising folders.

Pioneers of Precision Timekeeping, A Symposium (London: Antiquarian Horological Society, 196? [Monogr. No. 3]; 117 pp., illustrated). A collection of nicely illustrated articles on the details of timepieces of Harrison, Mudge, Le Roy, and others by craftsmen who have restored or otherwise studied intimately the pieces they write about.

Tardy, *Bibliographie Générale de la Mesure du Temps suivie d'un Essai de Classification Technique et Géographique* (Paris: Tardy, 1947). Over 5,000 entries, including books and articles on practical horology, chronometry, and the art of dialing. List of journals published in nineteen countries, pp. 262–266; subject index, pp. 267–332; geographic index, pp. 333–352. Main entries are alphabetical, by author. A companion volume to Baillie, preceding.

See also

For a convenient introduction and selected bibliography, see H. Alan Lloyd, "Mechanical Timekeepers," in **Singer,** *History of Technology* (p. 2), Vol. 3, pp. 648–675.

5. Scientific Instruments and Calculating Machines

Silvio A. Bedini, *Early American Scientific Instruments and their Makers* (Washington: Smithsonian Institution, 1964 [U.S.N.M., Bulletin 231], 184 pp.), nicely illustrated, has lists of instruments and makers. Bibliography, pp. 172–176. Reviewed by Brooke Hindle in *Science* (Jan. 1, 1965).

Reginald S. Clay and Thomas H. Court, *The History of the Microscope* (London: Griffin, 1932; 266 pp., 164 illustrations). Extensive list of makers. Still the best single work on the subject. [Note by Silvio A. Bedini.]

Maurice Daumas, *Les Instruments Scientifiques aux XVIIe et XVIIIe Siècles* (Paris: Presses universitaires de France, 1953; 420 pp. + 63 pls.). The author is concerned with the problems of making instruments. Bibliography, pp. 387–392.

Alfred N. Disney, Cyril F. Hill, and W. E. Watson, Eds., *Origins and Development of the Microscope* (London: Royal Microscopical Society, 1928; 303 pp., illustrated). Extensive annotated bibliography.

Brooke Hindle, *David Rittenhouse* (Princeton: Princeton University Press, 1964, 394 pp.). Detailed, definitive biography, in which has been assembled every available scrap of evidence regarding Rittenhouse's instrument-making. Bibliographical note, pp. 465–468. Reviewed by Norman B. Wilkinson in *Technology and Culture,* 6 (Summer 1965), 465–468.

Hebbel E. Hoff and L. A. Geddes, "The Beginnings of Graphic Recording" (*Isis,* 53 [Sept. 1962], 287–324), gives illustrations of recording devices from Leonardo through Leupold (1724).

Henry C. King, *The History of the Telescope* (London: Griffin, 1955, 456 pp.) is a painstaking work, well illustrated. Very numerous reference notes; no separate bibliography.

Francis Maddison, "Early Astronomical and Mathematical Instruments: A Brief Survey of Sources and Modern Studies," *History of Science,* 2 (1963), 17–50. A major survey of sources for the study of instruments of all ages and all places, including collections of instruments in and out of museums; photographic catalogues and inventories; libraries and manuscripts; and a review of scholarly work completed and of needs and opportunities. Impressively thorough and carefully constructed. Bibliography, pp. 34–50.

William E. K. Middleton, *A History of the Thermometer and Its Use in Meteorology* (Baltimore: The Johns Hopkins Press, 1966; 249 pp., illustrated). Bibliographical footnotes. This scholarly work is the first monograph on the thermometer since Henry C. Bolton's *Evolution of the Thermometer, 1592–1743* (Easton, Pa., 1900, 98 pp.).

W. E. Knowles Middleton, *The History of the Barometer* (Baltimore: The Johns Hopkins Press, 1964, 489 pp.), is the first full-length study of this important instrument. Reviewed by A. Rupert Hall in *Technology and Culture*, 6 (Winter 1965), 130–131.

Robert P. Multhauf, "The Introduction of Self-Registering Meteorological Instruments" (Washington, 1961 [U.S.N.M., *Bulletin 228*], paper 23), pp. 95–116. Although techniques were available earlier, self-recording instruments were not introduced into meteorology until about 1850.

Howard C. Rice, Jr., *The Rittenhouse Orrery* (Princeton: Princeton University Library, 1954; 88 pp., illustrated) describes the Princeton instrument, built by David Rittenhouse before 1800.

Charles E. Smart, *The Makers of Surveying Instruments in America since 1700* (Troy, N.Y.: Regnal Art Press, 1962; 182 pp., 60 illustrations). Reviewed by Silvio A. Bedini in *Technology and Culture*, 5 (Spring 1964), 260–261. The author is chairman of the Board of W. & L. E. Gurley, makers of surveying instruments.

University of Virginia, *Catalogue of the Adolph Lomb Optical Library at the University of Virginia* (Charlottesville, 1947 [University of Virginia Bibliographic Series, No. 7], 203 pp.). About 2,500 photoprints of author catalogue cards. The collection emphasizes the history and development of optical science. Alphabetically arranged; no subject index.

Ernst Zinner, *Deutsche und Niederländische Astronomische Instrumente des 11–18. Jahrhunderts* (Munich: C. H. Beck'sche, 1956; 679 pp., 80 pls.). Bibliography, 284 entries. One of the most important works on instrumentation ever published, or to be published. [Note by Silvio A. Bedini.]

Particular collections of instruments are described in the following titles:

M. V. Brewington, *The Peabody Museum Collection of Navigating Instruments, with Notes on Their Makers* (Salem, Mass.: Peabody Museum, 1963; 154 pp., 57 pls.). Reviewed by Silvio A. Bedini in *Technology and Culture*, 5 (Fall 1964), 640–641.

I. Bernard Cohen, *Some Early Tools of American Science* (Cambridge, Mass.: Harvard University Press, 1950, 201 pp.). A history of scientific instruments at Harvard.

[C. H. Josten], *Scientific Instruments (13th–19th Century): The Collection of J. A. Billmeir* (Oxford, 1954).

[Francis R. Maddison], *A Supplement to a Catalogue of Scientific Instruments in the Collection of J. A. Billmeir* (London: Partridge, 1957; 95 pp., 37 pls.). Bibliography.

[Maria Luisa Bonelli], *Catalogo degli Strumenti del Museo di Storia della Scienza* (Florence: Olschki, 1954; 394 pp., 89 illustrations). Catalogue of the Medici collection.

Florian Cajori, *A History of the Logarithmic Slide Rule* (1909?; 116 pp., illustrated). Reprinted in *String Figures and Other Monographs* (New York: Chelsea, 1960). The history is well told from the origins in 1630 through the log-log duplex slide rules of about 1900. A chronological annotated list of 256 "Slide rules designed and used since 1800" is followed by a "Bibliography of the Slide Rule," pp. 107–121.

Philip and Emily Morrison, Eds., *Charles Babbage and His Calculating Engines* (New York: Dover, 1961; 400 pp., illustrated). A collection of papers by Babbage and others, to which has been added an introduction by the editors. Bibliography, pp. xxxvii–xxxviii; list of Babbage's published papers, pp. 372–377.

René Taton, "Sur l'invention de la machine arithmétique," *Revue d'histoire des sciences et de leurs applications*, 16, No. 2 (Apr.–June 1963), 139–160. On Pascal's calculating machine.

P. J. Booker, *A History of Engineering Drawing* (London: Chatto & Windus, 1963; 239 pp., illustrated). A specialist's work, largely for one who is familiar with drafting techniques. Unfortunately, the author has avoided any extensive treatment of instruments. This deficiency is repaired, however, in the article that follows. Bibliography, eighty-six entries. Reviewed by D. Chilton in *Technology and Culture*, 6 (Winter 1965), 128–130.

Henry W. Dickinson, "A Brief History of Draughtsmen's Instruments," Newcomen Society, *Transactions*, 27 (1949–1951), 73–84 + 8 pls. The principal sources are indicated in footnotes.

Franz Maria Feldhaus, *Geschichte des Technischen Zeichnens* (2nd ed., Wilhelmshaven: Franz Kuhlmann KG, 1959; 121 pp., 90 figs. and 5 colored pls.). A handsome and well-filled book on the history of mechanical drawing. Apparatus and instruments are treated adequately. In an informative review by R. S. Hartenberg in *Technology and Culture*, 2 (Winter 1961), 45–49, reference is made to a three-part treatise on the same subject in *Blätter für Technikgeschichte*, *19–21* (1957–1959), which includes a bibliography of nearly two hundred entries.

See also
Wolf, *History of Science* (pp. 14) illustrates and describes many instruments of the sixteenth through eighteenth centuries.
Child, *Tools of the Chemist* (p. 262).
For surveying and navigational instruments, see Parts XIV.B.5 and XIV.C.2.
For electronic computers, see Part XIV.E.3.

6. *Weights, Measures, and Standards*

Torsten K. W. Althin, *C. E. Johansson 1864–1943*, Cyril Marshall, Trans. (Stockholm: Privately Printed, 1948; 165 pp., illustrated). The emphasis of the author, former Director of Tekniska Museet, is upon the details of Johansson's contributions to precise measurement in the development of gauge blocks. "Sources Employed," pp. 164–165.

A. E. Berriman, *Historical Metrology* (London: Dent, 1953; 224 pp., illustrated). A summary of weights and measures from the earliest times to about 1900, based upon extensive sources. Bibliography, pp. 200–215.

William Burden, *66 Centuries of Measurement* (Dayton, Ohio: The Sheffield Corporation, 1960, 136 pp.). A paperback compilation of twenty-two short articles published in *The Sheffielder*, a house organ. A bibliography, pp. 129–136, lists a number of significant references on various aspects of metrology and measuring instruments.

Paul Burguburu, *Essai de bibliographie métrologique universelle* (Paris, 1932, 328 pp.). More than 4,000 titles, many in lesser-known journals, are listed. All historical periods are included. This work appeared serially in *Bibliographe Moderne*, Paris, 1926–1931, which is in several U.S. libraries.

Rexmond C. Cochrane, *Measures for Progress. A History of the National Bureau of Standards* (Washington: National Bureau of Standards, 1966; 703 pp., illustrated). The bureau was established in 1901. Earlier history of U.S. standards is touched lightly.

Adrien Favre, *Les Origines du Système Métrique* (Paris: Presses universitaires de France, 1931, 242 pp.) gives the French view. C. D. Hellman, whose review of the book is in *Isis, 16* (1931), 449–450, repairs omissions and approves of the bibliography.

Sami Hamarneh, "The First Recorded Appeal for Unification of Weight and Measure Standards in Arabic Medicine," *Physis*, 5, No. 3 (1963), 230–248. Of the thirteenth century, A.D.

Bruno Kisch, *Scales and Weights: A Historical Outline* (New Haven: Yale University Press, 1965; 297 pp., illustrated). A comprehensive outline of weighing and its instruments from the origins to modern times. There are fourteen pages of bibliographic references. Reviewed by Silvio A. Bedini in *Science, 150* (Oct. 8, 1965), 203–204.

Douglas McKie, "Origins of the Metric System," *Endeavour*, 22 (Jan. 1963), 24–26. Brief and lucid survey of the need for and results of reforms of standards.

Ralph W. Smith, *The Federal Basis for Weights and Measures* (Washington: National Bureau of Standards, 1958 [Circular 593], 23 pp.). A pamphlet for popular distribution, citing *Annals of Congress, American State Papers*, and other NBS publications. The significant and in many ways exciting developments (for example, Joseph Saxton's work in making the state standards) deserve much better treatment than this.

The metric-versus-English system controversy, a fit subject for the comic talents of Gilbert and Sullivan, has been treated with ax-grinding solemnity by an unbelievably large number of advocates of (1) the status quo and (2) "progress." The catalogue of a large library will reveal the clearly drawn battle lines in the contest which is still very much alive.

The very first paper published in American Society of Mechanical Engineers, *Transactions, 1* (1880), 1–28, was an argument by **William Sellers** showing the meager advantages to be gained and the great cost of conversion to the metric system. Perhaps less heat would be generated by each new demi-generation as it discovers the metric system if a careful scholar would review the whole comedy.

Edward F. Cox, "The Metric System: A Quarter-century of Acceptance (1851–1876)," *Osiris, 13* (1958), 358–379. This article places the author in the camp of progress.

Edward F. Cox, "The International Institute: First Organized Opposition to the Metric System" (*Ohio Historical Quarterly, No. 1* [1959]), is an article I have not seen.

Carl F. Kayan, Ed., *Systems of Units: National and International Aspects* (Washington: American Assn. for the Advancement of Science, 1959 [Pub. 57]). Although this is a polemical work, nearly devoid of scholarship, one paper, by Chauncey D. Leake, has a good basic bibliography of twenty-eight entries, 1821–1959, on the metric-versus-English campaign.

Bernard Semmel, "Parliament and the Metric System," *Isis, 54* (Mar. 1963), 125–133. Traces several unsuccessful movements, 1807–1957, to adopt the metric system in England.

The development of standards for screw threads, nail sizes, wire gauges, and finally bearings, shafting, and so forth, which began in the nineteenth century, appears to have been an enormously important factor in our increasing material affluence. Only a few articles have been written on the subject.

Henry W. Dickinson and Henry Rogers, "Origin of Gauges for Wire, Sheets and Strip," Newcomen Society, *Transactions, 21* (1940–1941), 87–98. This is the only article I have seen on this particular aspect of the subject.

"Standards in Industry," a collection of thirty-six articles in American Academy of Political and Social Science, *Annals, 137* (May 1928), 1–258. Papers on technical, economic, and social effects of standards were contributed by representatives of industry, labor, government, and the colleges.

George V. Thompson, "Intercompany Technical Standardization in the Early American Automobile Business," *Journal of Economic History, 14* (Winter 1954), 1–20. A much-cited pioneering article.

7. Mechanical Power Transmission, Bearings, and Lubrication

Although electrical power transmission was used to send the power of Niagara Falls to Buffalo in 1895, it was this project that brought out some of the bravest schemes for long-distance power transmission by air and, for shorter distances, by rope.

John T. Nicholson, "Transmission and Distribution of Power by Compressed Air" (Canadian Society of Civil Engineers, *Transactions, 7* [1893], 56–93), demonstrates "the fact" that an air system is more economical, convenient, and secure than any yet known.

A very instructive summary of thirteen alternative proposals for transmitting power from Niagara Falls to Buffalo is in **Engineering, 52** (1891), 468, 559–562, 589–591.

John J. Flather, *Rope-Driving: a Treatise on the Transmission of Power by Means of Fibrous Ropes* (New York: Wiley, 1897, 230 pp.). A manual of practice.

W. C. Unwin, "Power Distribution from Central Stations," *Electrical World*, 20 (Aug. 20, 27, 1892), 124–125, 136–137, gives examples of high-pressure water systems (London), compressed-air systems (Paris), and electrical systems (Lauffen to Frankfort).

R. K. Allan, *Rolling Bearings* (London: Pitman, 1945; 401 pp., illustrated). Chapter 1, pp. 1–28, is a chronology of bearing development. Bibliography, pp. 369–381, has six references before 1900, fifty before 1920. List of manufacturers, pp. 382–388. Useful, but author is centrally concerned with current practice, not history.

E. Heidebroek, "Die Entwicklung des Lagers seit Ende des vorigen Jahrhunderts," *Technikgeschichte*, 28 (1939), 44–49. On the development of bearings since the end of the last century.

M. D. Hersey, "Notes on the History of Lubrication," American Society of Naval Engineers, *Journal*, 45 (1933), 411–429; 46 (1934), 369–385. I have not seen these articles.

W. Jürgensmeyer, *Die Wälzlager* (Berlin: Springer, 1937, 512 pp.). History of roller bearings, pp. 1–84. Title listed in Newcomen Society, *Transactions*, 17, 221.

Hudson T. Morton, *Anti-Friction Bearings* (Ann Arbor, Mich.: the Author, 1954; 394 pp., illustrated). A book for bearing users and for the instruction of engineers. The historical chapters, pp. 1–24, 355–382, include a few very early bearings, patent drawings of nearly a hundred designs, ca. 1863–1944, and capsule histories of thirty-odd U.S. bearing manufacturers. Twenty-five reference notes. Useful but not definitive.

H. Naylor, "Bearings and Lubrication," *Chartered Mechanical Engineer*, 12 (Dec. 1965), 642–647, 649. A general survey from earliest times. Despite the key role of improved lubrication in making possible high-speed machinery, I know of no monographic treatment of the subject.

In Rees, *Cyclopaedia* (p. 55) bearings of about 1810 are discussed in the article "Machinery" and roller bearings in the article "Millwork." Feldhaus (p. 2), treats bearings in cols. 595–602.

8. *Cranes, Rigging, and the Moving of Heavy Objects*

A. Burford, "Heavy Transport in Classical Antiquity," *Economic History Review,* 2nd ser., *13,* No. 1 (Aug. 1960), 1–18. A discussion of the use of oxen and wagons in Greece, and other matters.

Robert F. Heizer, "Ancient Heavy Transport, Methods and Achievements," *Science, 153* (Aug. 19, 1966), 821–830, 98 refs. Illustrated review of methods of hauling large, monolithic loads, from Egyptian times onward. The author is writing a book on the subject.

F. R. Forbes Taylor, "Heavy Goods Handling Prior to the Nineteenth Century," Newcomen Society, *Transactions, 35* (1962–1963), 179–191. A survey which includes an interesting summary of forty-six cranes in use around 1600.

H. Musical Instruments

Extensive articles on technical and acoustical details of instruments will be found in *Grove's Dictionary of Music and Musicians,* **Eric Blom,** Ed. (5th ed., 10 vols., New York: St. Martin's, 1960–1961). Articles are supplemented by bibliographies or reference notes. The piano, for example, is treated historically in twenty pages; this is followed by a bibliography of thirty-two entries.

Cecil Clutton and Austin Niland, *The British Organ* (London: Batsford, 1963; 320 pp., illustrated). Selective bibliography, twelve entries.

Curt Sachs, *The History of Musical Instruments* (New York: W. W. Norton, 1940; 505 pp., illustrated). A standard work that appeared first in 1913. The instruments of the world are described, employing a historical approach. Topical bibliography, pp. 469–487.

Curt Sachs, *Real-Lexikon der Musik-Instrumente* (2nd ed., New York: Dover, 1964; 452 pp., 200 illustrations). A corrected and revised version of the previous title. New Introduction in English; text in German.

André Schaeffner, *Origine des instruments de musique* (Paris: Payot, 1936, 406 pp.). **Russo** (p. 17), p. 152, notes that this book has "importante bibliographie." On the same page, Russo lists three titles on organs, one on bells, and one on stringed instruments.

Peter Williams, *The European Organ 1450–1800* (London: Batsford, 1966; 336 pp., illustrated). A handsomely illustrated book on the technical details of individual organs.

Alexander Wood, *The Physics of Music* (6th ed., London: Methuen, 1962; 258 pp., illustrated). This standard work, first published in 1944, has been revised by J. M. Bowsher. It is written for "those whose main interest is on the musical side" of the "interesting borderline territory which lies between physics and music." Pitch, quality, intensity, and loudness, all are grist for the authors' mills. Characteristics of individual instruments are analyzed.

I. Military Technology and War

As pointed out to me by Brooke Hindle (in his *Technology in Early America*, p. 47), the technical history of war can be relatively brief because so much of it is buried under other rubrics. Bridges, surveying, ships, road vehicles, aircraft, the arts of communication, manufacturing of all kinds, automatic control — these and many more subjects contribute to and have been stimulated by the technology of war.

In this context, I cannot help but feel that Lewis Mumford's question, which he asks as he explores our nearly unlimited technical capabilities, is central (in our time, at least) to the study of the history of military technology: "Is this association," he asks, "of inordinate power and productivity and equally inordinate violence and destruction a purely accidental one?" (*Technology and Culture*, 7 [Summer 1966], 313).

R. C. Anderson, *Oared Fighting Ships from Classical Times to the Coming of Steam* (London: Percival Marshall, 1962; 102 pp., illustrated). By an acknowledged maritime authority.

Ralph Andreano, Ed., *The Economic Impact of the Civil War* (Cambridge, Mass.: Schenkman, 1962, 205 pp.). A collection of eleven articles on a generally ignored aspect of this war *and* of U.S. technical history. There is also a "Statistical Supplement . . . 1850–1880," pp. 169–203.

Robert V. Bruce, *Lincoln and the Tools of War* (Indianapolis: Bobbs-Merrill, 1956; 368 pp., illustrated). The President's relationship to design and procurement of breech-loading rifles, machine guns, artillery, and other armaments is set out in this book which was written from primary sources. While I think the author depends too much upon his information to tell its own story, this is the only serious work I know of that considers technical details of a considerable part of Civil War armament.

Charles Ffoulkes, *The Armourer and His Craft* (London: Methuen, 1912, 199 pp., illustrated). Reprinted Bronx, N.Y.: Blom, 1967. A survey by a leading authority of tools, techniques, and materials employed in the making of body armour. A polyglot glossary of terms, pp. 153–167; select bibliography, pp. xx–xxii; short biographies of armourers, pp. 131–146.

Sidney Forman, "Early American Military Engineering Books," *Military Engineer*, 46 (Mar.–Apr. 1954), 93–95, describes briefly about a dozen titles earlier than 1830. The author omits mention of a volume of plates to accompany Tousard's work.

J. F. Hayward, *The Art of the Gunmaker* (2 vols., London: Barrie, 1962–1963). An attractive work that emphasizes decorative aspects, 1500–1830, of hand firearms. The author is a keeper in Victoria and Albert Museum.

A. V. B. Norman and Don Pottinger, *A History of War and Weapons, 449 to 1660. English Warfare from the Anglo-Saxons to Cromwell* (New York: Crowell, 1966; 224 pp., illustrated). A nicely illustrated short summary of tactics, armor, and weapons, intended for the general reader. The sources, consisting of somewhat over a dozen titles, were carefully chosen, and the list tells the student where to go next.

Ralph Payne-Gallwey, *The Crossbow . . . with a Treatise on the Balista and Catapult . . . and the Turkish Bow* (London, 1903) has been reprinted (London: Holland, 1958; 328, 24, 23 pp.). An illustrated, exhaustive work by an enthusiast (in the English sense).

George C. Stone, *A Glossary of the Construction, Decoration, and Use of Arms and Armor in All Countries and in All Times* (Portland, Me.: Southworth, 1934; fo., 694 pp.). Profusely illustrated, unfortunately with indifferent reproductions of good pictures. Bibliography, pp. 687–694.

Paul Wahl and Donald R. Toppel, *The Gatling Gun* (New York: Arco, 1965; 4to, 168 pp., illustrated). An exhaustive treatise on this machine gun, from about 1860. Bibliography, thirty-two entries.

The following three titles were taken from **Warren G. Ogden, Jr.,** "A Bibliography of the Art of Making Artillery, Concerning Books with Accurate Drawings of Early Cannon, and their Carriages, Collected during Forty Years of Active Model Engineering," a dittoed annotated list of twenty-one titles. Mr. Ogden's address is 316 Johnson St., North Andover, Mass. 01845.

Charles Ffoulkes, *The Gun-Founders of England* (Cambridge: Cambridge University Press, 1937). A standard work.

Gaspard Monge, *Description de l'art de fabriquer les canons* (Paris, 1793). This heavily illustrated work describes the principles and practice of building artillery. It is an important source book.

Bernhard Rathgen, *Das Geschütz im Mittelalter* (Berlin: VDI-Verlag GmbH, 1928). There are several dimensioned drawings among the twelve plates on artillery, mostly bombards. There is a bibliography listing nearly 200 titles.

See also
Mayer, *Islamic Armourers* (p. 281)
Cockle, *Bibliography* (p. 40), for titles before 1650
Horst de la Croix, "Fortification in Renaissance Italy" (p. 40)
Higgins Armory, *Catalog of Armor* (p. 191)
Spaulding and Karpinski, *Early Military Books* (p. 43)

J. Industrial Organization

1. *General, Including the "Labor Problem"*

Edith Abbott, *Women in Industry* (New York: Appleton, 1910, 409 pp.). On the duties, wages, conditions, and attitudes of women operatives in the United States. The trends during the nineteenth century are traced. Bibliography, pp. 392–399.

Hugh G. J. Aitken, *Explorations in Enterprise* (Cambridge, Mass.: Harvard University Press, 1965, 420 pp.). Eighteen articles on the entrepreneur by Cole, Schumpeter, Cochran, Passer, Flinn, Habakkuk, and others prominent in this branch of economic and business history. Eleven articles appeared first in *Explorations in Entrepreneurial History;* the rest appeared in other periodicals. The editor's introduction and interlocutory passages alone make the book worth while.

Elizabeth F. Baker, *Technology and Woman's Work* (New York: Columbia University Press, 1964, 460 pp.). Women constitute one third of the labor force in the United States today. The reasons for the sharp increase of women working outside the home are explored. Bibliography, pp. 443–450.

William Miller, Ed., *Men in Business. Essays on the Historical Role of the Entrepreneur* (New York: Harper & Row, 1962, 389 pp.). A "Torchbook" paperback reissue of a book first published in 1952, to which have been added two chapters by the editor. There is a chapter on John Stevens (1749–1838) and one on Frank Julian Sprague

(1875–1934). This work was one of the results of Harvard's Research Center in Entrepreneurial History which, like the Accademia del Cimento, lasted for just ten years.

Sidney Pollard, *The Genesis of Modern Management. A Study of the Industrial Revolution in Great Britain* (London: Arnold, 1965, 328 pp.). A fresh study of the managerial, as distinguished from the entrepreneurial, function. The distinction between manager and entrepreneur is drawn in order to concentrate on the former. The author is interested in how a firm is run after the broad decisions, such as what will be made and at what rate, have been reached. Bibliographic information is in the notes, pp. 273–324.

Richard S. Rosenbloom, "Some 19th-Century Analyses of Mechanization," *Technology and Culture,* 5 (Fall 1964), 489–511. A fresh look at the writings of Babbage, Ure, Marx, and others. I hope the author will continue to work toward the "grand theory" of the effects of mechanization that he treats so sensibly in this article.

George Unwin, *Industrial Organization in the Sixteenth and Seventeenth Centuries* (Oxford: Clarendon, 1904, 277 pp.). A pioneering work. Bibliography: manuscripts, pp. 253–262, books and articles, pp. 263–270.

W. A. Young, "Works Organization in the 17th Century," Newcomen Society, *Transactions,* 4 (1923–1924), 73–101. A modified putting-out system at Winlanton, near Newcastle upon Tyne, is described. A precursor of the company union was also present.

See also
Freudenberger, *Waldstein Woolen Mill* (p. 273)

2. *"American System" of Manufacture; Assembly Line*

In view of the central significance of the "American system" of manufacture — so named by the British around 1851 — in the development of mass production industries, it is remarkable that no monograph on the subject has yet appeared. Articles describing and analyzing the technical contributions of Hall, Warner, North, and others are needed, perhaps as the first steps toward a balanced synthesis.

Only the articles by Sawyer and Woodbury are directly concerned with the "American system"; the ancestry of the methods is suggested in the remaining titles. The "American system" of manufacture involves a sequential series of operations carried out on successive special-purpose machines that produce interchangeable parts.

K. R. Gilbert, *The Portsmouth Blockmaking Machinery* (London: Her Majesty's Stationery Office, 1965 [A Science Museum Monograph]; 41 pp., illustrated). This is a fully illustrated description of the development and functioning of the system for making ships' (pulley-) blocks at Portsmouth dockyard. The machines were in use before 1810.

Sten Lundvall, "Christopher Polhems Skärmaskiner för Urhjul," *Daedalus* [Stockholm], (1949); pp. 51–62, 10 illustrations. Although in Swedish, this article on Polhem's gang-milling machine of 1729 for the cutting of gear teeth and other similar machines conveys through its pictures the essential facts about surviving machines in the Tekniska Museet, Stockholm. An article in English on the same subject is in the following title.

Christopher Polhem, the Father of Swedish Technology, William A. Johnson, Trans. (Hartford, Conn.: Trustees of Trinity College, 1963; 259 pp., illustrated). A collection of articles by various authors on Polhem's activities in manufacturing, mining, and building. His manufacturing innovations are described in Gustaf Sellergren, "Polhem's Contributions to Applied Mechanics," pp. 109–162. A handsome book.

Samuel Rezneck, "The Rise and Early Development of Industrial Consciousness in the United States, 1760–1830," *Journal of Economic and Business History,* 4, pt. 2 (1932), 784–811. Using different data, Rezneck arrives at substantially the same conclusions as Sawyer regarding the attitudes of Americans toward industrial pursuits.

John E. Sawyer, "The Social Basis of the American System of Manufacturing," *Journal of Economic History,* 14 (1954), 361–379. The cultural attributes of nineteenth-century Americans which encouraged industrialization and which were noted by European visitors to the United States were noted again after the Second World War by the British "productivity teams" that inspected U.S. industries. Very good article.

Robert S. Woodbury, "The Legend of Eli Whitney," *Technology and Culture,* 1 (Summer 1960), 235–251. The contributions of Whitney to the "American system" of manufacture are critically examined. Except for Whitney's ability to sell an undeveloped idea, little remains of his title as father of mass production.

No connected account of the development of assembly lines has yet appeared. Some inquiry into the subject has, however, been made.
Giedion, in his *Mechanization Takes Command* (p. 12), has thought

more carefully about the origins and implications of assembly lines than anyone else, so far as I know. See, in his index, entries under "Assembly line." A number of references occur also in **Nevins**, *Ford* (p. 235), Vols. 1 and 2. See the index in each volume.

A series of articles by **Fred Colvin** describing Henry Ford's Dearborn plant was published in *American Machinist* during May–September 1913.

Horace L. Arnold and Fay Faurote, *Ford Methods and the Ford Shops* (New York: The Engineering Magazine Co., 1915; 440 pp., illustrated). This well-illustrated account of a tour through Ford's Highland Park plant appeared originally as a series of articles in *Engineering Magazine*. Close, admiring attention is given to machines and methods. Here is one of the milder remarks about the employees, who were very well paid: "Wilful insubordination is, of course, absolutely intolerable, and Ford workers must be, first of all, docile."

3. *Scientific Management and Systems Analysis*

In a perceptive review of **Graham Adams, Jr.,** *Age of Industrial Violence 1910–15* (p. 102), David Montgomery points out the connections between labor-management conflict and the rise of scientific management, which the book's author apparently missed. Montgomery's review, in *Technology and Culture*, 8 (Apr. 1967), 234–237, specifies the nature of the intellectual climate of the period in a way that had not occurred to me. The book, incidentally, is based upon the proceedings of a federal government commission.

Hugh G. J. Aitken, *Taylorism at Watertown Arsenal; Scientific Management in Action, 1908–1915* (Cambridge, Mass.: Harvard University Press, 1960, 269 pp.). By far the best analysis that I have read of the Taylor system. The book is unusual in that while it is critical of the assumptions and methods of Taylor and his disciples, it manages to convey its message without becoming shrill.

A contemporary review, pro and con, of Taylorism will be found in *American Economic Review Supplement*, 2 (1912), 117–130.

Robert Boguslaw, *The New Utopians. A Study of System Design and Social Change* (Englewood Cliffs, N.J.: Prentice-Hall, 1965, 213 pp.). A penetrating analysis of the current practice of "systems engineering," usually computer-centered. Drawing upon utopian schemes from the time of Plato to B. F. Skinner, the author shows convincingly that current systems are not only *not* value-free but that they necessarily incorporate the values of the systems engineer, the computer manufacturer, and the programmer. The new system designers "receive their impetus from the newly discovered capabilities of computational equipment rather than from the fundamental moral, intellectual, or even physical requirements of mankind." Very good.

V. M. Brown, *Scientific Management, A List of References in the New York Public Library* (New York: New York Public Library, 1917, 81 pp.). This appeared first in New York Public Library *Bulletin, 21* (1917), 19–43, 83–136.

Raymond E. Callahan, *Education and the Cult of Efficiency* (Chicago: The University of Chicago Press, 1962; 273 pp.). The intrusion of business values into the public school systems in the United States between 1900 and 1930 and the influence of the concepts of scientific management are explored in this informative and disturbing book.

Jean Christie, "Morris Llewellyn Cooke: Progressive Engineer," *Dissertation Abstracts, 25* (Sept. 1964), 1,870–1,871. This Columbia University dissertation is unpublished. Cooke (1872–1960) was an early disciple of Taylor, but he outgrew the assumptions of the scientific managers and became finally a liberal Democrat, rendering public service in the administrations of Presidents Roosevelt and Truman.

Morris L. Cooke, *Academic and Industrial Efficiency* (New York, 1910 [Carnegie Fund for the Advancement of Teaching, Bulletin No. 5], 134 pp.). Cooke was in 1910 a follower of Frederick W. Taylor. The reasons for his study (of a physics department, not an engineering department) were "partly in the existence in the colleges of new and large problems and partly in the criticisms of American colleges and universities made during the past few years by business men."

Frank B. Copley, *Frederick Winslow Taylor* (2 vols., New York, 1923) is the standard biography of the key figure in scientific management.

Ernest Dale and Charles Meloy, "Hamilton MacFarland Barksdale and the DuPont Contributions to Systematic Management," *Business History Review, 36* (Summer 1962), 127–152. Barksdale (1861–1918) was influential during a period of rapid growth of the DuPont Company.

Fifty Years Progress in Management 1910–1960 (New York: American Society of Mechanical Engineers, 1960; 4to, 329 pp.). Five decennial reports, each entitled "Ten Years Progress in Management," and the 1912 report, "Present State of the Art of Industrial Management," are here compiled. The compiler apparently was vaguely dismayed to find that various, and "even conflicting," views appear in the book. A historical sketch by Lillian M. Gilbreth and W. J. Jaffe, "Management's Past — A Guide to its Future," pp. 5–15, has a bibliography of seventy-nine entries, from Babbage to Wiener.

Samuel Haber, *Efficiency and Uplift. Scientific Management in the Progressive Era 1890–1920* (Chicago: University of Chicago Press, 1964, 181 pp.). A critical, rather flippant review of the effects of F. W. Taylor, Cooke, and others on social reforms of the period under review. Very good bibliography, pp. 169–173, and wide-ranging footnote references.

William J. Jaffe, *L. P. Alford and the Evolution of Modern Industrial Management* (New York: New York University Press, 1957, 366 pp.). List of Alford's writings, pp. 328–336; bibliography, pp. 337–345.

Joseph A. Litterer, "Systematic Management: Design for Organizational Recoupling in American Manufacturing Firms," *Business History Review,* 37 (Winter 1963), 369–391. A review of ideas advanced during the period 1875–1900 by Towne, Metcalf, and others.

Harwood F. Merrill, *Classics in Management* (New York: American Management Assn., 1960, 446 pp.) has selections from the writings of Babbage, Towne, Taylor, Gilbreth, Follet, and others. A "Selected Bibliography," pp. 437–446, lists from two to twelve additional titles by each of the authors represented.

Milton J. Nadworny, *Scientific Management and the Unions 1900–1932. A Historical Analysis* (Cambridge, Mass.: Harvard University Press, 1955, 187 pp.). A study of labor-management relations in the light of new innovations in management. Bibliographic reference notes, pp. 155–181.

Frederick W. Taylor, *Scientific Management* (New York: Harper, 1947). Three books are conveniently placed in a single volume: Taylor's *Shop Management* (1903, 207 pp.), Taylor's *The Principles of Scientific Management* (1911, 144 pp.), and *Taylor's Testimony Before the Special House Committee* (1912, 287 pp.). The testimony before the "Special Committee of the [U.S.] House of Representatives to Investigate the Taylor and Other Systems of Shop Management" portrays more clearly than anything else I have seen just what kind of a mind conceived and promoted the Taylor system. Very revealing, even if not unexpected.

C. Bertrand Thompson, *The Theory and Practice of Scientific Management* (Boston: Houghton Mifflin, 1917, 319 pp.). As pointed out by Haber (above), Chapter 5, "The Literature of Scientific Management," pp. 173–269, is a very good chronological and topical review of the subject. In addition, there is a formal bibliography, pp. 271–308.

L. Urwick and E. F. L. Brech, *The Making of Scientific Management* (3 vols., London: Pitman, 1966; 12mo, 196, 241, 225 pp.). First published *ca.* 1944–1948. Vol. 1; "Thirteen Pioneers," includes biographical sketches of Babbage, Taylor, Fayol, Follett, Rowntree, Gantt, Rathenau, Le Châtelier, Fréminville, Dennison, Gilbreth, and Elbourne; Vol. 2: "Management in British Industry," historically treated; Vol. 3: "The Hawthorne Investigations" of Western Electric Company, carried out during the 1930's.

Norman B. Wilkinson, "In Anticipation of Frederick W. Taylor: a Study of Work by Lammot du Pont, 1872," *Technology and Culture,* 6 (Spring 1966), 208–221. DuPont's study of movements of ingredients in a powder plant may or may not have been a unique early "efficiency" study. Wilkinson suggests several intriguing lines of further research.

The **Carl G. Barth** papers and a carton of his slide rules are in Harvard Business School, Baker Library. See *National Union Catalog of Manuscript Collections* (p. 114), MS62–4827.

Gilbreth manuscripts are in the Goss Library at Purdue University. See *Manual of the Library* (p. 27), pp. 179–206, and *National Union Catalog of Manuscript Collections* (p. 114), MS61–3075. A microfilm edition of a part of these papers is described in **Gilbreth,** *Selected Papers* (p. 116).

Elizabeth G. Hayward, Comp., *A Classified Guide to the Frederick Winslow Taylor Collection* (Hoboken: Stevens Institute of Technology, 1951, 45 pp.) lists papers and reports but does not reveal the extent of the collection. **Hamer,** *Guide to Archives* (p. 113), p. 315, says collection contains "thousands of items," including European diaries, 1869–1870; manuscripts of Taylor's articles and correspondence concerning them; correspondence, 1887–1915, with many distinguished men; photographs and lantern slides.

4. Quality Control

Quality-control techniques pervade much of American industry. Because the idea conveyed to an unreflecting public is one of making sure that quality remains high, two observations are in order. First, control means to check or regulate, to keep within bounds. Birth control, for example, is not intended to increase births. Second, Walter Shewhart, who was the pioneer in establishing the techniques of statistical quality control, was very clear in defining "quality" as an attribute or characteristic that may be described objectively, or at least with arbitrary exactness. Yet Shewhart apparently did not object to the connotation of high quality that the advertising mind took hold of.

A guide to the chronology of quality-control methods is in **S. B. Littauer,** "Development of Statistical Quality Control in the United States," *American Statistician, 4* (Dec. 1950), 14–20. The author has been intimately involved in the developments that he describes.

K. "Engineering Sciences": Thermodynamics, Hydraulics, Aerodynamics, Strength of Materials, and Kinematics

The province of the engineering sciences lies between pure physical science — physics, mechanics, mathematics — on the one hand and the empirical and intuitive knowledge of the engineer on the other. This section has been set apart not in order to bring together all of the strands of science that form a part of technology but merely to provide a place for a few works in disciplines that impinge on more than one subject; for example, hydraulics is of interest in both civil and mechanical engineering.

Aleksei Bogoliubov, *Istoriia mekhaniki mashin* (Kiev: Naukova Dumka, 1964; 460 pp., illustrated). A history of kinematics of machines, pointed out to me by Richard S. Hartenberg.

James B. Conant, Ed., *Harvard Case Histories in Experimental Science* (2 vols., Cambridge, Mass.: Harvard University Press, 1959). Carefully developed arguments, using long quotations from original sources, have been prepared by the editor and others on Boyle's experiments in pneumatics, concepts of temperature and heat, atomic-molecular theories, studies of fermentation, the concept of electric charge. Biological studies are also included in these volumes.

A. Rupert Hall, *Ballistics in the Seventeenth Century* (Cambridge: Cambridge University Press, 1952, 186 pp.) is an important study of scientific and technical understanding of guns and gunnery. Bibliography, pp. 172–181.

Stanley B. Hamilton, "The Historical Development of Structural Theory" (Institution of Civil Engineering, *Proceedings, 1,* pt. 3 [1952], 374–419), treats the subject from the time of Galileo to the present. Bibliography has seventy-four references.

R. S. Hartenberg and Jacques Denavit, *Kinematic Synthesis of Linkages* (New York: McGraw-Hill, 1964), opens with a mature and comprehensive chapter on rational approaches to the design of machines and mechanisms, from the seventeenth century onward. This chapter augments and corrects the following title.

Eugene S. Ferguson, "Kinematics of Mechanisms from the Time of Watt," U.S.N.M., *Bulletin* 228, Paper 27 (1962), 185–230. A review of both academic and practical approaches to mechanism design over a 200-year period.

Erwin N. Hiebert, *Historical Roots of the Conservation of Energy* (Madison: State Historical Society of Wisconsin, 1962). On the ideas of the Greeks and later thinkers on work and energy, leading up to the statement of thermodynamic principles in the nineteenth century.

Douglas McKie and Neils H. Heathcote, *The Discovery of Specific and Latent Heats* (London: Arnold, 1935, 155 pp.). Details of the work of Joseph Black (1728–1799), Johan Carl Wilcke (1732–1796), and others.

Morton Mott-Smith, *The Concept of Energy Simply Explained (formerly titled: The Story of Energy)* (New York: Dover, 1964; 215 pp., illustrated). A clear and generally sound nonmathematical explanation is given of the development of the ideas of thermodynamics, heat engines, and heat pumps. The book was published originally in 1934.

Gunhard Orvas and Leslie McLean, "Historical Development of Energetical Principles in Elastomechanics; Pt. 1, From Heraclitos to Maxwell," *Applied Mechanics Reviews*, 19 (Aug. 1966); 647–658, about 80 refs. On the development of ideas on conservation of energy.

Franz Reuleaux, *The Kinematics of Machinery: Outlines of a Theory of Machines*, Alexander B. W. Kennedy, Trans. (New York: Dover, 1963; 622 pp., illustrated). A reprint edition of the 1876 work that established the modern discipline of kinematics of mechanisms. The author has much of interest on the development of ideas regarding machines in general. His remarks on the "simple machines," for example, are penetrating and accurate.

Hunter Rouse and Simon Ince, *History of Hydraulics* (New York: Dover, 1957, 269 pp.). A systematic account of the academic understanding of phenomena of hydrostatics and hydrodynamics from classical times to the present. Bibliographic references are given at the end of each chapter.

Isaac Todhunter, *A History of the Theory of Elasticity and of the Strength of Materials, from Galileo to the Present*, edited and compiled by Karl Pearson (2 vols. in 3, Cambridge: Cambridge University Press, 1886–1893). A classic work on development of ideas, reprinted by Dover Publications, Inc.

Theodore von Kármán, *Aerodynamics* (New York: McGraw-Hill, 1963, 203 pp.). A paperback reissue of the Messenger Lectures of 1953, delivered at Cornell University by von Kármán. A chapter on aerodynamic research before 1900 is followed by a historical and nonmathematical explanation of subsequent developments.

Felix Wankel, *Rotary Piston Engines,* R. F. Ansdale, Trans. and Ed. (London: Iliffe, 1965). An attractive taxonomy of rotary engines and compressors, consisting of a series of figures in color which identify hundreds of designs from 1800 to the present. The presence of several rotary pumps in Ramelli's *Machine* of 1588 is noted, but since the objective of the book is not historical, identification of origins is understandably sketchy. Published originally as *Einleitung der Rotations-Kolbenmaschinen* (Stuttgart, 1963).

See also
Wolf (p. 14). The author is interested in showing the applications of science to practice.
Timoshenko, *History of Strength of Materials* (p. 275) is the authority in its field.

L. Process of Invention and Innovation

S. C. Gilfillan, *Inventing the Ship; A Study of the Inventions Made in her History between Floating Log and Rotorship* (Chicago: Follett, 1935). A well-known work on the sociology of invention.

John Jewkes, David Sawers, and Richard Stillerman, *The Sources of Invention* (London: Macmillan; New York: St. Martin's, 1958, 428 pp.). The authors have set out to examine beliefs, many of which are widely and dogmatically held, regarding the nurture and care of inventors and innovators. For example, the stereotype which holds that the lone inventor of the nineteenth century has been replaced in the twentieth by the research laboratory "team" is shown to be distorted and misleading. The general purpose of the book appears to be to provide "planners" with data on which to base decisions. While the answers are mostly wide of the mark, many of the right questions have for the first time been asked.

Fifty case histories of economically important twentieth-century inventions are given in synoptic form, pp. 263–410. Further sources for each case are listed.

Joseph Rossman, *Industrial Creativity; the Psychology of the Inventor* (3rd ed., New Hyde Park, N.Y.: University Books, 1964; 252 pp., illustrated). Bibliography, pp. 237–246. First published in 1931. The author is a noted patent attorney.

INDEX

Boldface numbers indicate principal entry. Numbers in parentheses indicate more than one entry on a page.

Individual technical societies that are listed alphabetically within geographical groups on pp. 162–169 are not separately indexed.

Incidental subject references (see, for example, Clark, p. 130) in general and collected works, bibliographies, and periodicals have not in general been separately indexed.

Abbot, Henry L., on stream gauging, 99
Abbott, Edith, *Women in Industry*, 297
Abell, Leggat, and Ogden, 278, 39
Abell, Sydney G., *Art of Turning*, 278, 39, 282
Abhandlungen und Berichte, 147
abstracts of dissertations, 30
Académie des Sciences
 archives, publication of, 121
 biographies, 77
 Descriptions des Arts et Métiers, 58
 history, in Stemple, 97–98
 Machines et inventions, 57
Accademia del Cimento, in *Landmarks in Science*, 171
Acheson, Edward G., 265
Acres, R. d', 44
Adams, Edward Dean, *Niagara Power*, 244
Adams, Graham, Jr.
 David Montgomery on, 300
 Industrial Violence, 102
Adcock, Henry, *Pocket-Book*, 65
Adelswärd, Axel, 131
Adney, Edwin T., 228
Adolphus, Lalit, 142
aerodynamics, 306
Africa
 Cline, *Mining and Metallurgy*, 257

Thomas, *The Harmless People*, 127
Agricola, Georgius
 in Beck, 40
 De Re Metallica, 44
 in Russo, 18
Agricultural History, 147
agricultural machinery, **214–217**
 Arthur Young's writings, Amery, 136
 preliminary inventory, National Archives, 110
 Science Museum titles, 187
 in U.S.D.A. reports, 101
Agriculture Department
 reports, 100–101
 yearbook, 101
Ahrons, Ernest L., 239
Aiken, John, *Country round Manchester*, 137
air conditioning, in Ingels, 249
aircraft, **240–243**
 A.I.A.A. papers, Library of Congress, 116
 Croydon (1935) airport film, 128
 illustrations and portraits, in Vanderbilt, 88, 124
 Lammot du Pont, Jr., collection, Eleutherian Mills, 116
 list of books in N.Y.P.L., 23
 preliminary inventory, National Archives, 110
 Science Museum titles, 187–188

307

Aitchison, Leslie, 256
Aitken, Hugh G. J.
  Exploration in Enterprise, 297
  Taylorism at Watertown, 300
  Welland Canal, 232
Alberti, Leon Battista, in Olschki, 11
Albion printing press, 151
Albion, Robert G.
  Naval and Maritime History, 228
  Rise of New York Port, 232
Aldred, Cyril, 8
Alembert, Jean d', *Encyclopédie*, 58
Alexander, E. P., *Iron Horses*, 239
Alexander, Patrick Y., *Aeronautics*, 241
Alford, L. P., and Modern Management, 302
Al-Jazari on Automata, 284
Allan, R. K., *Rolling Bearings*, 293
Allen, J. M., biography in Spalding, 83
Allen, James T., patent digests, 103–104
Allen, Zachariah
  Practical Tourist, 135
  Science of Mechanics, 62
Allgemeine Bauzeitung, 155
Alman, M., 120
Althin, Torsten K. W., *C. E. Johansson*, 290
American Assoc. for State and Local History, 35
American Assoc. for the Advancement of Science, bibliographies, 98
American Assoc. of Museums, *Directory*, 182
American Engineer, 155
American Historical Association
  bibliography of societies, 29
  Guide to Historical Literature, 18
  pamphlets, 18
  Writings on American History, 29
American Historical Review, 147
American Journal of Science, 155
American Mechanics' Magazine, 155
American Men of Science, 77
American Neptune, 147
American Philosophical Society
  history, 175
  manuscript guide, 117
American Quarterly, 147
American Railroad Journal, 156
American societies. See note at beginning of index.
American Society for Engineering Education, 179
American Society of Civil Engineers, library catalogue, 26
American State Papers, 100
American system of manufacture, 298–300
  in Tenth Census, 96
Amery, G. D., 136
Ames, John G., 93
Amman, Jost K.
  Ständebuch, 47
  in Klinckowstroem, 282
ancient and classical period
  general works, 8–10, 1–7
  Forbes, *Bibliographia antiqua*, 21
  Loeb Classical Library, 46
  Sarton, *History of Science*, 43
  Thorndike, *History of Science*, 44
agricultural technology
  Curwen, 215
  Anderson, 216
  Anderson, Bennett, Moritz, Storck, 217
chemical technology, Levey, Singer, 263
crafts and craftsmen, 280–281
energy conservation, Hiebert, Orvas, 305
engineering, civil
  Burr, 219
  Straub, 221
  Vitruvius, 48
glass, Duncan, 265
instruments, astronomical, Maddison, 287
lighting, Hough, O'Dea, Robins, Thwing, 250
masonry construction
  Atkinson, Blake, 221
  Edwards, Fakhry, 222
  Van Deman, 226
materials
  Lucas, Hodges, Smith, 255
  Weeks, 256
mechanisms and automata, 284
metals and mining
  Aitchison, 256
  Davies, 257
  Lavender, Rickard, Wertime, 258
military technology, 295–297
musical instruments, 294
oil, Forbes, 267
printing, Carter, 270
pumps, Ewbank, 244–245
roads, Forbes, 233
ships and navigation

Anderson, 229
Landstrom, 230
Taylor, 231
timekeepers, 285–286
tools, 276–277
transport, heavy, Burford, Heizer, 294
water supply, Herschel, Robins, 226
weights and measures, Berriman, Burden, Burguburu, 290, Kisch, 291
Ancient Peoples and Places (series), 8
Anderson, Joseph R., and Tredegar Iron Works, 259
Anderson, Oscar E., Refrigeration, 249
Anderson, R. C.
 early books on shipbuilding (2), 40
 Oared Fighting Ships, 295
 The Sailing-Ship, 229
Anderson, Russell H.
 Grain drills, 216
 Grain milling, 217
Andreano, Ralph L.
 American Petroleum Industry, 268
 Economic Impact of Civil War, 295
Annals of Electricity, 156
Annals of Science, 147
Annual of Scientific Discovery, 156
Annual Record of Science and Industry, 156
Anthony, L. J., 249
Appert, Nicholas, on canning, 217
Appleby, Ltd., Illustrated Handbook, 69
Applegath, Augustus, 151
Appleton's, American Biography, 74
Appleton's Cyclopaedia of Applied Mechanics, 65
Appleton's Dictionary of Machines, 63
Appleton's Mechanics' Magazine, 156
Appleton's Railroad and Steamboat Companion, 139
aqueduct, Los Angeles, 227
aqueduct, Roman
 in Blake, 221
 in Van Deman, 226
Arbman, Holger, 8
archaeology, ancient and classical periods, 8–10
archaeology, industrial, 191–192
 Industrial Archaeology, 150
archaeology, underwater, 8
archives, see manuscripts; National Archives
Archives Internationales d'Histoire des Sciences, 148

architecture, history of
 articles in Soc. of Architectural Historians, Journal, 153
 bibliography in Hitchcock, 222
 bibliography in R.I.B.A. Catalogue, 220
 Fletcher, 222
 Islamic Architects, 281
Argles, Michael, 176
Arkell and Douglas, catalogue, 69
Armengaud, J. E. and C. E., 196
Armed Forces Institute, in Kranzberg, 5–6
armor and arms, 295–297
 in Higgins Armory, 191
 Islamic, in Creswell, 20
 Islamic, in Mayer, 281
Armstrong, Edwin Howard, Man of High Fidelity, 254
Armytage, W. H. G.
 Civic Universities, 176
 Rise of the Technocrats, 204
 Social History of Engineering, 13
Arnold, Horace L., 300
Artz, Frederick B., 177
Ash, Lee, 29
Ashby, Eric
 "Education for Age of Technology," 176
 Technology and the Academics, 176
Ashton, Thomas S.
 Coal Industry of 18th Century, 266
 Iron and Steel, 258
Aspin, C., James Hargreaves, 272
assembly line, 299–300
Association. See note at beginning of index.
Atkinson, Frank, on horsepower, 243
Atkinson, R. J. C., Stonehenge, 221
Atlantic Cable, 253
Atlantic Monthly, 156
atomic energy, 249
Australia
 engineering profession, Lloyd, 178
 iron and steel, Hughes, 259
 technical education, Bruce, 178
autobiographies, 84–87
 guides, 79
automata, 284
 C.N.A.M. catalogue Z, 190
 in Diels, 9
automatic control, 284–285
automobile, 235–236
 Maxim, Horseless Carriage Days, 86
 Science Museum titles, 187–188
 U.S.N.M. catalogue, 191

automobile (*continued*)
  U.S. patents, Allen, 104
  see also internal-combustion engine
Automotive History Collection, Detroit, 235, 69
Ayres, Eugene, *Energy Sources*, 243

Babbage, Charles
  in Barlow, 63
  Calculating Engine, Morrison, 289
  memoir of, 98
Babcock, George H., 247
Bache family papers, 117
Bagrow, Leo, 227
Baier, Wolfgang, 264
Bailey, Erza, biography, in Spalding, 83
Baillie, Granville H.
  Britten's Old Clocks, 285
  Clocks and Watches, 285
Baines, Edward, Jr., *Cotton Manufacture*, 274
Baird, Spencer F.
  Annual Record, 156
  Iconographic Encylopaedia, 56–57
Baker, D. F., *Industrial Engineering*, 30
Baker, Elizabeth F., *Woman's Work*, 297
Baldwin Locomotive Works, papers, 114, 115
Bale, Manfred P., 275
Ballot, Charles, 13
Ballou's Pictorial, 159
Banks, John, 61
Bannister, Turpin, 221
Barba, Alvaro Alonso, in Russo, 17–18
Barksdale, Hamilton MacFarland, 301
Barlow, Peter, 63
Barnaby, K. C., 229
Barnard, John G., biography in NAS, 83
Barnes, Seth, biography in Spalding, 83–84
barometer, in Middleton, 288
Baron, Stanley, 218
Barr, E. Scott, 78
Barth, Carl G., 303
Bartol, B. H., 66
Barton, D. B.
  Cornish Beam Engine, 247
  Mines of East Cornwall and West Devon, 257, 192
  Mines of West Cornwall, 192
Barton, R. M., *China-Clay Industry*, 265
Barzun, Jacques, 204

BASF, Die, 148
Bass, George F., 8
Bassermann-Jordan, Ernst von, 285
Bate, John, in Ferguson, 15
Bates, Ralph S., 173
Bathe, Greville, publications, 170
Battison, Edwin A.
  "Eli Whitney and Milling Machine," 277
  reviews *Knaurs Geschichte*, 5
  "Screw-Thread Cutting," 278
bearings, antifriction, 293
Beaton, Kendall, 267
Beck, Ludwig, *Geschichte des Eisens*, 258
Beck, Theodor, *Geschichte des Maschinenbaues*, 40
Becker, Felix, *Lexikon*, 47
Becker, Max, engineering handbook, 64
Beckmann, Johann
  in Ferguson, 15
  History of Inventions, 16
Bedini, Silvio A.
  American Scientific Instruments, 287
  "Automata in History," 284
  "Compartmented Cylindrical Clepsydra," 285
  "Evolution of Science Museums," 184
  Mechanical Universe, 286
  notes by, 3, 10, 285, 288
  reviews Brewington, 288
  reviews Kisch, 291
  "Scent of Time," 285
beer, 218, 219
Beer, John J.
  German Dye Industry, 262
  reviews Siemens, 86–87
Beers, Henry P.
  Bibliographies in American History, 30
  Guide to Federal Archives of Civil War, 111
Begun, S. J., 253
Beiträge zur Geschichte der Technik und Industrie, 148
Belgium, travelers to U.S., 132
Belidor, Bernard, in Russo, 17–18
Bell, Thomas J., *Water Supply*, 225
Bell, Whitfield, Jr., A.P.S. collections, 117
Belt, Elmer
  Library of Vinciana, 50
  Manuscripts of Leonardo, 49
Bement, William B., in Scharf, 84

INDEX 311

Benedict and Burnham, in Van Slyck, 82
Benfield, Eric, 282
Benjamin, Park, 65
Bennett, R., *Corn Milling*, 217
Berger, Max, 129
Berkebile, Don H., 234
Berner, Richard C.
   manuscripts on microfilm, 119
   "Research in Forest History," 274–275
Bernstein, Henry T., 229
Bernt, Walter, 276
Berolzheimer, D. D., 263
Berriman, A. E., 290
Berry, W. Turner, 269
Bertrand, J. E., 59
Besson, Jacques
   in Beck, 40
   in Keller, 41
   in Russo, 17–18
   in Thorndike, 44
Besterman, Theodore
   World Bibliography, 18
   quoted, 213
Bettman Archive, address, 89
Bettman Portable Archive, 126
Beyern, Johann Matthias, 46
Bibliographic Index, 30
bibliographies, general, *see* Contents, Part II
bibliographies, subject matter, *see* Contents, Part XIV
Bibliothèque Nationale, portraits, 89
bicycles, *see* cycles
Bidwell, Percy W., 215
Bigelow, Erastus B., portrait, in Appleton's, 89
Bigelow, Jacob, *Elements of Technology*, 62
Bigmore, Edward C., 269
Billings, Charles E.
   in Spalding, 83–84
   in Van Slyck, 82
Billmeir, J. A., Collection of (2), 289
Bining, Arthur C., 258
biography, *see* Contents, Part VI
biological technology, in China, Needham, 2
Bion, Nicolas, in Russo, 17–18
Birembaut, Arthur, 18
Biringuccio, Vannocio
   in Beck, 40
   Pirotechnia, 47, 171
   in Russo, 17–18
Birkbeck, George, 176

Birr, Kendall, *Prelude to Point Four*, 200
Bishop, Elsie H., *Business History*, 23
Bishop, J. Leander
   American Manufactures, 256
   biographies in, 82
Bitting, A. W., 217
Bivins, Percy A., 115
Blackwood, Beatrice, 189
Blake, G. G., *Radio Telegraphy*, 253
Blake, Marion Elizabeth, *Roman Construction* (2), 221
Blake, Nelson M., *Water for the Cities*, 225
Bland, David, 269
Blätter für Technikgeschichte, 148
Blom, Eric, *Grove's Dictionary*, 294
Blum, André, on paper and printing (2), 269
Blümner, H., 8
Board for Testing Iron, etc., 101
Böckler, George Andreas
   in Reti, 42
   in Russo, 17–18
Bodmer, John G., travel diary, 137
Boehm, Eric H., 142
Bogoliubov, Aleksei, 304
Bogus, Allan G., 216
Boguslaw, Robert, 300
Bohatta, Hans, 19
boiler explosions
   government publications, in Poore, 92
   in *House Doc. 21*, 248
   in Thurston, 248
boilers, 247–248
Bolles, Albert S., 256
Bolton, Henry C.
   Bibliography of Chemistry, 98
   Scientific and Technical Periodicals, 143, 98
Bond, W. H., 49
Bondaroy, Fougeroux de, 59
Bonelli, Maria Luisa, 289
Boni, Albert, 264
Booker, P. J., 289
Borden, Gail, *Dairyman to a Nation*, 218
Borgnis, Giuseppe A.
   Traité Complet, 61
   in Reti, 42
Bossert, H. T., 51
Boston, Orlan W., 278
Bosworth, C. E., 49
Bouchu, M., *Art des Forges*, 48
Bowie, Theodore, 49
Bowsher, J. M., *Physics of Music*, 295

Boyd, Anne M., 92
Boyden, Seth, in Bishop, 82
Brace, Harold W., 217
Bradbury, John, 131
Bradford, Thomas L., 84
Bradstreet, J. M., 37
Bramah, Joseph, in Smiles, 81
Branca, Giovanni
  in Keller, 41
  in *Landmarks in Science*, 171
Brand, John, 138
Branner, John C., 266
brass and bronze
  in Aitchison, 256
  in Hull, 261
Brayer, Herbert O., 169
Brearley, Harry
  autobiography, 84
  Steel-Makers, 258
Brech, E. F. L., 303
Brewer, Griffith, 241
brewing
  in England, 219
  in U.S., 218
Brewington, Marion V.
  "Marine Museums," 183
  Navigating Instruments, 288
Brewster, David, 56
bridges, **223–225**
  in Condit, 219
  government publications, in Poore, 92
  Niagara railway suspension, 120, 134
  in R.I.B.A. *Catalogue*, 220
  U.S., 1810, in *Port Folio*, 160
  U.S., 1851, in Culmann, 132
Bridgewater, Derek, 219
Brigg, Thomas H., horse haulage, 244
Briggs, Martin S., *Building Crafts*, 280
Brigham, Clarence S., 169
Bright, Arthur A., Jr., 252
Brindley, James, in Smiles, 82
Brinkmann, Donald, 208
Britannica, Encyclopaedia, 57
British Journal for the History of Science, 148
British Museum
  Catalogue of Printed Books, 59
  portraits in, 88
British Technology Index, 144
British Union-Catalogue of Periodicals, 142
Britkin, A. S., 279
brittle failure, in Parker, Tipper, 260
Brockett, Paul, 240

Brockhaus
  early encyclopaedia, 56
  recent encyclopaedia, 53
Brocklehurst, H. J., 13
Broderick, John T., 85
Bronowski, J., (2), 204
Brookings Institution, 102
Brooklyn Bridge
  on building, Steinman, 224
  on rebuilding, A.S.C.E., 223
Brose, H. L., 10
Brown, John J., "Museums Census, Canada," 183
Brown, Joseph R., in Van Slyck, 82
Brown, Karl
  Guide to Collections, 23
  Parsons collection, 42
Brown, Lloyd A., *Story of Maps*, 227
Brown, Ralph R., *Mirror for Americans*, 139
Brown, V. M., *Scientific Management*, 301
Brown, William H., *First Locomotive*, 239
Brown Brothers, 89
Bruce, Alfred W., *Locomotive in America*, 240
Bruce, John L., education in N.S.W., 178
Bruce, Robert V., *Lincoln and Tools of War*, 295
Brunelleschi, Filippo
  his inventions, 49
  in Promis, 82–83
Brunet, J. C., 40
Bruno, Giordano, in Olschki, 11
Brunswig, H., 263
Brunton, John, 85
Bryant and May Museum, 186
Bryant, Lynwood
  "Four-Stroke Cycle," 245
  reviews Neren, 236
  reviews Sass, 246
  "Silent Otto," 245
  mentioned, 235
Buchanan, R. A., bibliography, 19
Buchanan, Robertson, *Mill Work*, 62
Bucks County Historical Society, 186
buildings and similar structures, **221–223**
  Cass Gilbert papers, Smithsonian, 115
  in Condit (2), 219
  in *Industrial Chicago*, 126
  in R.I.B.A., *Catalogue*, 220
Bulletin Signalétique, 19
Bullock, Alan, 19

INDEX   313

Burchard, John, 209
Burden, Henry, waterwheel, 120
Burden, William, on measurement, 290
Bureau of Railway Economics, library, 236
Bureau of Standards, history,
  in Cochrane, 290
  in Weber, 102
Burford, A., 294
Burguburu, Paul, 290
Burke, Edmund, *List of Patents*, 103
Burke, John G., *Technology and Human Values*, 205
Burlingame, Roger
  Backgrounds of Power, 205
  Engines of Democracy, 11
  March of the Iron Men, 11
Burlington Archives, Guide, 119, 236
Burlington Route, 238
Burr, Theodore, bridge of, illustrated, 55
Burr, William H., *Ancient and Modern Engineering*, 219
Busch, Gabriel, in Klinckowstroem, 15
business history
  articles in *Tradition*, 154–155
  British "Sources of Industrial History," 120
  Larson's *Guide*, 23
  manuscripts in Merrimack Valley Textile Museum, 115
  National Archives resources, in Fishbein, 111
"Business History," pamphlet, 22
Business History Review, 148
Butterworth, Benjamin, 104
Byrn, Edward W., *Progress of Invention*, 12
Byrne, Oliver
  American Engineer, 63
  Model Calculator, 66

Cajori, Florian, 289
calculating machines
  electronic, 253–254
  mechanical, 289
C.N.A.M. catalogue A, 190
Calhoun, Daniel H., 219, 232
California, University of, at Davis, Higgins collection, 116
Callahan, Raymond E., 301
Calvert, Monte A., 173
Camus, C., in Russo, 17–18
Canadian Pacific railroad, in Stevens, 87

canals, **232–233**
  government publications, Poore, 92–93
  S. H. Long on canals and railways, *Port Folio*, 160
  in Parsons collection, 42
  Pennsylvania works, in Nicklin, 134
  in Rees, *Cyclopaedia*, 55
  state documents, in Hasse, 94
  in travel accounts, 132–138
  see also Panama Canal
Candler, Isaac, 131
canning of food, 217–218
carborundum, in Acheson, 265
Cardano, Girolamo, in Beck, 40
  in Thorndike, 44
Carder, Frederick, 150
Cardwell, D. S. L.
  "Dyes and Dyeing," 263
  Organisation of Science, 176
  Steam Power in 18th Century, 247
Carlson, Robert E., 136
Carman, Harry J.
  Manuscripts, 117
  Printed Materials, 30
Carnegie Library of Pittsburgh
  biographies, 79
  Review of Iron and Steel Literature, 259
Carnot, Sadi, in Leprince-Ringuet, 76
Carpenter, Marjorie, 180
Carpenters Company papers, 117
Carr, J. C., 259
carriages and wagons, 234
Carriages at Shelburne Museum, 190
Carrier, Willis Haviland, Father of Air Conditioning, 249
Carter, Harry, Moxon's *Art of Printing*, 271, 47
Carter, Thomas F., *Invention of Printing*, 270
cartography, 227
  in Albion, 228
Cassier's Magazine, **156–157**
  biographies in, 78
  Subject Index, 157
catalogues, trade, 67–69
cathedrals
  construction, in Fitchen, 222
  in R.I.B.A., *Catalogue*, 220
  restoration, in Fox, 85
Cato the Elder, in Beck, 40
Cattell, Jacques, 77
Cauntner, C. F., 234
Caus, Salomon de
  in Beck, 40
  in Russo, 17–18

cement and concrete, **266**
  reinforced, in Condit, 222
  reinforced, in Vierendeel, 223
  Roman, in Blake, 221
Census Bureau, schedules in National Archives, 110–112
census records, 95–96
Centennial Exhibition, Philadelphia, 196–197
Centre de Documentation du C.N.R.S., 19
d'Histoire des Techniques, 19
ceramics, **265–266**
  ancient, in Forbes, 9
  C.N.A.M. catalogue S, 190
Chaloner, W. H.
  Industry and Technology, 13
  introduction to Jackman, 228
Chamberlin, Mary W., 89
Chambers, Ephraim, 55
Chandler, Dean, 268
Chanute, Octave
  bibliography of, 80
  on progress in flying, 156
Chapelle, Howard I.
  American Sailing Ships, 229
  Bark Canoes, 228
  National Watercraft Collection, 191
Chapman, S. D., 272
Chapuis, Alfred (2), 284
Checklist of U.S. Public Documents, 93
  on international exhibitions, 193–194
Chemical Abstracts, on biography in, 80
chemical industries, **262–265**
  articles in *Die BASF*, 148
  in China, Needham, 2
  Discovery of the Elements, 256
  in Ferguson, 40, 41
  in Patent Office Library, 27–28
  Science Museum titles, 187
  in Smithsonian *Reports*, Stemple, 97–98
Chesapeake and Ohio Canal, 233
Cheshire, Esther, 259
Chevalier, Michel, 132
Child, Ernest, *Tools of the Chemist*, 262
Childe, V. Gordon, *Man Makes Himself*, 8
Chilton, D., 289
China
  Bedini, "Scent of Time," 285
  Hommel, *China at Work*, 280
  iron and steel, in Needham, 260
  modern railroad, in *Cassier's*, 240
  Needham, *Heavenly Clockwork*, 286
  Needham, *Science and Civilisation*, 2
  steam engine precursors, in Needham, 248
  Sung, *Technology in 17th Century*, 48
  Wang Cheng, *Western Machines*, 48
Chittenden, Russell H., 180
Christensen, T. P., 218
Christian, Gerard Joseph, 61
Christie, Alexander G., autobiography, 85
Christie, Jean, life of Morris L. Cooke, 301
Ciba Review, 148
Cippola, Carlo M., 205
Cist, Charles, 139
cities
  Historian and the City, 209
  Streetcar Suburbs, 209
Civil Engineer and Architects' Journal, 157
civil engineering, *see* engineering, civil
Civil War, U.S.
  in Andreano, Bruce, 295
  records in National Archives, 111
Clark, Alvan, telescope of, 161
Clark, Daniel K.
  Machinery of 1862, 195–196
  Manual for Engineers, 66
Clark, George H., collection on radio, 115
Clark, George N., *Age of Newton*, 13
Clark, Ronald H.
  English Steam Wagon, 234
  English Traction Engine, 234
Clark, Thomas D., *Travels in Old and New South*, 130
Clark, Victor S.
  History of Manufactures, 256
  travel works used by, 129
Clarke, Somers, *Egyptian Masonry*, 280
classical period, *see* ancient and classical period
Clausius, Rudolf, in Leprince-Ringuet, 76
Clay, Reginald S., 287
Clement, Joseph, in Smiles, 81
Clerk, Dugald, 245

INDEX

Cleveland Public Library, history card file, 31
Cline, Walter, 257
Clockmakers of London, Museum, 186–187
clocks and watches, *see* timekeepers
Clough, Shepard, in AHA *Guide*, 18
Clow, Archibald and Nan L.
  Chemical Revolution, 262
  "Timber Famine," 210
Clutton, Cecil
  British Organ, 294
  Britten's Old Clocks, 285
Clymer, George, and the Columbian Press, 270
coal, 266–267
  coke-making, in *Harper's*, 158
  government publications, in Poore, 92–93
  mines described, 1853, in Lambert, 134
  in Rees, *Cyclopaedia*, 55
  in *Special Consular Reports*, 100
Cochrane, Rexmond C., 290
Cockle, Maurice J. D., 40
Coetlogon, Dennis de, 15
Coghlan, H. H.
  Early Iron, 189
  Prehistoric Metallurgy of Copper, 189
Cohen, I. Bernard
  Early Tools of American Science, 289
  Landmarks in Science, 171
Cole, Arthur H.
  Handicrafts of France, 58
  Wool Manufacture, 273
Coleman, D. C., *British Paper Industry*, 270
Coleman, Earle E., on 1853 New York exhibition, 195
Coleman, Laurence V., *Museum in America*, 183
Colles, Christopher, 132
Collins, Samuel W., papers, 118
Collison, Robert
  Bibliographies, 20
  Dictionaries of Foreign Languages, 70
  Encyclopaedias, 54
Colt, Samuel, in Poore, 92
Columbia University, Parsons railroad prints, 124
Columbian College, 173
Columbian Exposition, 197–198
Columbian Institute, 173

Columbian printing press, illustrated, 55
Columbian steam engine, illustrated, 56
Comenius, Johann, 45
Commerce and Labor Department, 100
communication, electrical 253–255
  bibliography in *Bulletin Signalétique*, 19
computers
  C.N.A.M. catalogue A, 190
  electronic, 253–254
  mechanical, 289
Conant, James B., 304
concrete, *see* cement
Condit, Carl
  American Building Art (2), 219
  Chicago School of Architecture, 221
  "Reinforced-Concrete Skyscraper," 222
  reviews *Engineering in History*, 5
  reviews Oliver, 12
Conestoga wagon, in Berkebile, Shumway, 234
Conservatoire national des Arts et Métiers
  bibliographies of exhibitions, in Mandell, 198
  catalogues of collections, 189–190
Consular Reports, 100
control, automatic, 284–285
Conway, H. G., 284
Cooke, Morris L.
  Academic and Industrial Efficiency, 301
  dissertation on, 301
Cooley, Mortimer E., 85
Coomaraswamy, Amanda K., 284
Cooper Union Library
  History of Engineering, 20
  Humanistic-Social Stem, 180
Copeland, Melvin T., *Cotton Manufacturing*, 273
Copley, Frank B., 301
copper
  in Aitchison, 256
  ancient, in Forbes, 9
  Michigan Copper, 261
  prehistoric, in Coghlan, 189
  in Tooker, 261
Copperthwaite, William C., 225
Corliss, George
  Centennial engine of, 57
  papers, 114
  in Van Slyck, 82

Corning Museum of Glass, library, 265
Cort, Henry, in Smiles, 81
Cossons, Neil, 192
cotton manufacture, *see* textiles
Court, Thomas H., 287
Cox, Edward F. (2), 291
crafts and craftsmen, 280–284
 articles in E.A.I.A., *Chronicle*, 149
 Chinese, in Vanderbilt, 124
 East Indian and others, Kress Library, 124
 Islamic, in Creswell, 20
 proposed film series, Smithsonian, 127
 Shell film series, 128
 U.S., in Dow, Gottesman, Prime, 170
 U.S., in Smith, Tunis, 278
Cramp, William, shipbuilding papers, 115
Crane, Walter R., 257
cranes, 294
Crerar, John, Library
 bibliography of magnetic recording, Wilson, 255
 catalogue, 26
Creswell, K. A. C., 20
Cresy, Edward, 63
Crick, B. R., 120
Critical Bibliography, in *Isis*, 22
Croix, Horst de la, 40
Crombie, A. C.
 edits *History of Science*, 150
 Medieval and Early Modern Science, 10
Croton reservoir, in N.Y.P.L., 23
Crowe, Sylvia, 209
Cugnot, Nicolas, in Leprince-Ringuet, 76
Culmann, Karl, 132–133
Culver Service, 89
Cummings, Hubertis, Pennsylvania canal records, 118, 232
Cummings, Richard O., *American Ice Harvests*, 249
Cunliffe, Marcus, 194
current meters, stream flow, 225, 99
Curti, Merle
 "America at World's Fairs," 194
 Prelude to Point Four, 200
Curtis, Carolyn, *Burlington Archives*, 119, 236
Curwen, E. Cecil, 215
Cutler, Carl C., 229
cycles
 in Cauntner, 234

in Duncan, 236
Science Museum titles, 187–188
U.S.N.M. catalogue, 191
U.S. patents, in Allen, 103–104

d'Acres, R., 44
Daedalus, Tekniska Museet, 149
Daguerre, Louis J. M.
 in Leprince-Ringuet, 76
 portrait, in Appleton's, 89
Dahl, Jürgen, 58
Dahlgren, John
 in N.A.S., 83
 papers, 116
dairy industry, 218
Dalby, W. E., 179
Dale, Ernest, 301
Daly, Charles P., 176
dams, 227
Daniel, Glyn, on Childe, 8
Danniels, Lorna M., 20
Darling, Arthur B., 5
Darmstaedter, Ernst, mineral industry bibliography, 257, 40
Darmstaedter, Ludwig, *Handbuch*, 1
Daum, Arnold R., 268
Daumas, Maurice
 Histoire générale, 1
 Instruments Scientifiques, 287
 reviews Rousseau, 6
Dautry, Jean, 59
Davies, Olivier, *Roman Mines*, 257
Davies, R. E. G., *World's Airlines*, 241
da Vinci, *see* Leonardo da Vinci
Davis, E. W., *Pioneering with Taconite*, 257
Davis, Herbert, Moxon's *Art of Printing*, 271, 47
Davis, Pearce, *American Glass Industry*, 265
Davis, R. H. C., *Medieval European History*, 22
Davison, C. St. C. B., *Steam Road Vehicles*, 234
Davy, M. J. B., 188
Dawson, Philip, 209
De Bow's Review, 157
De Forest, Lee, collection, 115
De Golyer Foundation, manuscripts, 115–116
Del Mar, William A., 67
Deman, van, *see* Van Deman
Denavit, Jacques
 Kinematic Synthesis, 304
 "Men and Machines," 41

INDEX

Derry, T. K., *Short History*, 3
Desaguliers, John T., 45
Deschamps, P., 40
Descriptions des Arts et Métiers, 58–59
Detroit Public Library, *Automotive History*, 235, 69
Deutsches Museum
 Abhandlungen und Berichte, 147
 history of, 189
 Magdeburg hemispheres in, 45
 portraits, 88
developing nations, 37
Dew, Charles B., 259
Dewhurst, P. C., 237
Diamond, History and Use of, 265
diaries, guides to, 79
Dibner, Bern
 Atlantic Cable, 253
 Early Electrical Machines, 252
 Heralds of Science, 40
 monographs of, 170–171
 Moving the Obelisks, 45
Dick, Otto, *Die Feile*, 276
Dickens, Charles, *American Notes*, 133
Dickinson, Henry W.
 on British and continental technology, 200
 "Draughtsmen's Instruments," 289
 Garret Workshop of James Watt, 188
 "Gauges for Wire, Sheets," 292
 James Watt and the Steam Engine, 247
 Short History of Steam Engine, 247
 "Water-Supply of London," 226
dictionaries, multilingual, 70–71
dictionaries of technology, 61–65
Dictionary of American Biography, 74
Dictionary of Canadian Biography, 74–75
Dictionary of National Biography, 75
Dictionary of Scientific Biography, 74
Dictionnaire Archéologique des Techniques, 8
Dictionnaire de l'Industrie, 59
Dictionnaire Technologique, 61–62
Diderot, Denis, *Encyclopédie*, 58
Diels, Hermann, *Antike Technik*, 9, 8
Diesel, Eugen, 245
Digges, Thomas, 201
Dijksterhuis, E. J., 4
Dilke, Charles W.
 on London 1851 Exhibition, 194
 on New York 1853 Exhibition, 133
Dingle, Herbert, 48

Dingler's Polytechnisches Journal, 157
directories, 35–38
 city directories, 84
dirigible, 243
Diringer, David, 8
Disney, Alfred N., 287
Dissertation Abstracts, 30
Dissertations in History, 31–32
Disston, Henry, in Scharf, 84
Dixon, Thomas, 66
Dodge, Grenville M.
 papers in Iowa, 114
 Union Pacific, 85
Donkin, Bryan, in Sellers, 86
Donkin, Bryan, Jr., *Gas, Oil, and Air Engines*, 245
Donndorf, J. A., in Klinckowstroem, 15
Doorman, G., 108
Dow, George F., 170
Drachmann, A. G.
 Ktesibios, Philon, and Heron, 284, 49
 Mechanical Technology of Antiquity, 284, 49
drawing, mechanical, 289–290
Dredge, James, 197
Drinker, Henry S.
 Explosive Compounds, 264
 Tunneling, 225
Droz, Edmond, 284
Drubba, H., 43
Dubester, Henry J., 96
Dublin International Exhibition of 1865, 196
Ducassé, P., 4
Du Cros, Arthur, 235
Due, John F.
 Electric Railways in Canada, 237
 Interurbans in America, 238
Dufton, A. F., 249
Duncan, George S., *Bibliography of Glass*, 265
Duncan, Herbert O., *World on Wheels*, 236
Du Pont, E. I., company papers, 116
Du Pont, Lammot, anticipates F. W. Taylor, 303
Du Pont, Lammot, Jr., collection of aeronautics, 116
Dupree, A. Hunter, 101
Dürer, Albrecht, in Olschki, 11
Durfee, W. F., in *Popular Science Monthly*, 160
Durrell, Edward, 234
Durrenberger, Joseph A., 233
dyes and dyeing, 262–263

dyes and dyeing (*continued*)
  in Lawrie, 273
  in Ploss, 273–274

Eads, James B., in N.A.S., 83
Eads Bridge, 225
Early American Industries Association, *Chronicle*, 149
early works, *see* Contents, Part IV
Eason, C. M., papers, 116
East Germany
  Dresden *Informationsdienst*, 145
  Herlitzius, "Technik und Philosophie," 208
  technical museums, list, 183–184
Eastin, Roy B., 92
Eavenson, Howard N., 266
Ebhardt, Bodo, 48
Eckhardt, George, 286
Eco, U., *Inventions*, 4
Ecole Nationale des Ponts et Chaussées, catalogue of library, 29
Ecole Polytechnique
  Livre du centenaire, 180
  in Taton, 177–178
ecology, human, 209–211
Economic History, 149
Economic History Review, 149
Economy and History, 149
Edgar Allen News, 149
Edinburgh Encyclopaedia, 56
education, engineering, 176–181
education, technical, *see* Contents, Part XII
  in Smithsonian *Reports*, Stemple, 97–98
Edwards, Everett E., *Bibliography of Agriculture*, 214
Edwards, I. E. S., *Pyramids of Egypt*, 222
Eiffel Tower, elevator systems, Vogel, 223
Eighty Years' Progress, 64
Eisenman Memorial Collection, 43
electric power, *see* power generation and transmission
electric railways
  in *Cassier's*, special issue, 156–157
  interurbans, Due, 237, Hilton, 238
  Sprague on electric locomotives, 98
electrical engineering, *see* engineering, electrical
electricity and electronics, *see* Contents Part XIV.E
electronics, 250–251, 253–255
electrostatic precipitator, White, 253

Eleutherian Mills-Hagley Foundation, 22
Eleutherian Mills Historical Library
  Guttman explosives collection, 264
  manuscripts in, 116
elevators
  in Eiffel Tower, Vogel, 223
  in Soc. of Architectural Historians, 153
Ellis, Cuthbert Hamilton
  British Railway History, 237–238
  Railway Carriages, 240
Ells, S. C., 85
Ellsworth, Lucius F., "Directory of Artifacts," 183, 22
Ellul, Jacques, 205
Elton, J., *Corn Milling*, 217
Emme, Eugene M.
  Aeronautics Chronology, 241
  Rocket Technology, 241
Emmet, William L., 85
Emporium of the Arts and Sciences, 157
encyclopaedias, *see* Contents, Part V
Encyclopedia of Associations, 35
Encyclopédie, 58
  in *Revue d'Histoire*, 153
Encyclopédie méthodique, 59
Endeavour, 149
Energy and Man, 210
energy conversion, *see* Contents, Part XIV.D
  in Patent Office Library, 27–28
  U.S., in *Tenth Census*, 96
energy resources, 243
Engelbach, Reginald
  Ancient Egyptian Masonry, 280
  in *Legacy of Egypt*, 10
  Problem of the Obelisks, 280
Engelbrecht, Martin, on early trades, 283
Engelmann, Wilhelm, bibliography of 1843, 21
Engineer, The, 157
  Index, 157–158
Engineering, 158
engineering, civil, general, **219–221**
  bibliography, *Bulletin Signalétique*, 19
  Corps of Engineers publications, 99
  descriptions of public works in travel accounts, 129–139
engineering, electrical, **250–255**
  bibliography, Higgins, 250
  bibliography, Maynard, 251
  list of books, Zischka, 54
  in Patent Office Library, 27–28

INDEX

Siemens, autobiography, 86–87
synopsis of events, Westcott, 25
Tesla papers, Library of Congress, 116
U.S. patents, Allen, 104
engineering, hydraulic, **225–227**
  bibliography, in Rowe, 232
  Corps of Engineers publications, 99
  see also hydro-power
engineering, industrial
  evolution of, 30
  see also Contents, Part XIV.J
engineering, mechanical, *see* Contents, Part XIV.G
  in China, Needham, 2
  in China, Wang Cheng, 48
  synopsis of events, Westcott, 25
  in U.S.N.M. catalogue, 191
Engineering, Story of, Finch, 4
Engineering and Western Civilization, Finch, 4, 219
Engineering Congress, 1893, 198
engineering education, 176–181
Engineering in History, Kirby et al, 5
Engineering Index, 144–145
Engineering Magazine, 158
Engineering News-Record, 158
engineering sciences, 304–306
Engineering Societies Library, catalogue, 26
Engineers, Amalgamated Society of, history, 30
Engineers Corps, U. S. Army, **99**
  history, 102
  letters in National Archives, 112
entrepreneurial history
  in Aitken, 297
  in Miller, 297–298
Ercker, Lazarus, 47
Ericsson, Henry, 85
Erie Canal 1792–1854, 233
Ersch, Johann, S., 60
Esper, Thomas, reviews Cippola, 205
Essayons Club, 99
Esterly, George, in Bishop, 82
Ewbank, Thomas, 244–245
Evans, Oliver, by Bathe, 170
Evans, Oliver
  Miller's Guide, 61
  in Poore, 92
  portrait, in Appleton's, 89
  Steam Engineer's Guide, 61
exhibition, international, **194–199**
  Belgian commissioners to U.S., in Smet, 132
  government publications, in Poore, 92
  preliminary inventory, National Archives, 110
  report of New York 1853, Lambert, 134
  reports of New York 1853, Parliamentary Papers, 133
exhibition, local and regional, 199–200
exhibition buildings
  in R.I.B.A., *Catalogue*, 220
  in Vierendeel, 223
explosives, chemical, **263–264**
  bibliography in Besterman, 18
  Du Pont works, 1818, in Scott, 134

Fahie, J. J. (2), 254
failures, structural
  of bridges, Finch, 223
  brittle failure in steel (2), 260
  in Feld, 222
  in Hammond, 220
fair, *see* exhibition
Fairbairn, William, in Smiles, 81
Fairbanks, Erastus and Moses, in Van Slyck, 82
Fakhry, Ahmed, 222
Falconer, John I., 215
Falk, Howard, reviews Ellul, 205
Fall River Line, 231
Farey, John, Jr.
  drawings in *Pantologia*, 56
  Treatise on the Steam Engine, 247
Faurote, Fay, 300
Favre, Adrien, 291
Faye, Helen, 123
Fearon, Henry, 131
Feld, Jacob, 222
Feldhaus, Franz Maria
  bibliography of early histories, 15
  Handbuch, 2
  history of mechanical drawing, 289–290
  pamphlet war with Matschoss, 76
  Ruhmesblätter, 4
  Die Säge, 276
  Technik der Antike, 9
Felibien, André, 282
Ferber, J. J., travels (5), 138
Ferguson, Eugene S.
  Cassier's Subject Index, 157
  George Escol Sellers, 86
  "Kinematics of Mechanisms," 305
  reviews Rolt, 279
  "Technical Museums," 184
Ferguson, John
  Bibliotheca Chemica, 40
  History of Inventions, 15

Ferguson, John (*continued*)
  on Polydore, 15
  on technological chemistry, 41
Ferguson, William, *America by River and Rail*, 133
Ferranti, S. Z., 189
Ferris, Herbert W., 222
Ffoulkes, Charles
  Armourer and His Craft, 295
  Gun-Founders of England, 296
Field, Joshua, travel diary, 137
Figuier, Louis, 126
Filarete
  in Olschki, 11
  in Spencer, 51
Finch, James Kip
  Engineering and Western Civilization, 4, 219
  School of Engineering, Columbia, 180
  Story of Engineering, 4
  "Wind Failures of Suspension Bridges," 223
Finn, Bernard S., 184
Fischer, Johann C., diary, 136
Fishbein, Meyer H., business history and census, 111
Fisher, Marvin, *Workshops in Wilderness*, 130
Fitchen, John 222
Fite, Gilbert C., 216
Flather, John J., 293
Fleming, Arthur P. M., *History of Engineering*, 13
Fleming, J. A., *Fifty Years of Electricity*, 85
Fletcher, Banister, 222
Flexner, James T., 229
Flinn, Michael W., 130
Flint, James, 131
flour milling, *see* grain milling
Fontana, Domenico
  in Beck, 40
  in Dibner, 45
food production and preservation, 214–219
  ancient, in Forbes, 9
Forbes, Robert J.
  Ancient Roads, 233
  Bibliographia antiqua, 21, 8
  Early Petroleum History, 267
  History of Science and Technology, 4
  and Loeb Library, 46
  Man the Maker, 4
  Studies in Ancient Technology, 9

Ford, Guy Stanton, on fairs, 193
Ford, Percy and Grace, on Parliamentary Papers, 94–95
Ford, by Nevins and Hill, 235
Ford Methods and the Ford Shops, 300
Foreign Commerce Bureau, 100
Forest History Society, 271
forestry, *see* timber and wood industries
Forman, Sidney
  on early American military books, 296
  West Point, 180
Forster, J. G. A., 138
Forsyth, David P., 143
Forti, Umberto, 10
Fourneyron, Benoit
  in Hunter, 244
  in Leprince-Ringuet, 76
Fox, James, in Smiles, 81
Fox, Francis
  63 Years of Engineering, 85
  on tunnels, 98
Franc, Georges le, 238
Francis, Clarence, *History of Food*, 218
Francis, George W., handbook, 63
Francis, James B.
  in Hunter, 244
  Lowell Hydraulic Experiments, 244
Frank, Edgar B., *Old French Ironwork*, 280
Franklin, Benjamin, papers, 117
Franklin Institute
  exhibitions in, 200
  history of (2), 175
  Journal, 165
  Journal, patents in, 105
  library catalogues (2), 26
  manuscript collections, 115
  trade catalogues, 68
Frantz, Joe B., 218
Frazier, Arthur H., on stream-flow meters (2), 225
Freedley, Edwin T., 125
Freese, Stanley, 282
Freitag, Ruth S., 142
Fremont, Charles
  articles on tools (4), 276
  bibliography, in Sarton, 255
  on materials testing, 275
Freudenberger, Herman, 273
Frey, Howard C., 234
Frick Art Reference Library, 88
Fritz, John, 85

Frontinus, Sextus Julius
  in Beck, 40
  Water Supply **226**, 45
Frumkin, M., 109
fuel cell, in Peattie, 246
Fueter, E., 7
Fulton, Robert
  in Parsons collections, 42
  in Poore, 92
Furman, Franklin, 180
Furniture of Greeks, Etruscans, and Romans, 281
Fussell, G. E.
  English Dairy Farmer, 218
  Farmer's Tools, 216
Fyrth, H. J., 4

gages, wire and sheet, in Dickinson, 292
Galilei, Galileo
  in Olschki, 11
  Works, in *Landmarks in Science*, 171
Galloupe, Francis E., 145
Gamble, William H., 241
Gannon, William L., 234
Garratt, G. R. M., 188
Garvan, Anthony N. B., 125
gas, illuminating, industry, **268**
  in *Archives Internationales*, 148
  in Stewart, *Town Gas*, 189
gas turbine, see turbine, gas
Gates, Paul W., *The Farmer's Age*, 215
Gates, W. B., *Michigan Copper*, 261
Gatling, R. J., in Spalding, 83–84
Gatling Gun, 296
gauges, wire and sheet, in Dickinson, 292
Geddes, L. A., 287
Geitel, Max, 5
Gelb, Arthur, 207
Gelis, Edmund, 284
Genêt, Citizen, Diplomat and Inventor, 170
German Engineering Society, see Verein Deutscher Ingenieure
Germany, Patent Office, catalogue, 27, 65
Gerstner, Franz Anton, Ritter von, 133
Geschichtsblätter für technik, 150
Gesner, Abraham, in Beaton, Butt, 267
Ghega, Carlo de, 133
Ghiberti, Buonaccorso
  notebooks of, 49
  in Olschki, 11

Gibb, George S., *Saco-Lowell Shops*, 273
Gibb, Hugh R., notes by, 238, 239, 240
Gibbons, Chester H., *Materials Testing Machines*, 275
Gibbs-Smith, Charles H.
  The Aeroplane, 242
  George Cayley's Aeronautics, 242
  Great Exhibition of 1851, 194
Giddens, P. H., 267
Giedion, Sigfried, 12
Gilbert, Cass, construction photos, 115
Gilbert, Gilbert H., *Subways and Tunnels*, 225
Gilbert, K. R., *Portsmouth Blockmaking*, **299**, 188
Gilbreth, Frank B., papers on microfilm, **116–117**, 303
Gilbreth, Lillian M., history of management, 301–302
Gilfillan, S. C.
  Inventing the Ship, 306
  on patents, 105
Gille, Bertrand
  Engineers of the Renaissance, 10
  mentioned, 39
Gill's Technical Repository, 158
Gillispie, Charles C.
  Diderot Encyclopedia, 58
  Landmarks in Science, 171
Gilpin, Joshua
  British tour, in Hancock, 136
  papermaking, in Hancock, 269
  Pennsylvania tour, 133
Gilpin, Thomas, papermaking, in Hancock, 269
glass, **265**
  ancient, in Forbes, 9
  Islamic, in Creswell, 20
Glass Technology, 150
Glazebrook, G. P. de T., 228
Gleason's Pictorial, 159
Gloag, John, 219
Gnudi, Martha Teach
  translates Biringuccio, 47
  mentioned, 39
Goguet, Antoine, in Ferguson, 15
Goldbeck, Gustav
  Engines to Autos, 245
  Siegfried Marcus, 246
Goldmann, K., in Treue, *Hausbuch*, 51–52
Goldsmith, Maurice
  Science, History and Technology, 4
  Science of Science, 205
  Society and Science, 2

Gomme, A. A., 200
Goode, George Browne
    on technical institutions, 173–174
    Smithsonian Institution, 98
Gooding, P., 266
Goodman, Gordon T., *Ecology and Industrial Society*, 209
Goodman, W. L.
    Dictionary of Hand Tools, 277
    History of Woodworking Tools, 276–277
    Woodwork, 277
Goodrich, Carter, on American canals, 232
Goodrich, L. Carrington, *Invention of Printing*, 270
Goodwin, Jack
    "Trade Literature Collection," 68
    notes by, 8, 214, 264
Goodyear, Charles, portrait, in Appleton's, 89
Gorman, Mel, 252
Gottesman, Rita S., 170
government publications, *see* Contents, Part VII
Graff, Frederick, drawings, 115
grain milling, 217
    Dutch patents, Doorman, 108–109
    miller's guides, 61–62
Gramme, Zenobe, in Leprince-Ringuet, 76
Grand Central Terminal, papers in N.Y.P.L., 23
Grässe, Johann G. T., 41
Gray, Lewis C., *Agriculture in Southern U.S.*, 215
Gray, R. B., *Agricultural Tractor*, 216
Gray, Solomon S., in Bishop, 82
Great Britain, National Register of Archives, 120
Great Britain, *Parliamentary Papers*, 94–95, 133
Great Britain, Patent office
    Abridgment of Specifications, in A.S.C.E., *Catalogue*, 26
    biographical data, 80
    indexes, 107–108
    Library, catalogues and guides, 27–28
    portrait gallery, 88
    records, in Besterman, 18
Great Exhibition, 194–195
    illustrated, 123
Great Industries of the United States, 64
Great Northern railroad, in Stevens, 87

Greathead, James H., tunnel shield, 225
Greeley, Horace, 64
Green, Charles, *Sutton Hoo*, 229
Green, E. R. R., *Industrial Archaeology*, 191
Greene, Arthur M., Jr.
    A.S.M.E. Boiler Code, 247
    Pumping Machinery, 245
Greene, Evarts B., *Sources for American History*, 30
Gregory, Olinthus
    Pantologia, 56
    Treatise of Mechanics, 61
Gregory, Winifred
    American Newspapers, 169
    Foreign Government Serials, 95
    International Congresses, 35
    Union List of Serials, now Titus, 142
ground-effect machines, 242
Grove's Dictionary of Music, 294
Gruber, J. G., 60
Guericke, Otto von
    Experiment nova, 45
    in *Landmarks in Science*, 171
Güldner, Hugo, 246
guns, 295–297
Gutenberg and Master of the Playing Cards, 270
Guttman, Oscar
    library of, 264
    Manufacture of Explosives, 264
Guyot, Edmund, 231

Haber, L. F., *Chemical Industry*, 262
Haber, Samuel, *Efficiency and Uplift*, 302
Hadfield, Charles, 232
Haferkorn, H. E., 21
Hale, Richard W., Jr., 31
Hall, A. Rupert
    Ballistics in 17th Century, 304
    History of Technology, 2–3
Hall, Basil, *Sketches in North America*, 134
Hall, Courtney R., *American Industrial Science*, 12
Hall, G. K. & Co.
    catalogues published by, 26, 119, 124, 230, 235, 271
Hall, Marie Boas, *History of Science*, 18
Hallidie, A. S., papers, 114
Halstead, P. E. (2), 266
Hamarneh, Sami, 291

INDEX   323

Hamer, Philip M., *Guide to Archives*, 113–114
Hamilton, Stanley B.
  "Continental and British Engineering," 200
  "Structural Theory," 304
Hammond, Rolt, *Structural Failures*, 220
Hancock, Harold B.
  "Gilpin Endless Paper Machine," 269
  Joshua Gilpin's travels, 136
  on National Archives, 111
  "Thomas and Joshua Gilpin," 269
hand tools, *see* tools
handbooks, 65–67
Handlin, Oscar
  Historian and the City, 209
  This Was America, 135
Handover, P. M., 270
Hansard, Thomas C., *Typographia*, 270
Hansard's Catalogue of Parliament, 94
harbors, 232–233
Harcup, Sara E., 36
Hardesty, Shortridge, 225
Harding, H., 108
Hargreaves, James, and the Spinning Jenny, 272
Harper, Josephine L., 118
Harper's Magazine, 158
Harper's Weekly, 159
Harris, J. R.
  Ancient Egyptian Materials, 255
  reviews Forbes, 9
Harris, John
  in *Landmarks in Science*, 171
  Lexicon technicum, 54
Harrison, John
  biography, Quill, 231
  in Smiles, 82
Harrison, Joseph, in Scharf, 84
Hartenberg, Richard S.
  Kinematic Synthesis, 304
  "Men and Machines," 41
  reviews *Blätter für Technikgeschichte*, 148
  reviews Feldhaus, 289–290
  reviews Timm, 7
Hartley, E. N., *Ironworks on the Saugus*, 259
Hartley, Harold, *Landmarks in Science*, 171
Harvard Guide to American History, 21

Haskell, Daniel C.
  European Railway Literature, 236
  periodical indexes, 143
Hasse, Adelaide R., 94
Hassler, Ferdinand, papers, 118
Haswell, Charles H., 65
Hatt, E. M., *Museums*, 185
Hatt, Gudmund, *Plough and Pasture*, 215
Haupt, Herman
  Bridge Construction, 224
  Periodical Engineering Literature, 143
Hausbuch
  Mendel, 51–52
  Nürnberg, 52
  Waldburg, 51
  unknown master, 49
Hawkins, Gerald S., 221
Hawthorne, John G.
  translates *Theophilus*, 47
  mentioned, 39
Hayes, Carlton J. H., 206
Haynes, Williams, 262
Hayward, Elizabeth G., *Guide to Taylor Collection*, 303
Hayward, J. F., *Art of the Gunmaker*, 296
Hayward, Leslie H., *Air Cushion Vehicles*, 242
Hazen, Edward, 284
Heathcote, Neils H., 305
heating, 249
  ancient, in Forbes, 9
  government publications, in Poore, 92–93
  in Hough, 250
Heaton, Herbert, 201
Heattman, Charles F., 271
Hebert, Luke, 62
Heeren, Friedrich, 63
Heidebroek, E., 293
Heise, Paul, 21
Heizer, Robert F., 294
Helmholtz, Hermann von, in Leprince-Ringuet, 76
Henderson, W. O., 136
Hendricks, Franz, 13
Herlitzius, E., 208
Hermann, F. B. W., 199
Hermelin, Samuel G., 134
Hero of Alexandria
  in Beck, 40
  in Drachmann, 49, 284
  edition of 1589, 45
  translation of *Pneumatics*, Woodcroft, 284

Heron, S. D., *Aviation Fuels*, 242
Herschel, Clemens, translates Frontinus, 226, 45
Hersey, M. D., 293
Heslin, James J., 68
Hidy, Ralph W., *Weyerhauser Story*, 275
Hiebert, Erwin N., 305
Higgins, F. Hal, Library, **116**, 214
  in Rasmussen, 68
Higgins, John Woodman, Armory, *Catalog*, 191
Higgins, Thomas James, bibliographies
  "Book-Length Biographies," 78
  Electric Engineering, 250
  "Electrical Engineers," 78
  mentioned, 81
highways, *see* roads
Hilken, T. J. N., 181
Hill, Cyril F., *Microscope*, 287
Hill, Frank E.
  Ford, 235
  Weyerhauser Story, 275
Hill, M. W., note by, 28–29
Hilton, George W., 238
Hindle, Brooke
  David Rittenhouse, 287
  on military technology, 295
  Pursuit of Science, 12
  Technology in Early America, 22
Hirschfeld, Charles, 195
Hislop, Codman, 229
Historic American Buildings Survey, 192
Historical Association, London, pamphlets, 22
historiography
  in Hindle, 22
  in Philosophy of Technology, 208
History of Science, periodical, 150
Hitchcock, Henry Russell, Jr., 222
Hobhouse, Christopher, 195
Hodges, Henry, *Artifacts*, 255
Hodgson, Adam, in Mesick, 131
Hoe, Richard M., portrait in Appleton's, 89
Hoefer, Ferdinand, 76
Hoff, Hebbel E., 287
Hoffman, Hester R., 31
Hogan, Donald W., 106
Hoglund, A. William, 210
Holland Tunnel papers, N.Y.P.L., 23
Holley, Alexander, drawings, 115
Hollunder, C. F., 138
Holmyard, E. J., *History of Technology*, 2–3

Holstein, Edwin J., 180
Holt, G. O., *Liverpool & Manchester Railway*, 238
Holt, William Stull, *Chief of Engineers*, 102
Holtzapffel, Charles and John Jacob, 279
Hommel, Rudolf P., 280
Honnecourt, Villard de, 49
Hoosac Tunnel, in *Scribner's*, 161
Hoover, Herbert and Lou Henry, *Agricola*, 44
Hoover, Herbert, Jr., in *Energy and Man*, 210
Hopkins, Joseph G. E., 74
horsepower and manpower, 243–244
Horwitz, H. T. (2), 52
Hoskin, M. A., *History of Science*, 150
Hospitalier, E., 67
hot air engine, *see* internal-combustion engine
Hotchkiss, B. B., in Bishop, 82
Hough, Walter
  "Fire-making Apparatus," 98
  "Heating and Lighting Utensils," 250
House Document 21, *Report on Steam Engines*, 248
Howard, A. A., translates Vitruvius, 48
Howard-White, F. B., *Nickel*, 261
Howe, Henry
  Eminent American Mechanics, 81
  mentioned in Carnegie, 79
Howe, James Lewis, *Bibliography of Platinum*, 261
Howitt, F. O., *Literature on Silk*, 273
Hubach, Robert R., 131
Hudson, Derek, *Royal Society of Arts*, 175
Hudson, Kenneth
  "Company Museums," 183
  edits periodical *Industrial Archaeology*, 150
  Industrial Archaeology, book, 191
  reviews Green, 191
Hudson River Day Line, 231
Hughes, Helen, *Australian Iron and Steel*, 259
Hughes, Thomas P.
  "British Electrical Industry," 252
  Development of Western Technology, 14
  Samuel Smiles, 82
Hull, Daniel R., 261
Hulme, E. Wyndham, 41
humanities in engineering education, 180

INDEX 325

Hunt, Frederick V., *Electroacoustics,* 254
Hunter, Dard, *Papermaking,* 270
Hunter, Louis C.
  "Iron Industry in Western Pennsylvania," 259
  "Origines des turbines," 244
  Steamboats on the Western Rivers, 230
  use of travel works, 129
Hunt's *Merchants' Magazine,* 158
Huntsman, Benjamin, in Smiles, 81
Hütte, handbook, 66
Hutton, Charles, portrait, in Appleton's, 89
Hyamson, Albert M., 73
hydraulic engineering, *see* engineering, hydraulic
hydraulics, in Rouse, 305
hydro-power, **244**
  ancient, in Forbes, 9
  Burden waterwheel ,Troy, N.Y., 120
  miller's guides, 61–62
  Monastic Watermills, 192
  at Niagara Falls, special issue, *Cassier's,* 156–157
  patent survey, tide motors, *Journal of P.O.S.,* 151
  Uriah Boyden papers, in Smithsonian, 115
  in U.S., *Tenth Census,* 96
  water power photos, *ca.* 1900, in Vanderbilt, 124
hydrometry, bibliography, Kolupaila, 226

ice trade
  in Cummings, 249
  state documents, Hasse, 94
  in U.S.D.A. reports, 100–101
Ickx, Jacques, 236
Iconographic Encyclopaedia, 56–57
Ilbert, C. A., *Britten's Old Clocks,* 285
Iles, George, 81
Illinois Central Archives, 119, 237
Illustrated London News, 158
Illustration, L', 158
  Histoire de la marine, 230
  Locomotion terrestre, 235
illustrations, 123–128
Imperial Chemical Industries
  in *Endeavour,* 149
  in *History of Technology,* 2–3
Imperial Cyclopaedia, 64
Ince, Simon, 305

India
  Bernstein, *Steamboats on Ganges,* 229
  civil engineer, Brunton, 85
  public works, in MacGeorge, 220
  White, "Tibet, India, and Malaya," 11
industrial archaeology, 191–192
Industrial Archaeology, periodical, 150
  on museums, 183
Industrial Arts Index (2), 145
Industrial Commission, U.S., 101
industrial engineering
  evolution of, 30
  see also Contents, Part XIV.J
industrial organization, 297–304
  see also Contents, Part XIV.J
Informationsdienst Geschichte der Technik, **145**
  on museums, 183–184
Ingels, Margaret, 249
ink, printing, Wiborg, 272
innovation, process of, 306
Institute. See note at beginning of index.
Institute of Early American History and Culture, 22
Institution of Civil Engineers, library catalogue, 29
instruments, chemical, in Child, 262
instruments, drawing
  in Dickinson, 289
  in Feldhaus, 289–290
instruments, musical, 294–295
instruments, scientific, **287–290**
  in Patent Office Library, 27–28
  Science Museum titles, 187–188
instruments, surveying, **227**
  in Bedini, 287
  in Kiely, 41
  in Smart, 288
interchangeable mechanisms, *see* American system
internal-combustion engines, **245–246**
  in aircraft, Schlaifer, 242
  early aircraft engines, Meyer, 242
  first Diesel in aircraft, *Annals of Flight,* 98
  Otto and Langen film, Klöckner, 127
International Congress on the History of Sciences, *Actes,* 150
international congresses, 35–38
international exhibitions, 194–199
invention, history of, *see* Contents, Part I

invention, process of, 306
Ireland, Norma O., 31
Iron Age, 159
iron and steel, **258–261**
   articles in *Revue d'Histoire de la Siderurgie,* 153
   barbed wire, manuscripts, 114
   cast iron in architecture, 219
   early iron, Coghlan, 189
   Holley drawings, Bessemer works, 115
   J. C. Fischer diary, Henderson, 136
   in Jars, travels, 136–137
   Mont Cenis ironworks, 1788, Ferber, 138
   Old French Ironwork, 280
   rolled shape dimensions, Ferris, 222
   Speedwell Iron Works, 1818, Scott, 134
   wiremakers, in Nutt, 84
Irwin, Raymond, 36
Isis, **150–151**
   "Critical Bibliography," 22

Jackman, W. T., 228
Jackson, Benjamin D., dates of Rees parts, 55
Jackson, Elizabeth C., *Burlington Archives,* 119, 236
Jacobson, Johann K. G., 42
Jaffe, William L.
   L. P. Alford, 302
   history of management, 301–302
Jakkula, Arne A., 224
James, Patricia, 136
Jamieson, Alexander, 62
Japan
   Bedini, "Scent of Time," 285
   students abroad, in Wantabe, 179
   Tuge, *Science and Technology,* 7, 178
Jars, Gabriel, 136–137
Jeffreys, Alan, 120
Jenkin, A. K. Hamilton, *Mines and Miners of Cornwall,* 257–258
Jenkins, Francis B., *Science Reference Sources,* 31
Jenkins, J. Geraint
   Country Craftsmen, 281
   Farm Wagon, 190
Jenkins, Rhys
   Collected Papers, 152
   R. d'Acres, 44
   Watt and the Steam Engine, 247
Jervis, John B., 86

Jesperson, Anders, 244
jet propulsion, *see* turbine, gas
Jewkes, John, 306
Johanssen, Otto, *Geschichte des Eisens,* 259
Johansson, C. E., biography of, 290
John Crerar Library
   bibliography of magnetic recording, Wilson, 255
   catalogue, 26
Johnson, Benjamin P., *Great Exhibition,* 195
Johnson, Cuthbert, agricultural cyclopaedia, 214–215
Johnson, John B., *Engineering Index,* 144
Johnson, William, *Cyclopaedia of Machinery,* 64
Johnson, William A., *Christopher Polhem,* 299
Jones, Samuel, *Pittsburgh,* 139
Jonval turbine, in Hunter, 244
Josephson, Askel G. S., 22
Josten, C. H., 289
Journal of American History, 151
Journal of Economic and Business History, 151
Journal of Economic History, 151
Journal of Science and Technology, 159
Journal of Transport History, 152
Juenger, Friedrich G., 206
Jürgensmeyer, W., 293

Kaempffert, Waldemar, 12
Kainen, Jacob, 270
Kaplan, Louis, 79
Kármán, Theodore von, 306
Karmarsch, Karl
   Geschichte der Technologie, 16
   Handbuch, 62
   Technisches Wörterbuch, 63
Karpinski, L. C., 43
Kastner, Richard H., 181
Katz, Herbert and Marjorie, 184
Kauffmann, M., 263
Kayan, Carl F., 292
Keller, Alex G.
   edits Turriano, 51
   "Renaissance Waterworks," 226
   reviews Forbes, *Man the Maker,* 4
   reviews Tartaglia, 48
   on Strada, 47
   Theatre of Machines, 41
   mentioned, 39

INDEX

Kellerman, Rudolf
   Kulturgeschichte der Schraube, 277
   in Treue, *Hausbuch,* 51–52
Kelley, Etna M., *Business Founding Dates,* 36
Kelly, Fred C., *Wright Brothers,* 242
Kelly, Thomas
   Adult Education, 177
   George Birkbeck, 176
Kent, William, 67
Kerker, Milton, 248
kerosene, in Beaton, Butt, 267
Kiely, Edmond R., 41
Kilgour, F. G., 5
Kimball, Dexter S., 86
kinematics, 304–306
   in Hartenberg, 41
King, Henry C., *Telescope,* 287
King, W. James, "Electrical Technology," 250–251
Kirby, Richard S.
   biographies, 79
   Early Years, Civil Engineering, 220
   Engineering in History, 5
   Engineers of New Haven, 82
Kirchheimer, Franz, on uranium, 261
Kirkland, Edward C., *Industry Comes of Age,* 13
Kisch, Bruno, 291
Klebs, Arnold C., 41
Klemm, Friedrich
   Handwerk und Technik, 126
   History of Western Technology, 5
   Kurze Geschichte, 5
   Technik, 5
   Technik der Neuzeit, 14
   in Treue, *Hausbuch,* 51–52
Klinckowstroem, Carl Graf von
   bibliographies of early histories (2), 15
   foreword, Rosenthal, 42–43
   illustrated books of trades, 282
   Knaurs Geschichte, 5
Klinckowström, Axel Leonhard, travels in America, 134
Klöckner-Humboldt-Deutz Co., 127
Klose, Gilbert C., 268
Knaurs Geschichte der Technik, 5
Knight, Cameron, *Mechanician and Constructor,* 279
Knight, Edward H., *Mechanical Dictionary,* 65
Knowles, L. J., in Bishop, 82
Koch, M., 42
Kolupaila, Stephen, 226
Körting, Johannes, 268
Kouwenhoven, John A., 206

Kranick, Frank N. G., papers, 116
Kranzberg, Melvin
   Technology in Western Civilization, 5–6, 206
   reviews Soulard, 7
Kress Library
   catalogue, 29
   pictures of crafts, 124
Krieg, M., 67
Kronick, David A., *Technical Periodicals,* 143
Krünitz, Johann G., 60
Kruzas, Anthony T., 36
Ktesibios, Drachmann on, 49, 284
Kuehl, Warren F., 31

labor and management, 297–304
   in Bolles, 256
   in Munn, 267
   U.S. Industrial Commission, 101
Laboulaye, Charles P. F. de, 64
Lacey, A. Douglas, 268
Lacroix, Eugene, 22
Lamb, Isaac W., in Bishop, 82
Lambert, Guillaume, 134
Lamme, Benjamin G., 86
Lampard, Eric E.
   Dairy Industry in Wisconsin, 218
   Industrial Revolution, 18
Landauer, Bella C., collection, 68
Landes, David S., in Cambridge History, 22–23
Landstrom, Bjorn, 230
Lane, Frederic C.
   "Economic Meaning of Compass," 231, 147
   Venetian Ships and Shipbuilders, 230
Langley, S. P., in N.A.S., 83
Larson, Esther E., *Swedish Commentators,* 131
Larson, Henrietta M., *Guide to Business History,* 23
laser, in Savin, 254
Latrobe, Charles J., in Mesick, 131
Laurson, F. G., 220
Lavender, David, 258
Lawrie, L. G., 273
Layton, Edwin T.
   "American Engineering Profession," 174
   note by, 175
   reviews Armytage, 204
leather
   ancient, in Forbes, 9
   gilt leather, in Doorman, 108–109
   tanning, in Welsh, 263

Leavitt, Erasmus, drawings of, 115
LeDuc, Thomas, on conservation, 210
Lee, Charles E., 238
LeGear, Clara E., 227
Legacy Series, 10
Leggat, John, *Art of Turning,* 278, 39
Lehmann-Haupt, Hellmut, 270
Leicester, Henry M., 256
Leighton, Albert C., 244
Lenoir, Etienne, in Leprince-Ringuet, 76
Lenthal collection, on shipbuilding, 115
Leon, Antoine, 177
Leonardo da Vinci, 49–51
  in Beck, 40
  gallery in Milan museum, 189
  in Gille, 10
  and Martini, 42
  in Olschki, 11
Leprince-Ringuet, Louis, 76
Leslie's Illustrated Weekly, 159
Lessing, Lawrence, 254
Leupold, Jacob
  bibliography, 42
  in Reti, 42
  Theatrum Machinarum, 45–46
LeVan, W. B., in Bishop, 82
Levey, Martin, 263
Lewis, Arthur O., *Of Men and Machines,* 206
Lewis, C. S., *Abolition of Man,* 206
Lewison, Paul, 111
library lists, 25–29
Library of Congress
  city directories, 84
  Directories in Science and Technology, 36
  Guide to Study of U.S., 32, 131
  international meetings, 36
  manuscripts in, 116
  National Union Catalog of Manuscript Collections, 114
  Newspapers on Microfilm, 169
  portrait files, 87, 88
  Reference Facilities in D.C., 36
Lieb, John W., library of Vinciana, 50
Lieber, Francis, 56
Lief, Alfred, 268
lighting, 250–253
  ancient, in Forbes, 9
  Bryant and May fire-making appliances, 186
  electric systems, tested 1883 in Cincinnati, 199
  gas industry, 268
  Hough on fire-making, 98

O'Dea, *Short History of Lighting,* 188
  patent review, *Journal* of P.O.S., 151
Lilius, Zacharias, in Feldhaus, 15
Lilley, S.
  Men, Machines, and History, 6
  "Nicholson's Journal," 160
Lin Hsien-Chou, on Wang Cheng, 48
Lincoln and the Tools of War, 295
Link, H. F. (2), 138
Lippman, E. O. von, 6
Litchfield, Paul W., 86
Littauer, S. B., "Quality Control," 304
Litterer, Joseph A., "Systematic Management," 302
Liverpool and Manchester Railway, in Holt, Marshall, 238
Liversidge, Archibald, 178
Lloyd, B. E., 178
locomotives, railroad, 236-240
  British Railway Locomotive, 189
  Science Museum titles, 188
  Sprague on electric locomotives, 98
  Van Name photos, in Vanderbilt, 124
Loeb Classical Library, 46
Lomb, Adolph, Optical Library, 288
Lombe, John, in Smiles, 82
Long, Stephen H., in *Port Folio,* 160
Loos, John L., 267
Lorini, Buonaiuto
  in Beck, 40
  in Promis, 82–83
Loudon, J. C., 214
Lovett, Robert W., 118
Lowe, Thaddeus S. C., papers, 116
lubrication, in Hersey, Naylor, 293
Lucas, A., 255
Luckhurst, David, *Monastic Watermills,* 192
Luckhurst, Kenneth
  Royal Society of Arts, 175
  Story of Exhibitions, 193
Ludwig, K.-H., 7
Luiken, Jan, 283
Lunar Society of Birmingham, 175
Lundvall, Sten, 299
Lunt, Edward C., 96
Luxon, Norval Neil, 160

Mabbott, Maureen C., 50
McAdam, Roger W., 231
McCormick reaper papers, 114
McCurdy, Edward, 50

INDEX

McDonald, Donald, *History of Platinum,* 261
MacGeorge, G. W., 220
McGivern, James G., 179
McGrath, E. J., 180
McGuire, J. D., on primitive drilling, 98
Machine, Le, periodical, 152
Machine in the Garden, Marx, 206
machine shops
  in Cassier's, special issue, 156–157
  European, in Sellers, 137
  Novelty Works, in *Harper's,* 158
  Novelty Works, in Hislop, 229
  Saco-Lowell Shops, 273
  Whitin Machine Works, 273
machine tools, *see* tools, hand and machine
machines, *see* mechanisms
McHugh, Jeanne, 259
Mackay, Alan
  Science of Science, 205
  Society and Science, 2
McKay, Donald, portrait, in Appleton's, 89
McKie, Douglas
  on John Harris, 54
  on metric system, 291
  Specific and Latent Heats, 305
McLane, Louis, 96
MacLaren, Malcolm, *Electrical Industry,* 252
Maclaurin, William R., *Radio Industry,* 254
McLean, Leslie, "Energetical Principles," 305
McMillen, James A., in *De Bow's,* 157
McMurtrie, Douglas C., bibliographies of printing, 271
Maddison, Francis
  "Astronomical and Mathematical Instruments," 287
  Billmeir Collection, 289
  Mechanical Universe, 286
Madrid, Leonardo manuscripts, 50–51
Magdeburg hemispheres, in Guericke, 45
Maier, Michael, in Klinckowstroem, 15
Malclès, Louis-Noëlle, 23
Malézieux, Emile, 134
Malthus, Thomas Robert, travels, 136
Man Makes Himself, 8
Man the Maker, 4
management, industrial, 297–304
Mandell, Richard D., *Paris 1900,* 198

Mandey, Venturus, *Mechanick-Powers,* 46
Mann, Charles R., 179
Manning, Gordon P., 190
Manning, Maxwell, and Moore, 69
Mantoux, Paul, 14
Manufactures Bureau, U.S., 100
manufacturing, 256
  state documents, Hasse, 94
  U.S., in *Tenth Census,* 96
manuscripts, 113–121
  early, 49–52
maps and cartography, 227
Marcus, Siegfried, biography of, 246
Marestier, Jean Baptiste, 134
Mariner's Mirror, 152
Mariner's Museum Library, catalogues, 230, 124
Marino, Samuel J., 131
Marks, Lionel, 67
Marland, E. A., *Electrical Communications,* 254
Marlowe, John, *Suez Canal,* 232
Marperger, Paul Jacob, in Klinckowstroem, 15
Marshall, C. F. Dendy, 238
Martin, Thomas (1813), *Circle of Mechanic Arts,* 283
Martin, Thomas (1964), "Royal Institution" (2), 175
Martineau, Harriet J., 131
Martini, Francesco di Giorgio
  in Olschki, 11
  in Reti, 42
Marx, E., on medieval manuscript, 52
Marx, Leo, *Machine in the Garden,* 206
Masiotti, A., edits Tartaglia, 48
masonry construction, 220–223
  in Clarke and Engelbach, 280
materials, strength of, 275–276, 304–305
materials and processes, *see* Contents, Part XIV.F
materials testing, 275–276
  of building materials, in Stemple, 97
  in C.N.A.M. catalogue B, 190
  in Sarton, "Fremont," 255
  Tatnall on Testing, 87
Mathias, Peter, *Brewing Industry,* 219
Matschoss, Conrad
  biographies, 76
  on steam engine, 248
Matthews, William, diaries and autobiographies, 79

Maudsley, Henry
  in Sellers, 86
  slide-rest described, in Gregory, 61
  in Smiles, 81
Maxim, Hiram P., reminiscences (2), 86
Maxim, Hudson, *Reminiscences*, 86
Maxson, John W., Jr.
  "Coleman Sellers," 269
  "Nathan Sellers," 269
Mayer, Julius Robert, in Leprince-Ringuet, 76
Mayer, L. A., *Islamic Armourers*, 281
Maynard, Katharine, 251
Mayr, Otto, reviews Feldhaus, 2
Mease, James
  Picture of Philadelphia, 139
  revises Brewster, 56
measurement, 290–292
  C.N.A.M. catalogue K, 190
mechanical dictionaries, 61–65
Mechanical Engineer in America, Calvert, 173
mechanical engineering, *see* engineering, mechanical
mechanical technology, *see* Contents, Part XIV.G
mechanics, theoretical and applied, 304–306
Mechanics' Magazine, Boston, London, New York, 159
mechanisms, 284–285
  bibliography, in Sotheran, 43
  C.N.A.M. catalogue C, 190
  in Ferguson, Reuleaux, 305
  in Hartenberg, 304
  in Wankel, 306
medieval and renaissance periods
  general works, 10–11, 1–9
  bibliography, Historical Association, 22
  Italian military biography, Promis, 82–83
  renaissance books and manuscripts, 39–52
  Sarton, *History of Science*, 43
  Thorndike, *History of Science*, 44
  White, "Historical Roots of Our Ecologic Crisis," 209
  White, "Middle Ages in American Wild West," 210
  agricultural technology
    Anderson, Bennett, Brace, Storck, 217
    Anderson, Fussell, 216
    Curwen, 215
  chemical technology, Singer, 263

crafts and craftsmen, 280–284, Salzmann, 255
energy conservation, Hiebert, Orvas, 305
engineering, civil, Straub, 221
glass, Duncan, 265
instruments, scientific
  Billmeir Collection, Bonelli, Dickinson, 289
  Maddison, 287
  Zinner, 288
lighting, Hough, O'Dea, Robins, Thwing, 250
masonry construction
  Fitchen, 222
  Shelby, 223
materials
  Smith, 255
  Weeks, 256
mechanisms and automata, 284–285
metals and mining
  Aitchison, 256
  Darmstaedter, 257
  Koch, 42
  Ress, 260
  Rickard, 258
  Schubert, 260
  Wertime, 261
military technology, 295–297
musical instruments, 294
oil, Forbes, 267
paper and printing
  Blum, 269
  Carter, Hunter, Lehmann-Haupt, 270
  McMurtrie, 271
pumps, Ewbank, 244–245
ships and navigation
  Anderson, Green, 229
  Landstrom, Lane, 230
  Lane, Taylor, 231
textiles, Ploss, 273
timekeepers, 285–286
tools, 276–278
transport, heavy, Heizer, Taylor, 294
water supply, Keller, Robins, 226
weights and measures
  Berriman, Burden, Burguburu, 290
  Hamarneh, Kisch, 291
Meigs, Montgomery C.
  construction photos, Library of Congress, 124
  in N.A.S., 83
  portrait, in Vanderbilt, 88
Meixell, Granville, 67
Melish, John, in Mesick, 131
Mellor, C. M., 263

INDEX

Meloy, Charles, 301
Mencken, August, *Railroad Passenger Car,* 240
Mendel *Hausbuch,* 51–52
Mercer, Henry C., *Ancient Carpenters' Tools,* 277, 186
Merdinger, Charles J., 220
Merrill, Harwood F., 302
Merrimack Valley Textile Museum
  business manuscripts, 115
  Wool Technology, 189
Mesick, Jane L., 131
metallography, history of, Smith, 256
metals: mining and metallurgy, 256–262
  ancient, in Forbes, 9
  bibliography
    in Darmstaedter, 40
    Seyfferts, Sotheran, 43
  Cassier's special issue, 156–157
  in Cornwall, Barton (2), 192
  early mining books
    Koch, 42
    Science Museum, 43
  important books, Smith, 43
  Leavitt drawings of equipment, Smithsonian, 115
  list of books, 19th century, Haferkorn, 21
  mineral industry education, Read, 181
  Mining for Metals in Wales, North, 186
  in Patent Office Library, 27–28
  Swedish observers in U.S.
    Hermelin, 134
    Larson, 131
  translated books, Smith, 47
  in travel accounts, Ferber, Hollander, Monnet, 138
  U.S. iron works, 1853, Lambert, 134
  U.S. mining, in *Tenth Census,* 96
Méthodique, Encyclopédie, 59
metric system
  controversy, 291–292
  origins, in Favre, McKie, 291
metrology, 290–292
  C.N.A.M. catalogue K, 190
Meyer, B. H., *Transportation in U.S.,* 228
Meyer, Hermann H. B., *Conservation of Resources,* 210
Meyer, Robert B., Jr., on early aircraft engines, 242
Mézières, military school, in Taton, 177–178

Michalowicz, Joseph C., 252
Michaud, Joseph and Louis, 75–76
microscope, in Clay, Disney, 287
Middleton, William E. K. (2), 288
Miley, D. G., 216
military technology, 295–297
  bibliography
    Bulletin Signalétique, 19
    Sotheran, 43
  in China, Needham, 2
  Dahlgren papers, U. of Syracuse, 116
  in Diels, 9
  early books, Cockle, 40
  early fortification, Croix, 40
  government publications, in Poore, 92
  Guns and Sails, Cippola, 205
  list of books, 19th century, Haferkorn, 21
  in Patent Office Library, 28
  renaissance biography, in Promis, 82–83
  rockets, 1819, in Rees, 56
Miller, Joseph A.
  bibliography in forest history, 274
  Pulp and Paper History, 271
Miller, William, *Men in Business,* 297–298
milling of grain, 217
  Dutch patents, Doorman, 108–109
  miller's guides, 61–62
Mingay, G. E., 215
mining, see metals
Minor, D. K., 156
Mint, U.S., records in National Archives, 112
mint machinery
  in Nepomucene, 137
  in Sellers, 86
Mississippi Valley Historical Review, 151
Mittelalterliche Hausbuch, 51
Moen, Philip L., in Nutt, 84
Mohr, Carolyn Curtis, 119, 237
Moléon, M. de, 256
Moles, Antoine, 223
Moll, F., 230
Monaghan, Frank, 131
Monceau, Duhamel du, 59
Moné, Frederick, 63
Monge, Gaspard, on cannon-making, 297
Monnet, A. G., 138
Montaigne, Michel de, 137
Montgomery, David, reviews Adams, 300

Montgomery, James, *Cotton Manufacture,* 274
Moody, Ernest A., on Leonardo, 50
Moorat, S. A. J., 51
Moore, C. K., *Electronics,* 253
Moorehead, Alan, 201
Moran, James, 271
Morgan, Morris H., translates Vitruvius, 48, 221
Morison, Elting E., *Men, Machines, and Modern Times,* 206–207
Morison, Samuel E.
  "History as a Literary Art," 21
  Ropemakers of Plymouth, 274
Moritz, L. A., 217
Morris, Richard B., *Sources for American History,* 30
Morrison, Philip and Emily, *Babbage,* 289
Morse, Dean, 208
Morton, Hudson T., *Anti-Friction Bearings,* 293
Morton, John C., agricultural cyclopaedia, 214
Morton Memorial, Stevens Inst. of Technology, 180
motion pictures, 127–128
motorcycles, *see* cycles
motors, electric, 250–253
  Siemens autobiography, 86–87
Mott-Smith, Morton, 305
Mottelay, Paul F., 251
Moxon, Joseph
  Art of Printing, 271, 47
  Mechanick Exercises, 46
  Mechanick-Powers, 46
Müller, Johannes, 143
Multhauf, Robert P.
  and encyclopaedias, 60
  "European Science Museums," 184
  "Museum Case History," 184
  "Self-Registering Instruments," 288
  and travel works, 137
Multhauf, Mrs. Robert, note by, 45
Mumford, Lewis
  Myth of the Machine, 207
  Technics and Civilization, 6
  quoted, 295
mummification, in Forbes, 9
Munden, Kenneth W., 111
Munn, Orson, edits *Scientific American,* 161
Munn, Robert F., *Coal Industry in America,* 267
Munroe, Charles E., 264
Murphey, Robert W., 32
Murra, Katherine O., 37

Murray, Matthew, in Smiles, 81
Museo Nazionale della Scienza e della Tecnica, 189
Museum of English Rural Life, 190
museums, *see* Contents, Part XII.C
Museums Directory, 182
Mushet, David, in Smiles, 81
musical instruments, 294–295
Musson, A. E.
  "Early Growth of Steam Power," 248
  Industry and Technology, 13
  introduction to Dickinson, 247
Mystic Seaport, manuscripts in, 119

Nadworny, Milton J., 302
Nartov, A. K., Russian machine builder, 279
Nasmyth, James, in Buchanan, 62
National Academy of Sciences
  biographies, 83
  scientific and technical societies, 37
National Advisory Committee on Aeronautics, *Bibliography,* 240–241
National Agricultural Library, catalogue, 214
National Archives, **109–112**
  aeronautical holdings, in Sunderman, 241
  on international exhibitions, 194
  patent records, 105
National Electrical Manufacturers Association, 251
National Museum, *see* Smithsonian Institution
National Reference Library of Science and Invention, 28
National Science Foundation, *Current Projects,* 207
Nature, **159**
  biography in, 78
Nature, La, 159
navigation and charting, **231**
  aerial navigation, in Nayler, 242
  instruments, in Brewington, 288
Navin, Thomas R., *Whitin Machine Works,* 273
Nayler, J. L., *Aviation: Technical Development,* 242
Naylor, H., "Bearings and Lubrication," 293
Near East
  Bosworth, Arabic encyclopedia, 49
  Creswell, *Arts and Crafts,* 20
  Mayer, *Islamic Armourers,* 281
  standard measures, in Hamarneh, 291

INDEX

steam engine precursors, in Needham, 248
Wulff, *Crafts of Persia*, 281
see also ancient and classical period
Needham, Joseph
"East and West," 2
Heavenly Clockwork, 286
Iron and Steel Technology in China, 260
in *Legacy of China*, 10
on Lynn White (2), 11
"Pre-Natal History of Steam Engine," 248
Science and Civilisation in China, 2
Nef, John U., *British Coal Industry*, 267
Neiderhauser, Clodaugh M., 274
Nepomucene, Sister St. John, 137
Nerén, John, 236
Nettels, Curtis P., 13
Neu, Irene D., *Railroad Network*, 239
Neu, John, "History of Science," 23
Neuberger, A., 10
Neudeck, G., 6
Nevins, Alan
British travelers, 132
energy symposium, 210
Ford, 235
Weyerhauser Story, 275
New York Public Library
bibliographies
aeronautics, 241
American interoceanic canals, 232
automobile tires, 235
bridges and viaducts, 224
electricity, 251
hydraulic engineering, 226
illumination, 250
scientific management, 301
Catalogue of Manuscripts, 119
Guide to the Reference Collections, 23
patent and trade mark publications, 107
portrait file, 87
technology-biography index, 31, 81
New York Times Index, 81
Newberry Library
History of Printing, 271
railroad archives, 119, 236, 237
Newcomen, Thomas, 248
Newcomen Bulletin, 153
Newcomen Society
Extra publications, 152–153
General Index, 152
Transactions, **152-153**, 24

Newcomen Society in North America, 153
Newhall, Beaumont, reviews Baier, 264
Newton's London Journal, 160
New Zealand, technical education, in Nicol, 178
Niagara Falls Power Company, 244
Nicholson, John, *Operative Mechanic*, 243
Nicholson, John T., on compressed air, 292
Nicholson, Peter, portrait, in Appleton's, 89
Nicholson, William, encyclopaedia, 56
Nicholson's Journal, 160
Nickel, Howard-White, 261
Nicklin, Philip H., travel account, 134
Nicol, John, *Technical Schools of New Zealand*, 178
Nielson, James Beaumont, in Smiles, 81
Niland, Austin, *British Organ*, 294
Niles' Weekly Register, 160
Norman, A. V. B., 296
North, F. J., 186
North American Review, 160
Novelty, S. S., in Hislop, 229
Novelty Iron Works, in *Harper's*, 158
nuclear power, in Anthony, 249
Nunn, G. W. A., British picture sources, 123
Nutt, Charles, *History of Worcester*, 84
Nystrom, John W., 66

Oakley, Kenneth P., 277
O'Dea, William T.
Short History of Lighting, 188
Social History of Lighting, 250
Ogden, Warren G., Jr.
Art of Turning, 278
and encyclopaedias, 60
notes by, 282, 283, 296–297
reviews Britkin, 279
oil, **267–268**
aircraft fuels, in Schlaifer, 242
ancient, in Forbes, 9
Canadian oil sands, Ells, 85
in Patent Office Library, 28
refining of, in Harper's, 158
in Smithsonian *Report*, 1861, Stemple, 97
in *Tenth Census*, 96
Old South Association, 21

Oliver, John W., *American Technology*, 12
Oliver, S. H., *Automobiles and Motorcycles*, 191
Olschki, Leonardo, 11
Opperman, C. A., 196
Ordnance Bureau, U.S., 101
Ornstein, Martha, *Scientific Societies*, 174
Orvas, Gunhard, 305
Osiris, 153
Osmawa, N. I., 7
Ostwald's Klassiker, 171
Ottley, George, British railway bibliography, 237
Otto, Nicolas A., in Bryant, Diesel, 245
Otto and Langen engine, on film, 127
Overton, Richard C., 238
Ower, E., 242

Palissy, Bernard, in Thorndike, 44
Panama Canal
    in Burr, 219
    history, in Smith, 102
    in "Interoceanic Canals," 232
    preliminary inventories, National Archives, 110
    in Stevens, 87
Panciroli, Guido, 15
Pannell, J. P. M.
    Civil Engineering, 220
    Industrial Archaeology, 191
Pantologia, A New Cyclopaedia, 56
paper and printing, *see* printing and paper
Pappus of Alexandria, in Beck, 40
Paris 1900, 198
Paris Universal Exposition, 1867, 196
Parker, Earl R., *Brittle Behavior*, 260
Parliamentary Papers, 94–95
Parsons, Charles A.
    Parsons Steam Turbine, 248
    in Smithsonian *Report*, Stemple, 97–98
Parsons, R. H., *Power Station Industry*, 253
Parsons, William B.
    collection of, 42
    Engineering in the Renaissance, 11
    railroad prints, 124
    mentioned, 39
Passer, Harold C., 253
Patent Office Society, *Journal*, 151
patents
    articles in *Journal of the Patent Office Society*, 151
    foreign, 107–109
    history, in *Archives Internationales*, 148
    U.S., 102–107
Payen, Jacques, 7
Payne-Gallwey, Ralph, *The Crossbow*, 296
Peabody Museum, Harvard, film, 127
Peale, Charles Willson, portraits, 88
Peale, Franklin, European visit, 137
Peale family papers, in Bell, 117
Pearce, William, in Pursell, 201
Pearson, Karl, edits Todhunter, 305
Peattie, C. Gordon, "Fuel Cell," 246
Peddie, R. A., *Subject Index of Books*, 32
Pelton, L. A., in Hunter, 244
Pender, Harold, electrical handbook, 67
Pennock, Samuel, in Bishop, 82
Penny Cyclopaedia, 56
Penrose, Boies, illustrations, 124–125
periodicals and serials, *see* Contents, Part XI
Perkins, Jacob
    biography, in Bathe, 170
    portrait, in Appleton's, 89
    ship pump illustrated, 55
perpetual motion, Dutch, in Doorman, 108–109
Perrot, Paul N., on Corning Museum library, 265
Perry, J. W., *One Hundred Books*, 42
Pertuch, Walter A. R., *Horological Books*, 286
Peters, Harry T., *America on Stone*, 125
Peterson, Clarence S., *County Histories*, 84
petroleum, *see* oil
Pfeiffer, L., in Feldhaus, 15
Philip, Alex J., 185
Philon, in Drachmann, 49
philosophy of technology, 208
photography, **264–265**
    C.N.A.M. catalogue L, 190
    The First Negatives, 188
Physis, 153
pipelines, oil and gas, 267–268
Piper, David, 88
Pirtle, T. R., *Dairy Industry*, 218
Pitt Rivers Museum, 189
platinum, in Howe, McDonald, 261
Pledge, H. T., 188
Ploss, Emil Ernst, old textile colors, 273–274

INDEX   335

Pluche, Abbé Noël, encyclopaedia, 283
Plummer, Osgood, in Nutt, 84
Poggendorf, Johann Christian, 75
Polhem, Christopher, Father of Swedish Technology, 299
Pollard, Sidney, *Modern Management*, 298
Polydore Vergil
  De Inventoribus, 15
  in Feldhaus, 15
Polytechnic Institution of London, history, 175
Polytechnique, Ecole
  Livre du centenaire, 180
  in Taton, 177–178
Poole, H. Edmund, *Annals of Printing*, 269
Poole, Mary Elizabeth, history in *Industrial Arts Index*, 145
Poole, William F., *Index to Periodicals*, 145
Poore, Ben Perley, index of government publications, 92
Pope, Franklin L., on electric motor, 252
Poppe, Johann von, history of technology, 16
Popular Science Monthly, 160
Port Folio, 160
Porta, G. della
  in Beck, 40
  in Ferguson, 15
  Natural Magick, 47
Porter, Charles T., *Engineering Reminiscences*, 86
Porter, Rufus, in *Scientific American*, 161
Portland cement, *see* cement and concrete
portraits, 87–89
Portsmouth Blockmaking Machinery, 299
Posner, Ernst, 117
Pottinger, Don, 296
Poussin, Guillaume Tell, 135
Power, Tyrone, travel account, 131
power generation and transmission, electrical, 252–253
  power failure, New England, in Rosenthal, 207
  Science Museum titles, 187
power transmission, mechanical, 292–293
Prager, Frank D., on Brunelleschi, 49
Pratt, F. A., in Spalding, 83–84
Prechtl, Joh. Jos., 62

Prescott, Samuel C., *Boston Tech*, 181
Prévost, Michel, *Dictionnaire de Biographie*, 76
Price, Derek J. de Solla
  "Automata in History," 284
  Heavenly Clockwork, 286
  introduces Porta, 47
  "Science of Science," 205
Prime, Alfred C., 170
Princeton Index of Christian Art, 126
printing and paper, **269–272**
  articles in *Journal of the Printing History Society*, 151
  bibliography, in *Bulletin Signalétique*, 19
  Dutch patents, in Doorman, 108–109
  paper, in *Special Consular Reports*, 100
  patent survey in *Journal* of P.O.S., 151
  U.S. papermaking, in Sellers, 86
Printing History Society, *Journal*, 151
Promis, Carlo, *Biografie*, 82–83
pumps, 244–245
Pursell, Carroll W., Jr.
  agricultural bibliography, 214
  "American Tin-Plate Industry," 260
  on Digges and Pearce, 201
  Technology in Western Civilization, **5–6**, 206
pyramids of Egypt, 222

quality control, 303–304
Que sais-je?
  No. 126, history of technology, 4
  No. 938, technical education, 177
Quill, Humphrey, 231

Rabb, Theodore K., 47
radio, **253–254**
  De Forest collection, 115
  George H. Clark collection, 115
  Science Museum titles, 188
Rae, John B.
  American Automobile, 235
  Automobile Manufacturers, 235
railroad track
  standard gauge, in Taylor, 239
  in Watkins, 239
railroads and rolling stock, **236–240**
  articles in *American Railroad Journal*, 156
  articles in *Railway Age*, 160
  in Bathe, 170
  Baldwin and Vauclain papers, in DeGolyer, 115–116

railroads and rolling stock (*continued*)
  government publications, in Poore, 92–93
  S. H. Long, in *Port Folio,* 160
  in Parsons collection, 42
  state documents, in Hasse, 94
  in travel accounts, 132–139
  see also locomotives, electric railways
Railway Age, 160
Railway Locomotives and Cars, 160
Railway and Locomotive Historical Society, 237
Ramelli, Agostino
  in Beck, 40
  in Keller, 41
  in Promis, 82–83
Randall, Merle, 103
Rankine, William J. M., *Useful Tables,* 66
Rasmussen, Wayne D.
  on Higgins collection, 68, 116, 214
  Readings in Agriculture, 215
Rathgen, Bernhard, 297
Reader's Guide to Periodical Literature, 146
Réaumur, R. A. F. de
  and *Descriptions,* 59
  Steel and Iron, 47
recording, acoustic
  in Begun, 253
  patents, in *Journal* of P.O.S., 151
  in Wilson, 255
recording, graphic
  in *Archives Internationales,* 148
  in Hoff, 287
  in Multhauf, 288
Redwood, Boverton, 268
Rees, Abraham, *Cyclopaedia,* 55–56
Reeves, Dorothea D., 32
refrigeration, 249
  ancient, in Forbes, 9
  patent survey, in *Journal* of P.O.S., 151
  Science Museum titles, 187
  in *Special Consular Reports,* 100
  see also ice trade
Reingold, Nathan
  Conference on Science Manuscripts, 113
  Coast and Geodetic Survey records, 118–119
  on Library of Congress, 118
  on National Archives, 109
  note by, 22
  on Patent Office records, 105

reviews Armytage, 204
reviews Kronick, 143
renaissance, *see* medieval and renaissance period
Rennie, George, *Illustrations of Mill Work,* 62
Rennie, John, in Smiles, 82
Rensselaer Polytechnic Institute
  history, 181
  theses in library, 119–120
Repertory of Arts, 160
ReQua, Eloise G., 37
Resources for the Future, Inc., 210
Ress, Franz M., 260
Reti, Ladislao
  Leonardo's steam engines, 50
  Madrid codices, 50–51
  Martini's plagiarists, 42
  on Turriano, 51
  mentioned, 39
Reuleaux, Franz
  Briefe aus Philadelphia, 197
  Kinematics of Machinery, 305
Reuss, Jeremias D., 146
Revue d'Histoire de la Sidérurgie, 153
Revue d'Histoire des Sciences, 153
Reynolds, James, portrait, in Appleton's, 89
Reznek, Samuel, 299
Rhees, William J., 97
Ricci, Seymour di, 51
Rice, Howard C., Jr., 288
Richards, Francis H., in Spalding, 83–84
Richards, John, *Wood-working Machines,* 275
Richards, Joseph W., *Aluminum,* 261
Riches, Phyllis M., *Collected Biography,* 81
Richeson, A. W., *English Land Measuring,* 227, 170
Richey, M. W., *Geometrical Seaman,* 231
Richner, Alfred, *Impact of Science,* 208
Richter, G. M. A., on classical furniture, 281
Rickard, Thomas A., *Man and Metals,* 258
Ricketts, Palmer C., history of Rensselaer, 181
Ridding, Arthur, S. Z. *Ferranti,* 189
Rider, K. J., bibliography, history of technology, 24
Riepe, Dale, reviews Morison, 207

rigging and moving heavy objects, 294
Ringwald, Donald C., 231
Rips, Rae E., 92
Ristow, Walter W.
  Guide to Historical Cartography, 227
  Survey of Roads, by Colles, 132, 234
Rittenhouse, David, 287
Rittenhouse, Jack D., carriage bibliography, 234
rivers and harbors, 232–233
roads, 233–234
  ancient, in Forbes, 9
  in Condit, 219
  government publications, in Poore, 92
  National Road, in *Port Folio,* 160
  patent survey, in *Journal* of P.O.S., 151
  preliminary inventory, National Archives, 110
  state documents, in Hasse, 94
Robbins, Michael, *Railway Age,* 238
Roberts, A. D., *Guide to Technical Literature,* 143–144
Robins, F. W.
  Story of the Lamp, 250
  Water Supply, 226
Robinson, E., "Growth of Steam Power," 248
Robinson, T. H., in *Dissertation Abstracts,* 30
Rochas, Beau de, 19
Roe, Joseph W., *Tool Builders,* 279
Roebling, John
  in *Builders of the Bridge,* 224
  handbook, 67
  Niagara bridge, 120, 134
Roebling, Washington A.
  in *Builders of the Bridge,* 224
  thesis, 120
Roebuck, John, in Smiles, 81
Rogers, Earl M., agricultural bibliography, 214
Rogers, Henry, on wire gauges, 292
Rogin, Leo, 216
Roller, Duane H. D., 171
Rolt, L. T. C.
  The Aeronauts, on ballooning, 242
  Horseless Carriage, 235
  Machine Tools, 279
  Thomas Newcomen, 248
Romaine, Lawrence B., 68
Ronalds, Francis, 251

rope driving
  in *Engineering,* 292
  in Flather, 293
ropemaking, in Morison, 274
Roper, Stephen, 66
Rose, Albert C., *Public Roads,* 233
Rose, Walter, *Village Carpenter,* 282
Rosen, Edward, *Landmarks in Science,* 171
Rosenberg, Nathan, on Allen, 62
Rosenbloom, Richard S., on 19th-century management, 298
Rosenthal, A. M., on power failure, 207
Rosenthal, Gottfried, *Literatur der Technologie,* 42–43
Ross, Earle D., *Democracy's College,* 179
Rossman, Joseph, *Industrial Creativity,* 306
Roubo, André Jacob, on cabinetmaking, 58
Rouse, Hunter, 305
Rousseau, P., 6
Rowe, Robert S., 232
Royal Aeronautical Society, 168
Royal Institute of British Architects
  Catalogue, 220
  publications, 168
Royal Institution
  history (3), 175
  manuscripts, 121
  publications, 168
Royal Society of Arts
  Lectures on the Exhibition, 195
  materials on America, 117
  publications, 168
Royal Society of London
  biographies, in *Subject Index,* 80–81
  biographies of members, 83
  Catalogue of Scientific Papers, 146
  International Catalogue, 146
  publications, 168
  Subject Index, 146
Ruggles, John, 102
Rumsey, James, in Poore, 92
Russell, John Scott, *Technical Education,* 177
Russia
  Nartov, *Machine Builder,* 279
  Timoshenko, *Engineering Education,* 178
  Westwood, *Russian Railways,* 239
Russo, Francois, 17
Rydberg, Sven, 131

Sachs, Curt, *Musical Instruments*, 294
Sachs, Hans, *Ständebuch*, 47
Sahlin, Carl, translates Triewald, 48
St. Bride Institute, library of printing, 271
St. Hardouin, F. P. H. Tarbé de, 82
Salaman, R. A., *Dictionary of Hand Tools*, 277
Salzmann, L. F., *English Industries*, 255
Sanders, Ralph, *Project Plowshare*, 209
Sanderlin, Walter S., *Chesepeake and Ohio Canal*, 233
Sandström, Gosta E., *Tunneling*, 225
San Gallo, Giuliano da, 52
Sarton, George
  "Charles Fremont, Historien," 255
  Guide to the History of Science, 17
  Introduction to the History of Science, 43
  quoted, 213
Sass, Friedrich, 246
Saveney, E., workmen visit 1867 exposition, 196
Savery, Thomas, engine illustrated, 55
Savin, J., on laser, 254
Sawers, David, *Sources of Invention*, 306
Sawyer, John E., "American System," 299
Sawyer, R. Tom, *Gas Turbine*, 246
Sawyer, Sylvanus, in Bishop, 82
Saxton, Joseph, in N.A.S., 83
Scaglia, Giustina, on Brunelleschi, 49
Scappi, Bartolomeo, in Keller, 41
Scarlott, Charles A., 243
Schaeffner, André, on musical instruments, 294
Scheffler, J. E., in Leupold, 45
Scharf, J. Thomas, *Philadelphia*, 84
Schildberger, Friedrich, *Engines to Autos*, 245
Schlaifer, Robert, *Aircraft Engines*, 242
Schlatter, Hugo, *Explosives Industry*, 264
Schlomann, Alfred, polyglot dictionary, 70
Schmeckebier, Lawrence F., 92
Schnabel, Fritz, on engineering colleges, 178
Schneider, K., in Treue, *Hausbuch*, 51–52
Schoen, Henri, on German higher education, 178

Schofield, Robert
  Case Archives, 113
  "Histories of Societies," 174
  Lunar Society, 175
  reviews Munden and Beers, 111
Schott, Caspar, in Ferguson, 15
Schubert, H. R., *British Iron and Steel*, 260
Schuchardin, S. W. (2), 7
Schultz, Charles R., manuscripts at Mystic (2), 119
Schultz-Wittuhn, Gerhard, on automobile industry, 236
Schussele's engraving, *Scientists and Inventors*, 171
Science Library
  Bibliographical Series, 24
  technical glossaries, 69
Science Museum
  First Hundred Years, 188
  historic books (2), 43
  photographs in, 125
  portraits in, 88
  publications of, 187–189
Science Library bibliographies, 24
Science of Science, 205
Scientific American
  founded 1845, 161
  since 1948, 153
Scientific American Supplement, 161
scientific instruments, 287–290
  in Patent Office Library, 27–28
  Science Museum titles, 187–189
scientific management, 300–303
Scientific Monthly, biographies in, 78
Scott, David, *Machinist's Assistant*, 63
Scott, Franklin D., *Klinckowström's America*, 134
Scovill, J. M. L. and W. H., in Van Slyck, 82
Scoville, Warren G., on glass industry (2), 265
Scribner's Engineers' Companion, 66
Scribner's Monthly, 161
Scudder, Samuel H., 144
Seeger and Guernsey, catalogue, 37
Seiffert, Christian E., 43
Sellergren, Gustaf, 299
Sellers, Coleman, in Maxson, 269
Sellers, George Escol, *Reminiscences*, 86
Sellers, Nathan, in Maxson, 269
Sellers, William
  in N.A.S., 83
  in Scharf, 84
  tool drawings, in Franklin Institute, 115

Sellers family papers, in Bell, 117
Semenov, S. A., *Prehistoric Technology*, 277
Semmel, Bernard, on metric system, 292
Senefelder, Alois, portrait, in Appleton's, 89
serials, *see* Contents, Part XI
Serrel, R., 254
Service Center for Teachers of History, 18
servo mechanisms, in Conway, 284
Severance, Belknap, 107
sewage disposal
    in Paris and London, 227
    in R.I.B.A. *Catalogue*, 220
    state documents, in Hasse, 94
sewing machines, patent review, in *Journal* of P.O.S., 151
Seyfferts, Christian E., 43
Shaler, Nathan S., 211
Shannon, Fred A., 215
Sharlin, Harold I., 251
Sharp, Paul, 185
Shaw, Ronald E., *Erie Canal*, 233
Shaw, Thomas, "Wire Telephony," 254
Shelburne Museum publications, 190, 278
Shelby, Lonnie R., 223
Shell International Petroleum Co., films, 127–128
Shelton, William V., 66
Shepard's Federal Reporter, 104
ships and boats, 228–231
    articles in *American Neptune*, 147, *Mariner's Mirror*, 152
    in Bathe, 170
    bibliography, in Sotheran, 43
    Cassier's, special issue, 156–157
    government publications, in Poore, 92
    in *Harper's*, 158
    Inventing the Ship, 306
    Mystic Seaport manuscripts, Schultz, 119
    National Archives, preliminary inventories, 110
    nineteenth-century books, in Haferkorn, 21
    Oared Fighting Ships, 295
    Science Museum titles, 187–189
    U.S., in *Tenth Census*, 96
    U.S. steamboats, in Marestier, 134
    U.S.N.M. catalogue, 191
Shirref, Patrick, in Mesick, 131
Shumway, George, 234

Siemens, Werner von
    autobiography, 86–87
    in Leprince-Ringuet, 76
silk, bibliography in Howitt, 273
Silliman's *Journal*, 155
Simmons, Jack, *Railways of Britain*, 239
Simpson, Bernice, *List of Glossaries*, 70
Sinclair, Angus, *Locomotive Engine*, 240
Sinclair, Joseph Bruce, on Franklin Institute, 175
Singer, Charles
    Earliest Chemical Industry, 263
    History of Technology, 2–3
Singer, Dorothea W., translates Klemm, 5
Sisco, Anneliese G., translates metals books (3), in Smith, 47
Skempton, A. W., "Portland Cements," 266
Skinner, I. L., patent journal, 105
Skipper, Otis C., on *DeBow's Review*, 157
Slater, Samuel
    memoir of, in White, 274
    portrait, in Appleton's, 89
slide rule, in Cajori, 289
Sloane, Eric, *American Tools*, 277
Slocum, Robert B., on biographical works, 74
Slyck, J. D. Van, 82
Smart, Charles E., *Surveying Instruments*, 288
Smeaton, John
    Catalogue of Engineering Designs, 152
    travel diary, 137, 152
Smet, Antoine de, 132
Smiles, Samuel (4), 81–82
Smith, Cyril Stanley
    "Discovery of Carbon in Steel," 260
    History of Metallography, 256
    "Important Books," 43
    "Materials and Civilization," 255–256
    Sorby Centennial Symposium, 257
    translations of early works, 47
Smith, Darrell H., *Panama Canal*, 102
Smith, David, *Industrial Archaeology*, 192
Smith, Edgar C.
    biographical index, 80
    Marine Engineering, 230
    Pioneers of Refrigeration, 249

Smith, H. P., on cotton harvesting, 216
Smith, H. R. Bradley, *Blacksmiths' Tools*, 278
Smith, H. Shirley, *Great Bridges*, 224
Smith, James, mechanical dictionary, 61
Smith, Murphy D., guide to A.P.S. archives, 117
Smith, Ralph W., "Weights and Measures," 291
Smith, Richard K., on dirigibles *Akron* and *Macon*, 243
Smithsonian Annals of Flight, 98
Smithsonian Institution, 97–99
  catalogues of collections, 191
  Descriptions des Arts et Métiers in, 58
  industrial films planned, 127
  manuscript materials in, 115
  photographs in, 125
  portrait file, engineering division, 87
  trade catalogues, 67, 69
Smithsonian Journal of History, 153
societies, technical and scientific, see Contents, Part XI, XII.A
Society and Science, 2
Society for the History of Technology monograph series, 170
  Technology and Culture, 154
Society for the Promotion of Engineering Education, 179
Society of, see note at beginning of index.
Society of Architectural Historians, *Journal*, 153
Sorby, Henry C., centennial symposium, in Smith, 257, 43
Sotheran, Henry, *Bibliotheca Chemico-Mathematica*, 24, 43
Soulard, R., 7
Southern, R. W., on Lynn White, Jr., 11
spacecraft, 241
Spackman, Charles, 266
Spalding, J. A., *Biography of Connecticut*, 83–84
Spaulding, Thomas H., *Military Books*, 43
Spear, Dorothea N., 84
Spence, Clark C., *God Speed the Plow*, 216
Spencer, Christopher M., in Spalding, 83–84
Spencer, K. J., *Electronics*, 253
Spencer, John R., on Filarete, 51
Spon's Dictionary of Engineering, 64

Sprague, Frank Julian, in Miller, 297–298
Spratt, H. P.
  Steam Navigation, 189
  synopsis of events, mechanical and electrical, 25, 189
Springfield Armory, in *Harper's*, 158
standards, weights and measures, 290–292
  A.S.M.E. Boiler Code, 247
Standards Bureau, U.S.
  in Cochrane, 290
  in Smith, 291
  in Weber, 102
Standish, Philander M., papers, 116
Statham, Jane, 37
Staveley, Ronald, 36
steam engines, turbines, boilers, 247–249
  Cornish engine films, in Shell, 127
  Leonardo's, in Reti, 50
  nineteenth-century, in Porter, 86
  Parsons on turbines, in Stemple, 97–98
  in Rees, *Cyclopaedia*, 55–56
  tested, 1874, in Cincinnati, 199
steam road vehicles, see automobile
steel, see iron and steel
Steen, Herman, 217
Steers, George, portrait, in Appleton's, 89
Steinitz, Kate Trauman, 49
Steinman, David B., Roebling biography, 224
Stemple, Ruth M., index of Smithsonian *Reports*, 97–98
Stephenson, Robert, & Co., 240
Steuben Glass Works, 150
Stevens, Isaac I., railroad papers, 119
Stevens, John
  family papers, 118
  in Miller, 297
Stevens, John F., *Engineer's Recollections*, 87
Stevenson, David, *Civil Engineering of North America*, 135
Stewart, E. G., *Town Gas*, 189
Stewart, Robert E., Jr., *Adolph Sutro*, 258
Stillerman, Richard, *Sources of Invention*, 306
Stillwell, Margaret B., *Incunabula and Americana*, 44
Stone, George C., *Arms and Armor*, 296
Stone, J. F. S., in Ancient Peoples, 8
Storck, John, on flour milling, 217

INDEX 341

Storck, W. F., *Hausbuch*, 51
Stotz, Louis, 268
Stover, John F., *American Railroads*, 239
Stowers, A., "Water Power," 244
Strada, Jacob de
  in Keller, 41
  Künstlicher Abriss, 47
  in Reti, 42
Stradanus, Johannes, *New Discoveries*, 47
Straet, Jan van der, 47
Straub, Hans, 221
strength of materials, 304–306
Strickland, William, *Reports on Canals*, 137
Strong, George Templeton, 87
Stromer, W. V., in Treue, *Hausbuch*, 51–52
Struik, Dirk, 12
Stuart, Charles B., 81
  in Carnegie, 79
Stucklé, Henri, 135
Sturgeon, William, *Annals of Electricity*, 161
Sturt, George, *Wheelwright's Shop*, 282
Sturtevant, B. F., in Van Slyck, 82
Subira, Oriol Valls i, 272
submarine telegraph
  Atlantic Cable, 253
  One Hundred Years, 188
Suez Canal
  in Marlowe, 232
  in Vernon-Harcourt, 233
Sun, E-Tu Zen and Shiou-Chuan, 48
Sunderman, James F., 241
Sung Ying-Hsing, 48
Suplee, Henry H.
  Engineer's Reference, 67
  Gas Turbine, 246
surveying and instruments, 227
  in Bedini, 287
  in Kiely, 41
  in Smart, 288
Sutro, Adolph
  biography, by Stewart, 258
  papers, in Hamer, 113–114
Svinin, Paul, 135
Swank, James M., 260
Swanson, Edward B., 268
Swedenborg, Emanuel, 48
Switzer, Stephen, in Reti, 42
Sworykin, A. A., 7
systems analysis, in Boguslaw, 300

Taccola, Marianus Jacobus, in Thorndike, 51
Tacoma Narrows Bridge failure
  on film, 127
  in Finch, 223
Taconite, in Davis, 257
Taft, Robert, *Photography*, 265
Tainter, Charles S., collection, 115
Taisnier, Jean, in Thorndike, 44
Tallis, John, 195
Tanner, H. S., 139
tanning of leather, in Welsh, 263
Taplin, W., 259
Tarbé de St. Hardouin, F. P. H., 82
Tardy, bibliography of timekeeping, 286
Tartaglia, Nicoló
  in Olschki, 11
  Quesiti et Inventioni, 48
Tasks of Economic History, 154
Tatnall, Frank G., *Tatnall on Testing*, 87
Taton, René
  archives of Paris Académie, 121
  "La machine arithmétique," 289
  on scientific and technical education, 177
Taylor, A. J. P., European history bibliography, 19
Taylor, E. G. R., on navigation and instruments, 231
Taylor, Frank A., *Engineering Collections*, catalogue, 191
Taylor, Frederick W.
  papers of, 303
  Scientific Management, 302
Taylor, F. R. Forbes, "Heavy Goods Handling," 294
Taylor, F. Sherwood
  Industrial Chemistry, 263
  on museums, 185
Taylor, George R.
  American Railroad Network, 239
  Transportation Revolution, 13, 228
Teague, W. D., 217
Teani, Renato, 45
Technical Book Review Index, 32
Technics and Civilization, 6
Technikgeschichte, 154
Techniques et Civilisations, 154
technology, history of, see Contents, Part I
technology and culture, see Contents, Part XIII
Technology and Culture, 154

Tekniska Museet
   Daedalus, yearbook, 149
   history card file, 32
telegraph, **253–254**
   in King, 250–251
   in Sharlin, 251
   in Smithsonian *Reports*, Stemple, 97
telegraph, submarine
   Atlantic Cable, 253
   One Hundred Years, 188
telephone, **253–254**
   in King, 250–251
   in Sharlin, 251
telescope, **287–289**
   Dutch patents, in Doorman, 108–109
Telford, Thomas, in Smiles, 82
Temin, Peter, *Iron and Steel*, 260
Templeton, William, 66
Tesla, Nikola, papers, 116
Testing Board, for iron, etc., 101
testing of materials, *see* materials testing
Textile History, 154
textiles, **272–274**
   ancient, in Forbes, 9
   articles in *Ciba Review*, 148
   bibliography, *Bulletin Signalétique*, 19
   bibliography, Zischka, 54
   cotton manufactures, in *Harper's*, 158
   Dutch patents, in Doorman, 108–109
   Islamic, in Creswell, 20
   loom patent survey, *Journal* of P.O.S., 151
   Mamluk Costume, 281
   in Patent Office Library, 27–28
   in Rees, *Cyclopaedia*, 55–56
   in *Special Consular Reports*, 100
   U.S., in Parliamentary Papers, 94
   see also Merrimack Valley Textile Museum
Theophilus (12th Century)
   in Ferguson, 15
   On Divers Arts, 47
thermodynamics, **305**
   on heat radiation, 1859, in Stemple, 97
   in Kerker, 248
thermometer
   in Middleton, 288
   in *Revue d'Histoire*, 153
Thieme, Ulrich, *Lexikon*, 47
Thomas, D. B., on photography, 188

Thomas, Elizabeth M., *The Harmless People*, 127
Thomas, Isaiah, *Printing in America*, 271–272
Thomas Register, 37
Thomen, Harold O., *Checklist of Hearings*, 93
Thompson, Arthur W.
   Manuscripts, 117
   Printed Materials, 30
Thompson, C. Bertrand, *Scientific Management*, 302
Thompson, Daniel V., Jr., "Medieval Craftsmanship," 51
Thompson, George V., on standardization, 292
Thompson, J. David, *Learned Societies*, 174
Thompson, John S., *Composing Machines*, 272
Thompson, Robert L., *Wiring a Continent*, 254
Thomson, Thomas R., *American Railroads*, 237
Thorndike, Lynn
   History of Science, 44
   on Taccola, 51
Thornton, John L., 32
Thurston, Robert H.
   Boiler Explosions, 248
   "Technical Education," 179
   on Vienna 1873 exhibition, 196
Thwing, Leroy, 250
Ticonderoga, Story of the, 190
timber and wood industries, **274–275**
   patent review, in *Journal* of P.O.S., 151
   Puget Sound area manuscripts, 119
timekeepers, **285–286**
   C.N.A.M. catalogue JB, 190
   Dutch patents, in Doorman, 108–109
   early, in Diels, 9
   list of books, in Zischka, 54
   London Clockmaker's museum, 186–187
   machine-made watches, in *Harper's*, 158
   in Patent Office Library, 28
   Science Museum titles, 188–189
Times, New York, Index, 81
Times, The, indexes, 81
Timm, Albrecht, *Kleine Geschichte*, 7
Timoshenko, Stephen P.
   Engineering Education in Russia, 178

INDEX 343

History of Strength of Materials, 275
tin plate
 Historic Tinned Foods, 218
 in Pursell, 260
 in Tooker, 261
Tin Research and Development Council, 218
Tipper, Constance F., 260
tires, 235–236
Titus, Edna Brown, *Union List,* 142
Todd, A. C., 201
Todhunter, Isaac, 305
Tolansky, S., 265
Tomlinson, Charles, 63
Tooker, Elva, 261
Tooley, R. V., 227
tools, hand and machine, 276–280
 articles in EAIA *Chronicle,* 149
 bibliography, *Bulletin Signalétique,* 19
 British exports, in Parliamentary Papers, 94
 Nasmyth on, in Buchanan, 62
 Science Museum titles, 187–188
 stone tools, in Pitt Rivers, 189
 see also machine shops
tools, woodworking machine, 275
Toppel, Donald R., 296
Towne, Henry R., in Spalding, 83–84
tractor, farm, 216–217
trade catalogues, 67–69
Tradition, 154–155
transduction, in Hunt, 254
transfer of technology, 200–201
Transport History, 155
transportation, general, 228
 articles in *American Railroad Journal,* 156
 articles in *Journal of Transport History,* 152
 articles in *Transport History,* 155
 bibliography, *Bulletin Signalétique,* 19
 C.N.A.M. catalogues, 190
 in *Special Consular Reports,* 100
 in travel accounts, 132–139
 U.S., in *Tenth Census,* 96
Trautwine, John C., 66
travel and description, *see* Contents, Part X
Tredegar Iron Works, 259
Trenton bridge, illustrated, 56
Treue, Wilhelm
 Kulturgeschichte der Schraube, 277
 Mendel *Hausbuch,* 51–52

Trevethick, James, in Leprince-Ringuet, 76
Triewald, Marten, *Steam Engine,* 48
Trotter, Nathan, *Philadelphia Merchant,* 261
Tschernyschew, W. I., 7
Tsiolkovskiy, K. E., 243
Tuckerman, Alfred, thermodynamics, 98
Tudor, Frederic, in Cummings, 249
Tuge, Hidomei, 7
Tully, R. I. J., 32
tunnels, 225
 Fox, on Alpine tunnels, in Stemple, 97–98
 Simplon Tunnel, in Fox, 85
 Sutro drainage tunnel, 258
 U.S., 1853, in Lambert, 134
Tunis, Edwin, 278
turbine, gas
 in aircraft, Whittle, 243
 bibliography, A.S.M.E., 246
 bibliography, Besterman, 18
 in Sawyer, Suplee, 246
turbines, hydraulic, *see* hydro-power
turbines, steam, *see* steam engine
Turgan, Julien F., *Grandes Usines,* 64, 126
Turriano, Juanelo
 in Beck, 40
 in Reti, 51
Tylecote, Mabel, *Mechanics' Institutes,* 177
Tyler, David B.
 American Clyde, 230
 Steam Conquers the Atlantic, 230
typesetting machines
 in Moran, 271
 in Thompson, 272
typewriters, in Science Museum, 188
Tyrell, Henry G., 224

Uccelli, Arturo
 Enciclopedia, 3
 on Leonardo (2), 50
 Scienza e Tecnica, 3
 Storia della Tecnica, 7
Udall, Stewart L., 209
underdeveloped nations, in ReQua, 37
underwater archaeology, 8
UNESCO
 glossaries, in Wüster, 70
 interlingual dictionaries, 71
 "Museums of Science and Technology," 184
Union des Associations Internationales, 37–38

Union List of Serials, 142
United States
  government departments, *see* name of department or bureau
  government publications, *see* Contents, Part VII
  Historical Statistics of, 96
United States Magazine of Science, Art, 161
Unwin, George, *Industrial Organization*, 298
Unwin, W. C., "Power Distribution," 293
uranium
  in Kirchheimer, 261
  in Wilhelm, 262
Ure, Andrew, 63
Urwick, L., 303
Usher, A. P., *History of Mechanical Inventions*, 7, 44
Utica Steam Engine Co., 66

Van Deman, Esther B., *Roman Aqueducts*, 226
Van Gelder, Arthur P., *Explosives Industry*, 264
Van Melsen, Andrew G., *Science and Technology*, 207
Van Nostrand's Magazine, 161
Van Slyck, J. D., *New England Manufacturers*, 82
van der Straet, Jan, in Stradanus, 47
Vance, Lucile E., *Illustration Index*, 124
Vanderbilt, Paul, *Guide to Prints and Photographs*, 124, 88
Vaucanson, Jacques, in Leprince-Ringuet, 76
Vauclain, Samuel, papers, 115
Verantius, Faustus
  in Beck, 40
  in Keller, 41
  Machinae novae, 48
Verein Deutscher Ingenieure
  museums directory, 185
  Rundbrief, 25
Vergil, Polydore, *see* Polydore
Vernon-Harcourt, L. F., 233
Vernon, K. D. C.
  Royal Institution history, 175
  Royal Institution manuscripts, 121
Vidinov, S. S., 279
Vierendeel, Arthur
  Construction Architecturale, 223
  Esquisse d'un Histoire, 8
Villedeuill, Pierre-C. L. de, 237
Vinci, *see* Leonardo da Vinci

Vitruvius
  in Beck, 40
  Ten Books, 48, 221
Vogel, Robert M.
  elevators in Eiffel Tower, 223
  "Hall of Civil Engineering," 184
  on Wendel Bollman, 224
Vollbeding, Joh. Christoph, in Klinckowstroem, 15
von Kármán, Theodore, 306
von Klinkowstroem, *see* Klinckowstroem
von Lippman, *see* Lippman

Waddell, J. A. L., 179
wagons, *see* carriages
Wahl, Paul, 296
Wailes, Rex, "Windmill Fallacies," 244
Waldburg-Wolfegg, Johannes Graf, 51
Waldburg-Wolfegg-Waldsee, 51
Wales, National Museum, 186
Walford, A. J.
  foreign dictionaries, 71
  Guide to Reference Material, 33
Walker, Charles R., on technology and civilization, 207
Walker, Oakley S., in Nutt, 84
Wallas, Graham
  The Great Society, 208
  quoted, 203
Wallis, George, on New York 1853 exhibition, 133
Wang Cheng, *Western Machines*, 48
Wang Ling, *Heavenly Clockwork*, 286
Wankel, Felix, 306
Wansey, Henry, in Mesick, 131
Wantabe, Minoru, 179
war, *see* military technology
Warner, Aaron W., *Impact of Science on Technology*, 208
Warner, Sam B., Jr.
  bibliography of cities, 209
  Streetcar Suburbs, 209
Warren, J. G. H., *Robert Stephenson & Co.*, 240
Washburn, Ichabod, in Nutt, 84
Washburn and Moen Co., papers, 118
Washington, University of, manuscripts, 119
Washington Mill Co. papers, 119
water supply, 225–227
  state documents, in Hasse, 94
Waters, D. W., 231
waterwheels, *see* hydro-power
Watkin, Edward W., travel account, 135

INDEX

Watkins, J. Elfreth, "American Rail and Track," 239, 98
Watson, Elkanah, travel account, 135
Watson, Evelyn B., *U.S. Patent Numbers,* 103
Watson, W. E., *Microscope,* 287
Watson, W. J., *Bridge Architecture,* 224
Watt, James, and the Steam Engine, 247
Watt, James, in Leprince-Ringuet, 76
Watts, George B.
  Encyclopédie méthodique, 59
  Handicrafts of France, 58-59
  on Swiss editions, *Encyclopédie,* 58
Weale, John, 123
Weatherford, Willis D., 157
Weaver, William D., *Wheeler Gift,* 251
Weber, Gustavas A., 102
Webster's Biographical Dictionary, 74
Weeks, Mary E., 256
Wegmann, Edward
  Dams, 227
  Water-Supply of New York, 226
Weigel, Christoff
  Abbildung, 283, 48
  in Klinckowstroem, 282
weights and measures, 290-292
  C.N.A.M. catalogue K, 190
Weinberg, Meyer, note by, 133
Weissenborn, Gustavus, 64
Welch, Frank J., on cotton harvesting, 216
Welland Canal, 232
Wellcome Historical Medical Library
  Manuscripts, 51
  Printed Books, 44
Wells, Horace, portrait, in Appleton's, 89
Welsh, Peter C.
  notes by, 130
  on patents, 106
  Tanning in the U.S., 263
  Victorian American, 125
  "Woodworking Tools," 278
Wernwag, Louis, bridge illustrated, in Rees, 55
Wertime, Theodore
  Coming of the Age of Steel, 261
  "First Encounters with Metallurgy," 258
West, Clarence J.
  bibliography of pulp and paper, 272
  on chemical technology, 263

Westcott, G. F.
  British Railway Locomotive, 189
  synopsis of events, mechanical and electrical, 25, 189
Westcott, Thompson, *Philadelphia,* 84
West Point, by Forman, 180
Westwood, J. N., 239
Weyerhauser Story, 275
Wheeler, Charles (Philadelphia), in Scharf, 84
Wheeler, Nathaniel
  portrait, in Appleton's, 89
  in Van Slyck, 82
Wheeler Gift, library on electricity, 251
Whitworth, Joseph, travel account, 133
Whipple, Squire, 224
Whistler, George W., portrait, in Appleton's, 89
White, Francis S., *History of Inventions,* 16
White, George S., *Memoir of Samuel Slater,* 274
White, Harry J., *Electrostatic Precipitation,* 253
White, J. T., *American Biography,* 74
White, John H., Cincinnati locomotives, 240
White, Lynn, Jr.
  "Ecologic Crisis," 209
  "Invention in the Middle Ages," 11
  Medieval Technology and Social Change, 11
  "Middle Ages in Wild West," 210
  on origin of technical history, 14
  reviews Daumas, *Histoire,* 1–2
  "Tibet, India, and Malaya," 11
White, W. Howard, "European Railways," 137
Whitin, John C., in Van Slyck, 82
Whitney, Amos, in Spalding, 83–84
Whitney, Charles S., *Bridges,* 224
Whitney, Eli, and milling machine, in Battison, 277
Whittle, Frank, on jet propulsion, 243
Wiborg, Frank B., 272
Wiener, Philip P., 185
Wik, Reynold M.
  Steam Power on the Farm, 217
  on tractors, 216
Wildung, Frank H., *Woodworking Tools,* 278, 190
Wilhelm, H. A., on uranium production, 262
Wilkin, W. J., Australian education, 178

Wilkinson, Norman B.
  "Anticipation of F. W. Taylor," 303
  "Gilpin Endless Paper Machine," 269
  Joshua Gilpin's travels, 136
  Pennsylvania bibliography, 25
  "Thomas and Joshua Gilpin," 269
  on transfer of technology, 201
Williams, Judith B., guide to English economic history, 25
Williams, J. F. L., *Inventions and Discoveries*, 16
Williams, Peter, *European Organ*, 294
Williams, Trevor I.
  History of Technology, 2–3
  Short History, 3
Williams, Wellington, travel guides, 139
Williamson, Harold F., *American Petroleum Industry*, 268
Willings Press Guide, 38
Willis, Robert
  in Buchanan, 62
  edits Honnecourt, 49
Wilson, Allen B., in Van Slyck, 82
Wilson, Carmen, *Magnetic Recording*, 255
Wilson, Joseph M., *Centennial Exhibition*, 197
Wilson, Mitchell, *American Science and Invention*, 12
Winchell, Constance M.
  in *A.H.A. Guide*, 18
  Guide to Reference Books, 33
windmills, 244
  ancient, in Forbes, 9
  in Bathe, 170
  Dutch patents, in Doorman, 108–109
Wisconsin State Historical Society, manuscripts, 118
Wissner, A., in Treue, *Hausbuch*, 51–52
Withington, Sidney
  Engineering in History, 5
  Marestier translation, 134
Wolf, Abraham (2), 14
Wood, Alexander, *Physics of Music*, 295
Wood, Ethel May, *History of the Polytechnic*, 175
Wood, Frederic J., *Turnpikes of New England*, 234
Wood, R. D., in Scharf, 84
wood and timber industries, 274–275
  patent reviews, in *Journal* of P.O.S., 151

Puget Sound area manuscripts, 119
Woodbury, C. J. H., *Bibliography of Cotton*, 274
Woodbury, Robert S.
  histories of gear-cutting, grinding, milling machines, 279
  history of lathe, 280
  "Legend of Eli Whitney," 299
  reviews Rolt, 279
Woodcroft, Bennet
  biographical data, 80
  patent indexes, 107–108
  Pneumatics of Hero, 284
  portrait gallery, in Patent Museum, 88
Woods, James L., collection, Franklin Institute, 115
Woodward, Calvin M., *St. Louis Bridge*, 225
woolen manufacturers, see textiles
World List of Scientific Periodicals, 142
World of Learning, 38
world's fairs, see exhibition, international
Worsam, W. S., 275
Worth, Fremont P., 211
Wright, Carroll D., U.S. census history, 96
Wright, Sydney L., *Franklin Institute*, 175
Wright brothers
  biography, by Kelly, 242
  papers, in Franklin Institute, 115
  photographs, in Vanderbilt, 88
Wroth, Lawrence, *History of Printed Book*, 272
Wulff, Hans E., *Crafts of Persia*, 281
Wüster, Eugen, 70
Wyman, C. W. H., 269

Yabauti, K., 48
Yarmolinsky, Avrahm, 135
Yarranton, Andrew, in Smiles, 81
Young, Arthur, writings of, 136
Young, Pearl I.
  Octave Chanute, 80
  mentioned, 135
Young, W. A., "Works Organization," 298
Yverdon Swiss encyclopaedia, 58

Zagorsky, F. N., 280
Zammatio, Carlo, 50
Zaunmüller, Wolfram, 71
Zedler, Johann H., 60

Zeising, Heinrich
  in Beck, 40
  in Reti, 42
Zinner, Ernst, 288
Zirnbauer, H., 52
Zischka, Gert A.
  on glossaries, 70

Index lexicorum, 54
  on multilingual dictionaries, 71
Zonca, Vittorio
  in Beck, 40
  in Keller, 41
  in Reti, 42
Zorzoli, G. B., *Inventions,* 4

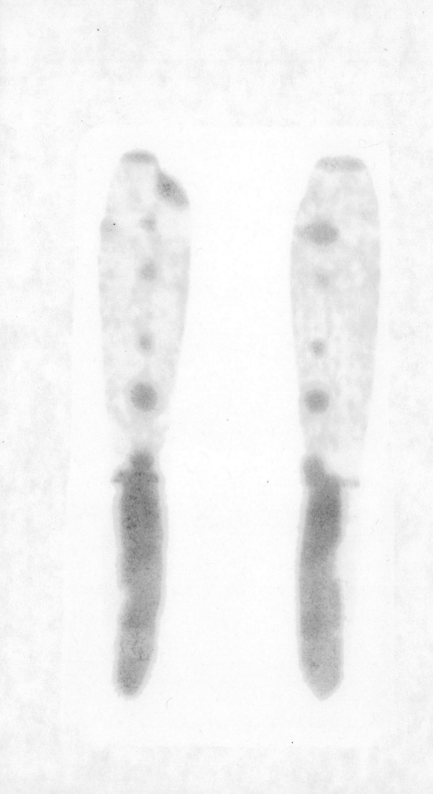

74104 Z
 7914
 H5
 F4

FERGUSON, EUGENE
    BIBLIOGRAPHY OF THE HISTORY OF
TECHNOLOGY.

DATE DUE

GAYLORD PRINTED IN U.S.A.